SUPRAMOLECULAR ORGANIZATION
AND MATERIALS DESIGN

To our families for their support

SUPRAMOLECULAR ORGANIZATION AND MATERIALS DESIGN

W. JONES AND C. N. R. RAO

Editors

with additional contributions by

LIA ADDADI
R. BALASUBRAMANIAN
ELIA BENIASH
SANTANU BHATTACHARYA
ANDREW D. BOND
NED BOWDEN
MARK E. DAVIS
WILHELM T. S. HUCK
IZUMI ICHINOSE
W. JONES
G. U. KULKARNI
TOYOKI KUNITAKE
SUENG-WOO LEE

UDAY MAITRA
SRINIVASAN NATARAJAN
STEVEN P. NEWMAN
GEOFFREY A. OZIN
C. N. R. RAO
FRANCISCO M. RAYMO
IVANA SOTEN
J. FRASER STODDART
P. JOHN THOMAS
JOE TIEN
PAUL WAGNER
STEVE WEINER
GEORGE M. WHITESIDES

CAMBRIDGE
UNIVERSITY PRESS

PUBLISHED BY THE PRESS SYNDICATE OF THE UNIVERSITY OF CAMBRIDGE
The Pitt Building, Trumpington Street, Cambridge, United Kingdom

CAMBRIDGE UNIVERSITY PRESS
The Edinburgh Building, Cambridge CB2 2RU, UK
40 West 20th Street, New York, NY 10011–4211, USA
477 Williamstown Road, Port Melbourne, VIC 3207, Australia
Ruiz de Alarcón 13, 28014 Madrid, Spain
Dock House, The Waterfront, Cape Town 8001, South Africa

http://www.cambridge.org

First published 2002

Printed in the United Kingdom at the University Press, Cambridge

Typeface Times NR MT 11/14pt *System* QuarkXPress™ [SE]

A catalogue record for this book is available from the British Library

Library of Congress Cataloguing in Publication data

Supramolecular organization and materials design / W. Jones and C. N. R. Rao, authors
and editors ; with additional contributions by Lia Addadi . . . [et al.].
p. cm.
Includes bibliogrpahical references and index.
ISBN 0 521 66240 0
1. Supramolecular chemistry.   2. Self-organizing systems.   I. Jones, William, 1949–   . II.
Rao, C. N. R. (Chintamani Nagesa Ramachandra), 1934–
QD878.S87   2001
547'.7–dc21   2001025398

ISBN 0 521 66240 0 hardback

Contents

List of contributors

Lia Addadi
Department of Structural Biology, Weizmann Institute of Science, 76100, Rehovot, Israel

S. Balasubramanian
Department of Organic Chemistry, Indian Institute of Science, Bangalore 560 012, India

Elia Beniash
Department of Structural Biology, Weizmann Institute of Science, 76100, Rehovot, Israel

Santanu Bhattacharya
Department of Organic Chemistry, Indian Institute of Science, Bangalore 560 012, India and
Chemical Biology Unit, Jawaharlal Nehru Centre for Advanced Scientific Research, Bangalore 560 012, India

Andrew D. Bond
Department of Chemistry, University of Cambridge, Lensfield Road, Cambridge CB2 1EW, UK

Ned Bowden
Department of Chemistry and Chemical Biology, Harvard University, 12 Oxford Street, Cambridge, MA 02138, USA

Mark E. Davis
Division of Chemistry and Chemical Engineering, California Institute of Technology, Pasadena, CA 91125, USA

Wilhelm T. S. Huck
Department of Chemistry and Chemical Biology, Harvard University, 12 Oxford Street, Cambridge, MA 02138, USA

Izumi Ichinose
Frontier Research System, The Institute of Physical and Chemical Research, Wako, 351-0198, Japan

William Jones
Department of Chemistry, University of Cambridge, Lensfield Road, Cambridge CB2 1EW, UK

G. U. Kulkarni
Chemistry and Physics of Materials Unit, Jawaharlal Nehru Centre for Advanced Scientific Research, Jakkur P. O., Bangalore 560 064, India

Toyoki Kunitake
Frontier Research System, The Institute of Physical and Chemical Research, Wako, 351-0198, Japan

Seung-Woo Lee
Frontier Research System, The Institute of Physical and Chemical Research, Wako, 351-0198, Japan

Uday Maitra
Department of Organic Chemistry, Indian Institute of Science, Bangalore 560 012, India

Srinivason Natarajan
Chemistry and Physics of Materials Unit and CSIR Centre of Excellence in Chemistry, Jawaharlal Nehru Centre for Advanced Scientific Research, Jakkur P.O., Bangalore 560 064, India

Steven P. Newman
Department of Chemistry, University of Cambridge, Lensfield Road, Cambridge CB2 1EW, UK

Geoffrey A. Ozin
Materials Chemistry Research Group, Chemistry Department, University of Toronto, Toronto, Ontario, Canada M5S 3H6

C. N. R. Rao
Chemistry and Physics of Materials Unit and CSIR Centre of Excellence in Chemistry, Jawaharlal Nehru Centre for Advanced Scientific Research, Jakkur P. O., Bangalore 560 064, India

Françisco M. Raymo
Department of Chemistry and Biochemistry, University of California, Los Angeles, 405 Hilgard Avenue, Los Angeles, CA 90095-1569, USA

Ivana Soten
Materials Chemistry Research Group, Chemistry Department, University of Toronto, Toronto, Ontario, Canada M5S 3H6

J. Fraser Stoddart
Department of Chemistry and Biochemistry, University of California, Los Angeles, 405 Hilgard Avenue, Los Angeles, CA 90095–1569, USA

P. John Thomas
Chemistry and Physics of Materials Unit, Jawaharlal Nehru Centre for Advanced Scientific Research, Jakkur P.O., Bangalore 560 064, India

Joe Tien
Department of Chemistry and Chemical Biology, Harvard University, 12 Oxford Street, Cambridge, MA 02138, USA

Paul Wagner
Division of Chemistry and Chemical Engineering, California Institute of Technology, Pasadena, CA 91125, USA

Steve Weiner
Department of Structural Biology, Weizmann Institute of Science, 76100, Rehovot, Israel

George M. Whitesides
Department of Chemistry and Chemical Biology, Harvard University, 12 Oxford Street, Cambridge, MA 02138, USA

Preface

The role of supramolecular organization in the design and synthesis of new molecular systems is well recognized and assumes an increasingly important role in modern materials design. There is good reason to believe that the future of materials science will be closely linked to the use of soft chemical methods and processes such as self-assembly. The intelligent use of pre-organized templates onto which novel materials may be grown will result in materials with unprecedented properties and applications. Self-organization at various length scales from the molecular to the macroscale will also create important materials. The application of these materials will range from biological systems and biomineralization through shape selective catalysis, organic solid state chemistry, molecular machines and mesoscale synthesis. These developments are indeed extremely broad and cover organic, inorganic, biological and metallic systems.

This book, with chapters written by international experts, highlights some current aspects of materials synthesis where supramolecular organization plays a significant role. We expect that the principal audience for the book will be scientists in industry and academia, although the book should be useful for graduate courses in Chemistry, Physics, Chemical Engineering and Materials Chemistry as well as in interface courses between Biology and Materials/ Chemistry.

The idea of this book came about while one of us (C. N. R.) was Linnett Visiting Professor at the Chemistry Department, University of Cambridge, and Sidney Sussex College. We are grateful to the University and the Master and Fellows of Sidney Sussex for their support.

W. Jones
C. N. R. Rao

1

Assembly and mineralization processes in biomineralization: strategies for forming biological composite materials

LIA ADDADI, ELIA BENIASH AND STEVE WEINER

1.1 Introduction

Many organisms produce materials with mechanical properties that are better than those of comparable synthetically produced materials, built from the same or similar basic components [1, 2]. The key to biological material production is, of course, the genetically programmed activities of cells, that reflect the accumulated experience of evolving by trial and error field testing over tens, if not hundreds, of millions of years. It is highly unlikely that the synthetic processes that humans are capable of performing will ever be able to reproduce or even mimic these cellular functions. Detailed studies of the structure of the formed material and the formation processes involved in producing the material can, however, provide an understanding of the basic design and assembly principles incorporated into the structure. Materials scientists may then be able to use some of this knowledge to improve the fabrication processes and the performance of novel synthetic materials.

The key components in all sophisticated biological materials are the macromolecules that the cells produce and subsequently incorporate into the material. These include proteins, glycoproteins, proteoglycans, lipid assemblies and polysaccharides. Many biological materials are composed almost entirely of these assembled macromolecules. Common examples are the cuticles of many insects, the skin and tendons of vertebrates or the silk of spider webs. A very widespread adaptation is to stiffen the material by the introduction of a mineral phase. Common mineralized biological materials include the shells of mollusks, the carapaces of crustaceans, and the bones and teeth of vertebrates. Many of them are composite materials and are known to possess remarkable mechanical properties, especially when taking into account that they form at ambient temperatures and pressures, and that their mineral components are often commonplace materials with rather poor natural mechanical properties [2, 3].

The major objective of this review is to describe some of the assembly and design principles used by organisms to form mineralized materials that perform mechanical functions. Understanding the mineralization processes in their biological setting is certainly a challenge that has boggled the minds of many biologists and materials scientists over the centuries [4, 5]. From the first elegant pioneering studies to this day, scientists have tried to characterize mineralized tissues, initially describing and dissecting the components making up their complex assemblies, and later attempting to unify the understanding of their modes of formation within a conceptual framework embodying general underlying principles and processes [1, 6, 7]. A general consensus has been reached on some basic issues concerning mineralized tissues formed under controlled conditions. These are as follows:

– The material is inevitably formed in a defined and specialized space that is usually isolated from the intracellular environment [7, 8]. We refer to this as the microenvironment of mineral formation.
– In almost every case where mineral is formed under controlled conditions, it is introduced into a preexisting macromolecular framework [9].
– Where crystalline materials are involved, a supersaturated solution is set up prior to the onset of crystallization, followed by controlled crystal nucleation, controlled crystal growth and finally controlled cessation of growth.

The types of control mechanism vary from case to case and are often insufficiently well understood. They are thus the subjects of ongoing debates. The basic concepts defined above, by and large, remain unchallenged. Based on recent knowledge from a variety of different mineralization processes, we feel that there is a need to reconsider whether or not this consensus is always justified, especially concerning the microenvironment of crystallization. We have noted that in many mineralization processes that have been studied in detail at different stages of their development, crystallization does not appear to occur from a supersaturated aqueous solution, but rather from structured colloidal phases. The locations, length scales and time scales of initial mineral deposition, followed by the transformation to the final product, are all under well-defined cellular control. We suspect that this phenomenon may be more widespread than was previously thought. This may have far-reaching consequences in our basic understanding of the underlying concepts involved in biomineralization, as well as providing inspiration from biology for the fabrication of new advanced materials.

The most convenient biological systems to study are those that are organized in such a way that the different stages of formation are separated in both space and time. Relatively few such mineralizing systems are known. Here we will first describe in some detail three such mineralizing processes, which are among the best documented to date. In the second part of this review, we shall discuss more general concepts, in terms of setting up the chemistry, the structure and in particular the microenvironment of mineral deposition.

1.2 Three biogenic materials

The three mineralized hard parts are the teeth of certain mollusks called chitons, the dentin component of vertebrate teeth, and the skeleton of the larvae of the sea urchin. They all possess very different material properties. Taxonomically they are also very different, with each being formed by organisms that belong to different phyla: the Mollusca, the Chordata and the Echinodermata respectively.

Figure 1.1 shows photographs of the three hard parts. Each material forms in spaces that are delineated differently from the rest of the organism, and each space is further subdivided by frameworks composed of different macromolecules and mineral components. The major differences are briefly as follows. Chiton teeth are made up of separate regions, each of which is composed of multicrystalline randomly oriented iron oxide (magnetite) crystals, amorphous iron hydroxide (lepidocrocite) and calcium phosphate (dahllite or carbonated apatite) crystals. The main framework component is the polysaccharide chitin. A vertebrate tooth is composed mostly of dentin, except for the outer working surface which is composed of very hard enamel. Dentin is composed of thousands, if not millions, of tiny crystals of calcium phosphate which grow inside a collagen framework. The crystals are oriented relative to the protein fibers. Finally, the sea urchin larva contains two or more spicules. Each spicule is composed of a single crystal of calcium carbonate (calcite). In contrast to the other two objects, the site of mineralization is exclusively delimited by a membrane. Furthermore, it is not completely preformed, but expands continuously while the mineral is deposited.

A comparison of the three materials thus offers an excellent opportunity to home in on some of the basic underlying strategies that are used in their fabrication. In the following sections we will briefly describe the relevant characteristics of the three mineralized tissues, using in each case the same categories. In this way we will list not only what is known, but can point out gaps in our knowledge. This section will then constitute the basis for a broader discussion of the assembly and mineralization processes of these and other mineralized tissues.

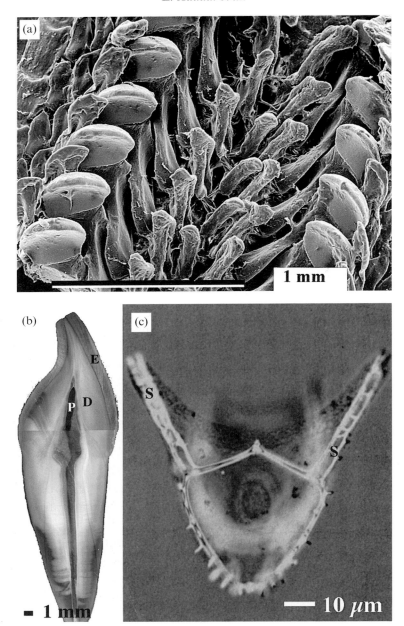

Figure 1.1. Views of three biogenic mineralized materials discussed in this chapter. (a) Scanning electron micrograph of the radula of the chiton *Acanthopleura haddoni*, showing five tooth rows. Two teeth per row (the larger, spoon-shaped objects) are mineralized. (b) Light micrograph of a longitudinal section through a whole human premolar. The external enamel layer (E) overlies the dentin (D). The central space is the pulp cavity (P). (c) Polarized light micrograph of a whole larva of *Anthocidaris crassispina*. The pair of spicules (S) are illuminated (modified from a picture of Okazaki [10]).

Figure 1.2. Light micrograph of the radula of the chiton *Acanthopleura haddoni*, showing 46 tooth rows. On the right hand side of the figure the first seven tooth rows are totally transparent (not mineralized), while in the next four rows the first mineral deposits appear. The dark color is due to the iron oxide mineral, magnetite.

1.2.1 The chiton tooth

The formation of the chiton tooth takes place on a tongue-like process, the radula, which comprises more than a hundred rows of teeth and is around 3 cm long. Two of the teeth in each row are mineralized (Fig. 1.2). The chiton uses its teeth to scrape the rocky substrate for algae and sponges living on the surface or just below the surface [11, 12]. Teeth are worn out at a rate of about a row every 12–48 h, and are continuously replaced [13, 14]. Thus the radula is in essence a conveyor belt, and every tooth row represents the product of 12–48 h worth of assembly and mineralization activities. The chiton radula is thus an ideal object to study assembly and mineralization. Its major drawback is that the teeth are small: 100 μm across [6].

Architecture during formation

Tooth formation takes place inside an organ called the radula sac (Fig. 1.3). Each nascent tooth forms in a cavity that is completely surrounded by cells; the so-called superior epithelial cells. These cells remain on the tooth surface and the whole tissue complex progresses together during the formation stages [15]. In the first 8–10 rows, the cells synthesize and secrete the organic matrix. This matrix undergoes a series of changes that have been visualized histochemically, and probably include cross-linking reactions [16]. X-ray diffraction patterns of the various rows show that the α-chitin becomes more ordered during the process [17]. As the cells are not in direct contact with the forming matrix, it can be assumed that this is largely a self-assembly process, although the macroscopic shape of the tooth, and thus of the macromolecular assembly, must be dictated by the cells at some stage. Interestingly, the matrix forms with different structures in the two major tooth layers, without the direct intervention of the cells. Before mineralization begins, the final shape and size of the mature tooth is achieved. We do not know if all the other macromolecular components are

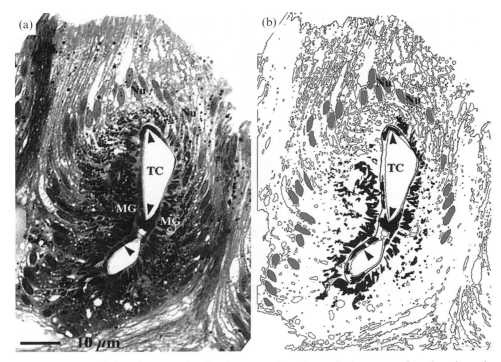

Figure 1.3. (a) Light micrograph of a tissue section through the zone of early mineralization in the radula sac of the chiton *L. hartwegi*. (b) A contour scan of the same figure which highlights some of the important features. The elongated cusp cells (cell nuclei (Nu) coloured gray in (b)) contain dark iron granules (MG) in their basal and apical poles. The central part (TC) is the tooth compartment, which is already partially mineralized (arrowheads). Reproduced with permission from Nesson [14].

already in place. Thus the cells supply the raw materials required for mineralization and orchestrate the process of mineralization while they are located at distances of tens to hundreds of micrometers away from the mineralization sites.

The organic matrix framework

The major component of the framework is α-chitin [16, 18]. All we know about its structural motif at the nanometer level is that the chitin chains of the whole tooth do have a preferred orientation. In another phylum (the Crustacea), chitin frameworks are organized into layers, in which the fibrils of each layer have a preferred orientation. These have been compared to liquid crystals [19]. The orientation direction changes from layer to layer, usually by discrete angles and in one direction. Thus a plywood-like structure is formed.

Towe and Lowenstam [20] showed that the magnetite crystals are located in a well-defined three-dimensional matrix framework. Van der Wal *et al.* [21] estimated that the organic content of the magnetite layer is about 3% by weight. The framework delineates elongated cylindrically shaped voids in which the mineral phase is located. Evans *et al.* [18] showed that in the demineralized magnetite layer the fibrils are rather thin and sparsely distributed, although a preferred orientation is apparent. The chitin framework of the carbonated apatite layer is more dense, the fibrils are thicker and they are also arranged to form tubules about 2 μm in diameter [18].

Protein comprises about 10% by weight of the organic matrix. The proteins are phosphorylated, and are enriched in aspartic acid, glutamic acid and glycine. Some of the proteins and the chitin are intimately associated. The carbonated apatite layer contains much more protein than the magnetite layer [22]. To date, nothing is known about the individual proteins within the framework.

The chiton tooth radula is perhaps the best documented example of the general principle in biomineralization, namely that the organic matrix is formed first, and only then are the spaces within the framework filled by the mineral.

Mineralization

The first formed mineralized deposits are Fe^{3+} containing granules present in the superior epithelial cells. These are composed of ferritin, which is a complex of the mineral, ferrihydrite ($5Fe_2O_3.9H_2O$), that forms inside a spherical shell of protein, apoferritin [14]. Some of the protein shells coalesce and larger ferrihydrite containing granules also form. They are observed in high concentrations within the cell in the zone adjacent to the matrix framework. These intracellular deposits can be detected in the cells that are located three rows prior to the first observable mineral deposits in the extracellular framework [15].

The first row of teeth with extracellular mineral in the framework has an ochre color. This is due to large accumulations of ferrihydrite granules. The iron is derived from the ferritin in the epithelial cells, but is transported out of the cells in the soluble reduced form [14]. Within a row or two the color darkens as the ferrihydrite precursor mineral converts to magnetite. All of the above is confined to the posterior tooth layer that is ultimately composed entirely of magnetite ($FeO.Fe_2O_3$) [11, 13].

The next stages of mineralization differ according to species. In some species, the remaining matrix framework is filled with an amorphous hydrous ferric phosphate [12]. In most species the process is more complex. Lepidocrocite (γ-FeO(OH)) forms in a thin layer just beneath the magnetite layer [23, 24] and

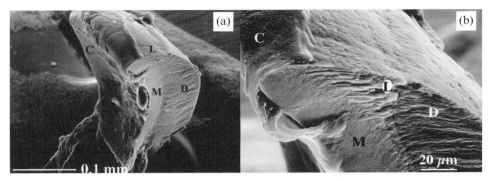

Figure 1.4. (a) Scanning electron micrograph of a fractured cross section of a mineralized tooth from the radula in Fig. 1.1. (b) Enlargement of the upper part of the same
fractured cross section. The magnetite layer (M) lines the interior surface. The thick
dahllite layer (D) is on the exterior. The thin lepidocrocite layer (L) is sandwiched
between the two. C, organic sheath and cell remnants.

then the remainder of the framework is filled in with amorphous calcium phosphate. After several weeks, the latter transforms into crystalline carbonated
apatite, and in some species fluoride is present as well. Significantly, the crystal
axes of the carbonated apatite have a preferred orientation and are aligned with
the surfaces of the matrix sheets, implying a possible role of the matrix in the
crystallization process [17].

Function and material design features

The function of the tooth is to scrape the rocky substrate. This is clearly a
mechanically very demanding task. The tooth has therefore been designed in a
unique manner. The layer that does the actual scraping is composed of a relatively hard mineral, magnetite [11, 16, 18] (Fig. 1.4). Beneath the magnetite
layer is a thicker layer in which the mineral phases are softer than the magnetite
and the matrix framework is better developed [18]. The design principles of
chiton teeth have been investigated [21]. The juxtaposition of a hard and stiff
working layer with a softer more pliant backing material is clearly important,
as it is in vertebrate teeth (see next section). The orientation of the magnetite
units is perpendicular to the cutting edge and about 60° to the leading surface.
This results in the tooth being a very efficient cutting device, as it is never
blunted or over-sharp.

1.2.2 *Dentin*

Dentin comprises the bulk of the vertebrate tooth (Fig. 1.5). During the formation of the tooth, dentin is produced by cells located in the pulp cavity [25, 26].

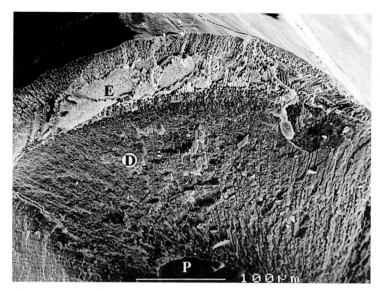

Figure 1.5. Transverse section through a mouse incisor tooth. E, enamel; D, dentin; P, pulp cavity. Note the dentin tubules, radiating out from the pulp cavity to the dentin/enamel interface.

The cells continuously withdraw as they form more dentin, and the pulp cavity decreases in size. The zone of formation between the cell surface and the final mineralized product is usually only a few tens of micrometers thick (Fig. 1.6). It is known as the predentin. Many rodents have continuously forming incisors. They, like the chiton teeth, are ground away at one end, and are continuously replaced. Thus all stages of tooth formation can be studied in a single rodent incisor. In the rat tooth the stages merge into each other, and are not separated, as is the case in the chiton tooth.

Dentin is composed primarily of type I collagen, the calcium phosphate mineral, carbonated apatite (also known as dahllite) and water (Fig. 1.7). The relative volume proportions of these major constituents are roughly 1:1 [27]. In this respect dentin is very similar to bone. In fact dentin is regarded as being a member of the bone family of materials [28]. All members of the family have as their basic building block, the mineralized collagen fibril (Fig. 1.7d). They differ in the manner in which the fibrils are organized at higher hierarchical structural levels.

Architecture during formation

The nascent tooth bud forms in a space between a layer of epithelial cells (ameloblasts) that will form the enamel, and a layer of mesenchymal cells (odontoblasts) that will form the dentin. The epithelial amelogenic cells signal to the underlying mesenchymal cells to differentiate into odontoblasts (i.e. to start forming dentin) [29, 30] and then the presence of the new dentin is a return signal for subsequent enamel formation [31, 32]. The result is that both layers form simultaneously, but the directions of growth are opposite. The space in which the two materials form is totally delineated by cells (Fig. 1.6).

During dentin growth, the odontoblast cells produce the type I collagen that constitutes the major framework component of dentin. The cells secrete both the collagen and other macromolecules into the predentin. They withdraw a little from the site of formation and secrete more macromolecules (Fig.1.6). The odontoblasts, however, do not completely withdraw, but each cell leaves an extension or process inside a tubule that is embedded in the dentinal tissue (Figs. 1.5, 1.6). By the time the dentin is fully formed, these processes extend from close to the dentin–enamel junction all the way to the cell in the pulp cavity; a distance of up to several millimeters (Fig. 1.5). Thus in dentin, the cells responsible for its formation are always in intimate contact with the forming mineral. These processes also fulfil an important sensor function during the mechanical operation of the mature tooth [26].

The organic matrix framework

The basic building block of the dentin matrix is the mineralized collagen fibril (Fig. 1.7d) [33]. The collagen polypeptide chains are synthesized within the cells, where they begin to self-assemble into triple helical molecules and subsequently fibril bundles within secretion vesicles [34–36]. After secretion into the extracellular environment they further self-assemble into bundles of aligned fibrils. The cells are presumably responsible for determining the orientations of the collagen bundles. The spatial organization of the bundles is such that almost all the collagen fiber axes are aligned in a plane parallel to the surface of the odontoblast cell layer, and are therefore perpendicular to the direction of the tubule long axes (Fig. 1.7c). However, within the plane, the fibril bundles are not well aligned, and in some cases show no preferred alignment (Figs. 1.6, 1.7c) [33]. A face-on view of these collagen layers can be seen in Fig. 1.7c, which is an unstained transmission electron microscope (TEM) image of the collagen bundles preserved in vitreous ice. The collagen is clearly closely packed.

The framework contains many other proteins, polysaccharides and lipids, in addition to collagen. These include noncollagenous proteins (NCPs) and proteoglycans that are also present in bone, lipids and proteins present in serum, as well as dentin-specific proteins [25, 27]. The dentin-specific NCPs are rich in aspartic acid and the amino acid phosphoserine. The sequences of some of these proteins have been determined, revealing large stretches of repeating

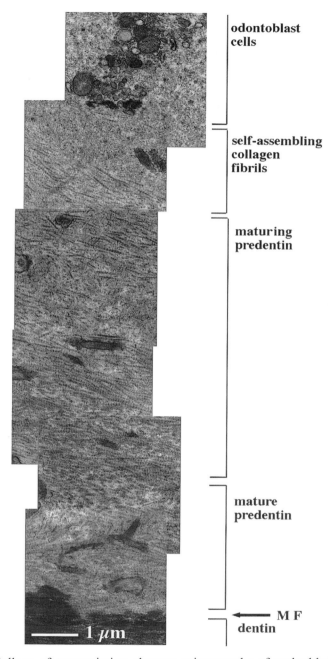

Figure 1.6. Collage of transmission electron micrographs of embedded and stained sections through rat predentin tissue, showing all the maturation stages from the cells to the mineralized material. The banded collagen fibers appear increasingly tightly packed through the predentin layers up to the mineralization front (MF), where they mineralize, see text.

short sequences of residues [37, 38]. Clearly, the abundance of negatively charged residues must be the predominant factor in determining the structure and conformation of these proteins. Not surprisingly, spectroscopic information (NMR, CD, IR) indicates that they appear to favor extended conformations [39, 40]. Unfortunately as the structures of these or other similarly highly charged proteins have not been determined at high resolution, structure predictions based on sequence are uncertain. The possibility of reliable and accurate structure determination is further compromised by the necessity of extracting them from their *in vivo* microenvironment. In proteins such as these, that presumably interact in a specific manner with a substrate, extraction may substantially modify their conformations.

Dentin formation, like many other mineralization processes, involves two steps: the formation of the framework and then the mineralization [27]. The first-formed collagen framework layers accumulate to a thickness of several tens of micrometers before mineralization begins. Pulse-chase studies using the continuously growing rat incisor showed that about 24 h elapse following synthesis and before the collagen framework begins to mineralize [41]. In this tooth, the odontoblasts are located about 20 μm away from the mineralization front. They do, however, maintain a 'short-cut' to the mineralization front in the form of the cell processes. Thus the raw materials for mineralization, if supplied through the processes, have to travel a distance of only 2 or 3 μm. The pulse-chase studies showed that the highly acidic dentin proteins, phosphophoryns, arrive at the mineralization front via the processes [36]. This occurs within 4 h after synthesis. On the other hand, the proteoglycans appear to be distributed throughout the entire predentin framework; however, their concentration is higher in the predentin closest to the odontoblasts [42].

Preparing the framework for mineralization involves not only the addition of macromolecules, but also the removal of specific components just prior to mineralization. One important example of removal of macromolecules, is the proteoglycans. They are broken down and removed just prior to the mineralization front [42]. It has been postulated that their breakdown must be a key step in the mineralization process, especially when bearing in mind their highly charged nature. Many have proposed that their function is to inhibit premature mineralization [25]. Proteoglycans are certainly good inhibitors of mineralization *in vitro* [43–45]. Their function *in vivo*, however, is not known. Another proposed reason for proteoglycan breakdown is that the concurrent decrease in charged groups associated with the proteoglycans, changes the local osmotic pressure. This in turn causes the collagen fibrils to expand. Volpi *et al.* [46] suggest that this conformation change may be important for initiating mineral formation.

Mineralization

Mineralization takes place in intimate association with the collagen framework, as well as in matrix vesicles that bud off the odontoblast processes. The very first dentin to mineralize (mantle dentin) contains a relatively large proportion of matrix vesicle derived mineral [31], whereas in the bulk of the dentin, most of the mineral is associated with collagen. The collagen fibril structure is therefore the key to understanding dentin structure and mineralization. Its main features are schematically depicted in Fig. 1.7d. We assume below that the basic characteristics of dentin collagen mineralization are the same as those for bone.

Most of the crystals of carbonated apatite that form at the predentin mineralization front are very thin (2–3 nm thick) and plate-shaped. Extracted crystals observed in the TEM have very irregular shapes, but are roughly tens of nanometers long and wide [48, 49] (Fig. 1.7a). The crystal length and width dimensions may be much larger *in vivo*, based on observations of dentin prepared by ion beam milling [50]. The dahllite crystals are, for the most part, nucleated within the collagen fibrils [51, 52]. The smallest crystals observed in the fibrils (of mineralizing tendon and bones) are needle-shaped, but soon after nucleation they adopt their characteristic plate shape [53]. In bone and in mineralizing tendon they have been observed to nucleate at a very specific location within the fibril (the e-band) [54]. It is thought that the nucleation event must be controlled, as the crystals inevitably have their *c*-axes aligned with the collagen fibril axis [55]. It is not known yet how this control occurs at the molecular level. The earliest formed crystal clusters are barely 20 nm in diameter and already have a lattice structure that is characteristic of apatite [56]. The paradigm of collagen mineralization for the past 30 years has been that at least some of the so-called noncollagenous proteins are involved in nucleation [25, 27]. In fact the first such proteins identified, the phosphophoryns, were discovered in dentin by Veis and Perry [57], and they are among the few dentin specific proteins known today [58]. It has recently been shown that peritubular dentin also contains proteins similar to the phosphophoryns present in bulk dentin [59].

The stages of crystal growth are best documented in the fibrils of mineralizing turkey tendon [53]. This tissue is particularly amenable to study because the various stages of mineralization can be observed. The collagen fibrils are all parallel, and in register with each other. The crystals grow rapidly along their *c*-axes to fill the gap region, and then spread laterally to fill the groove or channel. Finally they manage to push themselves into the overlap region. The crystals are organized in layers across the fibril [51, 52, 60]. This organization

L. Addadi et al.

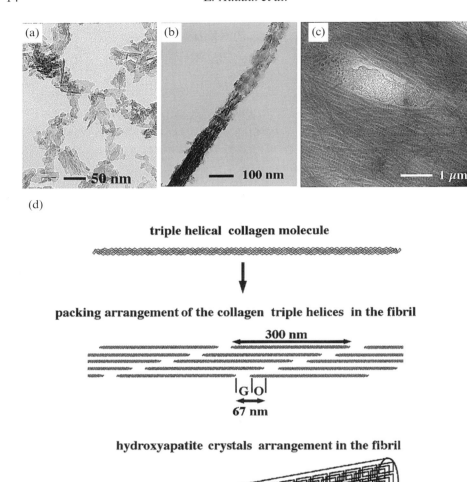

(a)
50 nm

(b)
100 nm

(c)
1 μm

(d)

triple helical collagen molecule

packing arrangement of the collagen triple helices in the fibril

300 nm

G O

67 nm

hydroxyapatite crystals arrangement in the fibril

Figure 1.7. (a) Plate-shaped carbonated apatite crystals extracted from rat bone. (b) Single mineralized collagen fibril disaggregated from human dentin. The fibril is twisted in the middle, such that the crystal plates are viewed edge on in the middle of the fibril, and face on in the upper part of the picture. (c) Unstained cryosection from the predentin layer of a rat incisor viewed in the TEM in a thin layer of vitrified ice. The section is cut parallel to the pulp cavity. In the center of the picture is a tubule, partially filled by an odontoblast process. The collagen fibrils lie randomly in the plane, but almost all are perpendicular to the tubule direction. (d) Schematic representations (top) of a triple helical collagen molecule; (middle) section one triple helical molecule thick through a fibril showing the staggered array [47] with gap (G) and overlap (O) regions; (bottom) a mineralized fibril showing layers of crystals.

essentially reflects the space delineated by the channels in the collagen fibril framework itself (Fig. 1.7d). In dentin, it has been observed that the orientations of the crystal layers in neighboring fibrils are not aligned, which is in contrast to the situation in bone and tendon, where they are aligned for at least several micrometers [61].

The second mode of crystal formation that occurs in dentin is via matrix vesicles. These are phospholipid delimited packages of specialized enzymes, macromolecular complexes and ions, that induce the precipitation of amorphous calcium phosphate. At some point the latter crystallizes into carbonated apatite crystals, that have no preferred orientation [62]. These appear smaller and denser than the crystals that form in the collagen framework.

In contrast to bone, the crystals in dentin grow to a certain size and density and then cease to grow, judging from the distribution of microhardness values measured on tooth sections. Hardness is mainly a function of crystal orientation and density [63]. Furthermore, this varies in a systematic manner throughout the dentin. Thus in dentin there must be a mechanism for finely controlling the cessation of crystal growth (perhaps the distribution of certain noncollagenous components?), whereas in bone, crystal growth continues until the process of remodeling removes the older bone and replaces it with newly formed bone.

Function and material design features

Dentin constitutes the bulk material of all vertebrate teeth. The outer working surface of a vertebrate tooth is composed of a much stiffer material called enamel, or in the case of fish, enameloid [6]. The two materials work together during mastication to provide the tooth with its functional properties [33]. In general the softer dentin functions in distributing and absorbing the compressive stress that is transmitted through the outer enamel layer [26].

Teeth fulfill very specific functions and it can be safely assumed that the tooth structure is finely tuned for this purpose. The dentin structure is clearly anisotropic, with all the mineralized fibrils being located on one plane, and all the tubules oriented perpendicular to this plane. It is therefore most surprising that in terms of microhardness, dentin is isotropic. This has recently been confirmed in a careful study in which root dentin microhardness was measured at the same precise location in three orthogonal directions [33].

The isotropic microhardness properties of dentin contrast markedly with the anisotropic microhardness properties of various types of bone [64]. Microhardness is strongly dependent on crystal orientation, and less dependent on fibril orientation. Wang and Weiner [33] therefore proposed that the isotropy arises from a combination of structural features related to crystal orientation. The crystal layers in adjacent fibrils within the same bundle, are not aligned

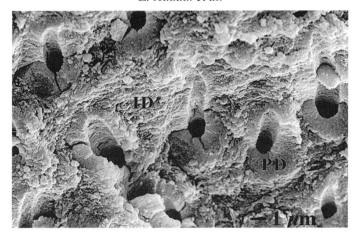

Figure 1.8. Scanning electron micrograph of a fracture surface through crown dentin of a human tooth. Various tubules are viewed in cross section. The dense envelope around the tubules is composed of peritubular dentin (PT). Intertubular dentin (ID) is located in the intervening spaces.

within the plane in which the fibril axes are aligned, there is little or no preferred orientation of fibril axes, and the matrix vesicle derived crystals are randomly oriented and presumably therefore have isotropic properties.

Crown dentin differs from root dentin in that about 20–30% by volume of the crown is composed of peritubular dentin (Fig. 1.8). In terms of crystal organization the peritubular dentin is very similar to the rest of the dentin, but differs in that it contains only small amounts of collagen framework [65]. This makes it much stiffer. Furthermore, the fact that the peritubular dentin surrounds the tubules, means that it forms a dense array of pipes all oriented more or less perpendicular to the enamel cap surface. It therefore appears to have a buttressing function designed to support the stiff enamel cap. Interestingly, there is a soft layer of 200 μm thickness between the enamel cap and the buttressed crown dentin, which Wang and Weiner [33] suggest may be designed to allow these very different materials to function together – a sort of gasket or cushion. This is needed not only to absorb the compressive stresses of mastication, but also to prevent differential expansion of the dentin layer due to thermal or humidity changes that may cause cracks in the enamel. The structure–mechanical function of this layer have not been studied.

Another unusual property of dentin is that its hardness properties vary continuously in three dimensions [33]. This is clearly under cellular control, and it was inferred that this design feature is part of the mechanism for distributing the applied compressive stress in such a way as to avoid cracks from developing. Unlike bone, teeth are generally not replaced (except once at puberty or thereabouts in humans), and do not undergo remodeling. The interesting exceptions are the continuously growing teeth of rodents, that form at the same rate as they are worn down at the working tip. The structural design of these teeth presumably also incorporates a self-sharpening feature. Another design strategy is adopted by certain teeth of bovids. They wear down over a life time, but are not replaced. The structural basis for the manner in which they wear down but remain functional, needs to be understood.

Vertebrate teeth, with all the morphological variations that are present in just one oral cavity, thus present a treasure trove of design features that can, and should be understood in terms of their structure–mechanical function relations.

1.2.3 Sea urchin larval skeleton

Sea urchin larvae are about a hundred micrometers in diameter. They have an internal skeleton that supports the soft tissues. The skeleton is composed of one or several pairs of intricately shaped spicules, the morphologies of which vary among different species. The spicule is composed of two different minerals, amorphous calcium carbonate and calcite [66]. The mineral phases in the adult skeleton are thought to be similar to those in the larval skeleton [67]. In fact almost the whole echinoderm phylum appears to use this type of material for constructing a large variety of skeletal elements.

The sea urchin larva is one of the most thoroughly investigated organisms in terms of its development. This is due to the ease with which synchronized cultures of larvae can be grown in the laboratory. Furthermore, methodology also exists for extracting the cells responsible for spicule formation and to culture them *in vitro* such that they produce spicules [68]. This greatly enhances the possibilities of investigating spicule formation processes. Wilt has written a review of spicule formation and mineralization [69].

Architecture during formation

Spicule formation takes place within the intracellular environment. The spicules, however, are much larger than individual cells. This is achieved by many cells fusing their membranes to enclose an extended space (called a syncytium) (Fig. 1.9). Spicule formation takes place inside a membrane delineated vacuole within this space [70, 71]. The size and the shape of the syncytium constantly increase and change during growth of the spicule [72, 73]. In fact, a freeze-fracture TEM study of the relation between the membrane and the growing spicule shows that the membrane is always juxtaposed to the spicule surface. There is thus no bulk solution within which the spicule forms [74]. Thus spicule

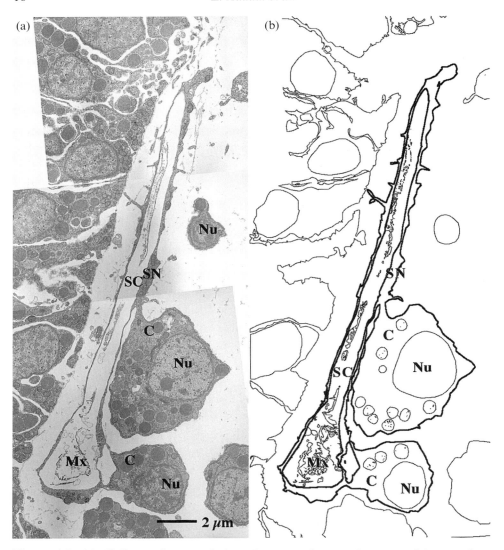

Figure 1.9. (a) Collage of transmission electron micrographs, comprising sections through the spiculogenic compartment during the early stages of spicule development in the sea urchin *Paracentrotus lividus* larva. (b) Contour profile, highlighting the syncytium. C, cells; SC, spiculogenic compartment; SN, syncytial envelope of the spicule; Nu, cell nuclei; Mx, organic matrix of the spicule.

formation is to a large extent controlled by the shape and properties of the syn-cytium, and the cells are in intimate contact with the forming spicule via the membrane.

The 'Organic Matrix Framework'

In essence, sea urchin larval spicule formation takes place in a preformed mem-brane framework that continuously changes. There is, however, also an organic matrix-like framework within the spiculogenic cavity. It is composed of poly-saccharides and proteins that remain insoluble after the mineral phase of the mature spicule is dissolved [75]. The framework forms concentric sheaths around the spicule long axis, and has radiating fibers that connect the sheaths laterally. It is not, however, known whether this matrix is preformed and func-tions as a framework to guide the mineral deposition, or whether it is deposited periodically as the mineral is introduced.

Most of the proteins that are present within the spicule are not part of the insoluble framework, but are dispersed within the mineral phase [76]. They are preferentially located on certain crystal planes that are more or less parallel to the spicule long axes [77]. These macromolecules are mainly glycoproteins that are not particularly acidic.

Mineralization

Mineralization is initiated at the point where three cellular extensions fuse to form a syncytium with a triradiate shape. The first mineral deposits form within a rounded spiculogenic vacuole [72]. Several such deposits form in the syncytium, but only two persist [73]. Furthermore, it has been observed by using light microscopy, that the deposits are birefringent and hence crystalline, and have the shape characteristic of a single crystal of calcite [66, 78]. This calcite crystal rapidly grows to produce three extensions (Fig. 1.10a). These extensions are aligned with the *a*-crystallographic axes of the calcite crystal, but their surfaces are smooth and do not express the stable faces of calcite [10]. Beniash *et al.* [66] showed that the additional mineral phase added to the growing spicule is actually amorphous calcium carbonate. With time, however, the amorphous calcium carbonate transforms into calcite by overgrowth of the initial crystal [66]. The spicule material, whether amorphous or crystalline, is directly juxtaposed to the envelope membrane, with no visible interstitial volume occupied by solution [74]. There also does not appear to be a well defined crystalline propagation front, as the presence of crystalline material can be detected all along the spicule by birefringence [74]. It was thus suggested that the crystal/amorphous phase interface forms a diffuse network, rather than a well-defined surface. At a later stage of spicule growth, one of the radii

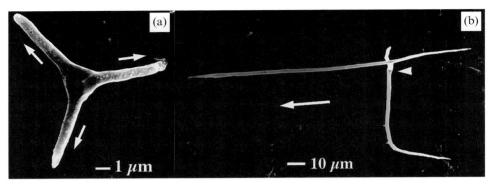

Figure 1.10. Larval spicules from the sea urchin *Paracentrotus lividus* at two stages of development. (a) Triradiate stage. The initially deposited single crystal is still visible in the center of the spicule. The direction of the crystallographic *a*-axes of calcite are marked. (b) Fully developed spicule. The direction of the crystallographic *c*-axis of calcite is marked by an arrow. The initially formed triradiate part can be seen (arrowhead).

changes its growth direction by 90° such that it grows along the crystallographic *c*-axis direction. There is thus an interesting interplay between morphological control and crystallographic orientation. In the light of these and other related findings, the concepts of what is the nature of this single crystal, how it grows, and properties associated with it, need to be considered.

The cells surrounding the spiculogenic cavity have granules that contain amorphous calcium carbonate [74, 79]. It is suspected that this mineral phase forms initially within the cells and is then transported as such to the spiculogenic cavity.

Function and material design features

One basic function of the larval spicular skeleton is probably to provide a rigid framework for the soft tissues. Interestingly, Thompson pointed out that the basic symmetry of the larval skeleton is related to the manner in which a sphere can be divided into equal parts [80]. Emlet [81] has measured the elastic modulus of the spicule material and found that its strength is higher and stiffness is lower than in synthetic calcite. He also compared the mechanical properties of a spicule comprising a single rod (Fig. 1.10b) with one composed of a number of parallel rods connected by struts and found that the stiffness of the latter type of spicule is three times higher than the former due to the structural differences. The fracture properties of the spicules also differ from calcite. The spicules do not cleave along the 'cleavage rhombohedron' planes characteristic of synthetic and geological calcite. The fracture surface of the spicule has a smooth appearance typical of amorphous rather than crystalline minerals

[71, 78]. Thus, although the spicules optically behave as single crystals of calcite, their mechanical and textural properties differ from those of synthetic and geological calcite.

There are several factors that may contribute to the mechanical properties of the spicules. Intracrystalline macromolecules are one such factor. It has been suggested that the presence of proteins occluded on crystal planes oblique to the cleavage planes, may have a crack-stopping function, by both dispersing and deviating the energy of the advancing cracks [82]. Another parameter which must be taken into account in understanding the mechanical properties of the spicules is the recently demonstrated presence of amorphous calcium carbonate [66]. The proportion of amorphous material reaches 80% in the developing spicules. Developed spicules are estimated to still contain approximately 20% amorphous material, which appears to be uniformly distributed along the whole spicule, with no evident phase separation boundaries. Although the exact mechanism of incorporation of the amorphous component into the crystalline matrix is not yet well understood, its presence may well account for the lower stiffness, higher strength and smooth fracture surfaces of the material compared to the corresponding crystalline material.

1.3 Discussion

The chiton tooth, dentin and the sea urchin larval spicule reflect the enormous diversity of the field of biomineralization. They differ with respect to the nature of their mineral and macromolecular components, as well as their structures. Few underlying common strategies can be recognized: the delineation of a dedicated space in which the mineralized tissue forms, the formation of mineral in a preformed framework within this space, and the precipitation of mineral from a supersaturated phase. In this section we will reexamine some of these underlying issues, focussing in particular on the microenvironment in which mineralization occurs.

1.3.1 Self assembly versus cell assembly

Mineralization is a biological process as cells are intimately involved in almost all aspects. The basic question that arises from the fact that the mineralization process itself occurs in a confined space is how remote the cellular control is from the active sites of mineralization. It can be assumed that the more remote the control is, the more know-how is built into the macromolecules that need to self-assemble and function, and the greater dependence there is on exerting control through the microenvironment of mineralization.

In the chiton tooth, the organic framework components are synthesized and secreted by the cells into the extracellular space, and there they self assemble. By the time mineralization is about to occur the cells are tens of micrometers away from many of the mineralization sites. They must therefore operate by remote control. The mineralization sites themselves are within a complex chitin framework, the dimensions of which are in the nanometer range. The sea urchin larval spicule represents the exact opposite situation. Mineralization occurs in a vacuole defined by a membrane, and the entire apparatus is within a consortium of fused cells (the syncytium). The membrane of the syncytium tightly surrounds the growing spicule [74]. Therefore, it has been proposed that the cells directly control spicule formation. The mineralization vacuole is subdivided by framework proteins. Nothing is known about the structure of the one nucleation site per spicule in the larvae, but in the adult a well-defined location, enclosed within a framework, has been identified as the nucleation site [83]. Dentin formation is intermediate between the two. It is an extracellular process, and the distances between cells or cell processes and mineralization sites are in the range of tens of micrometers or several micrometers respectively. Nucleation occurs within the fibril or at its surface and is associated with a site on the fibril surface some 7 or 8 nm wide [54]. The space available for crystal growth within the fibril is even smaller in one of the dimensions, namely 2 or 3 nm wide.

'Geometry', time scale and length scale are thus key factors in assessing possible modes of transport of the mineral phase components. The chemical state of the medium from which the mineral forms, and its nucleation and growth processes, are other key factors in determining the final properties of the mineral. These will be discussed below.

1.3.2 *The chemistry of the medium/solution from which minerals form*

Little is known with certainty about the exact chemical composition of the solution/medium from which mineral deposition occurs in the three systems addressed here, as well as in most, if not all, biomineralization processes. This is because it is extremely difficult to reliably extract or stabilize samples from the immediate environment of mineral deposition, let alone perform a chemical analysis on such small volumes. Spectroscopic techniques such as proton induced X-ray or γ-ray emission, can map the *in situ* distribution of some of the chemical elements, and monitor their changes in concentration with time, even before mineralization begins.

The concentrations of Fe, Zn, K, Ca, P, S, Cl, Na and F have been analysed in the forming teeth of the chiton radula [84]. The concentration of Fe increases steeply already in the second row of teeth, becoming constant after the fifth row. The first Fe mineral deposits occur in the seventh row. Ca and P concentrations increase simultaneously together with F starting in the seventh row, and stabilize after row 9, when the first deposition of amorphous calcium phosphate (ACP) appears. Na, S and Cl concentrations are constant in all rows, indicating that they are probably not directly involved in setting up the environment for mineral deposition. Zn concentrations increase between rows 3 and 7, and K concentrations increase slightly throughout the tooth development. Thus the introduction of the appropriate ions into the framework is clearly a well-controlled process. Furthermore, the Fe enters only into the part of the tooth where magnetite will be formed, and the calcium and phosphate where ACP is to be deposited.

In dentine mineralization, the concentrations of Ca, P, S, Na and K have been mapped by energy dispersive X-ray microanalysis in the transmission electron microscope [85]. Areas with high levels of K, P and S were found in mature predentine close to the primary zone of mineralization, in regions that are not yet mineralized. K appears to co-localize with P, suggesting a possible connection between the two ions in premineralization areas. How Ca and phosphate are transported to the mineralization site is not known exactly. There are probably several transport and transient storage mechanisms. Alkaline phosphatase, an enzyme that hydrolyses organic phosphate esters to phosphate ion, is abundant in all mineralizing phosphate tissues [6].

These observations clearly preclude the possibility that the solution from which mineral forms is essentially that of blood or other body fluids, or even sea water. Analyses of fluid extracted from the space in mollusks between the mantle, the organ that forms the shell, and the shell surface, show that it is different from sea water or the blood [86]. It is not, however, clear that this is the actual fluid from which the mineral precipitates. The trace element compositions of the mollusk shell minerals, as well as the shells of many other organisms, are also incompatible with crystallization from a medium similar to sea water [87], indicating that mineral deposition occurs from a specialized medium.

Another means of assessing the nature of the medium/solution from which the mineral forms is to analyse the very first mineral deposits precipitated. These must reflect the chemistry of the medium in which they formed. In most chiton teeth a poorly crystalline ferrihydrite is the first mineral deposited and it subsequently transforms into magnetite. In the anterior layer, amorphous calcium phosphate is first deposited, and only after several weeks does it begin to crystallize into dahllite or francolite [17]. In the sea urchin larva, calcite is the first mineral deposited, but subsequent growth of the spicule occurs through the prior deposition of amorphous calcium carbonate [66]. This then transforms into calcite. There is no evidence that the dahllite formed inside the collagen fibers in dentin is preceded by the deposition of amorphous calcium phosphate. The presence of amorphous calcium phosphate was shown, however, inside the matrix vesicles of mantle dentin [88, 89].

The presence of these amorphous precursor phases in the cases of the chitons and the sea urchin, are not unique. Lowenstam and Weiner [6] listed several other examples. We suspect that the phenomenon is much more common, as it is not easy to detect such transient phases, especially if they are present together with a crystalline phase. The first mineral deposit may not be at the site of mineralization itself, but rather within the cells that are responsible for the mineralization process. For example, in the sea urchin larvae the cells surrounding the mineralizing vacuole contain vesicles with amorphous calcium carbonate deposits [74]. It is not yet clearly established whether these are transported as such into the mineralizing space or whether they dissolve and are then reprecipitated. The former scenario is, however, more consistent with the knowledge existing to date. A similar situation exists in the case of the chiton tooth. The cells do contain abundant deposits of ferrihydrite, and this is also the first-formed mineral within the framework. Nesson [14], however, showed that the cellular deposits are transported into the framework in a solubilized form. Dentin mineralizes in two ways – mainly in association with the collagen framework, and to a lesser extent in vesicles that bud off from the cells. These so-called matrix vesicles have enzymes and other structured complexes, that allow them to induce the formation of amorphous calcium phosphate within the internal space [88, 89]. The vesicles migrate through the about-to-be mineralized tissue, and at some point the amorphous phase crystallizes into a spherical body of unoriented crystals of carbonate apatite. Thus even in dentin, some aspects of mineralization involve an amorphous precursor phase.

The deposition of unstable amorphous precursor phases requires a hydrated medium that has a very high ion concentration, and an inordinately high supersaturation relative to the corresponding crystalline phases. In order to stabilize, even transiently, such high supersaturations, specialized inhibitors of crystallization probably need to be present. Such high concentrations of ions cannot be regarded as a solution, but rather a structured colloidal phase. Concepts such as mechanisms of diffusion, levels of supersaturation and consequently kinetics of crystallization, must be reconsidered if crystallization does not occur from free solution.

1.3.3 Implications of a structured colloidal phase in biology

One generally valid observation is that organisms do not let chemical processes proceed uncontrolled down an energetically favored thermodynamic cascade. Such situations could too easily result in uncontrolled mineralization, sensitive to the whims of external parameters such as temperature, pressure, or even mechanical shocks. This is never observed in physiological processes of mineralization that are under biological control. Crystallization always occurs at well-defined points in time, location and development stages, following accurate 'blueprints', and the products formed by a given species do not vary among individuals. In the few known cases in which such variability does exist, they are clearly poorly controlled processes [6].

Highly metastable supersaturated solutions are not easily formed unless premature crystallization is inhibited either transiently or permanently [90]. In organisms, most of this complex control system is regulated through specialized macromolecules that interact directly with the mineral phases. A specialized family of glycoproteins inhibits the crystallization of calcite in the triradiate sponge spicules of *Clathrina* and in the so-called antler spicules of the ascidian *Pyura pachydermatina* [91]. The resultant amorphous calcium carbonate phase, although highly metastable in the absence of additives, is indefinitely stabilized at the loci of precipitation. A similar regulation of crystallization by inhibitory macromolecules may be operating in the stabilization of transient amorphous phases in the sea urchin larval spicule and in the chiton teeth. The inhibitory activity would have to be removed in a controlled manner to allow crystallization to occur with the correct timing.

An interesting analogous *in vitro* experiment was performed by Xu *et al.* [92] in which the presence of a polyelectrolyte in solution resulted in a layer of amorphous calcium carbonate forming under a structured monolayer. This subsequently transformed into a thin layer of polycrystalline calcite. The crystal growth in this *in vitro* system occurs by phase transition of amorphous calcium carbonate into calcite and not by dissolution of amorphous phase and reprecipitation of calcite crystals.

The role of Mg in the solution from which calcium carbonate minerals form presents an interesting enigma. Aragonite, the less stable polymorph of calcium carbonate, precipitates from sea water because of the presence of large concentrations of Mg in sea water. The Mg/Ca ratio is around 4. The Mg, with its strongly bound hydration sphere, inhibits the growth of calcite nucleii, and thus by default the denser aragonite forms. The paradox is that many marine organisms are able to form Mg-bearing calcite, with Mg contents of up to about 40 mole % [93, 94]. It has been shown that when Mg/Ca

ratios are very high, or when sea water is artificially supersaturated by diffusion of ammonium carbonate, then high Mg calcites can be formed *in vitro*
under ambient conditions. The problem with the experiments performed *in
vitro*, is the lack of reproducibility and control on the resulting phases. It is
also known that the presence of Mg or other inhibitors of nucleation such as
phosphate or charged macromolecules, induce the precipitation of a transient
amorphous phase of calcium carbonate [90]. It has recently been demonstrated *in vitro* that high-Mg calcite can form via such a transient amorphous
phase [93] under ambient conditions. The half-life of the amorphous phase is
further increased by the presence of matrix macromolecules or polyelectrolytes. We suspect that similar processes may be occurring *in vivo*, and that
high-Mg calcite, and possibly other minerals, may actually be forming from
structured colloidal phases. If the relation between high-Mg contents and
transient amorphous colloidal phases were established in biogenic calcites,
then the large number of very different calcitic skeletal elements with Mg contents > 10% would be, itself, a tell-tale sign of how widespread is the use of
these transient phases.

1.3.4 Diffusion

When distances of micrometers are involved, and the ions are in solution, diffusion becomes very important. There are several indications that one biological
strategy in diffusion control is to separate temporally, and maybe even spatially,
the introduction of the cation and anion components into the microenvironment of mineralization. Several observations point to the fact that many of the
charged matrix macromolecules that are components of the preformed matrix
framework, do not adopt a regular conformation unless calcium is present as a
counterion [39, 95]. This implies that the framework loads up first with the
cation, and then only when the anion is introduced does mineralization begin.
This may well be the reason why the enzymes that control the concentrations of
the relevant forms of calcium carbonate and calcium phosphate (carbonic
anhydrase and alkaline phosphatase respectively) are so intimately associated
with the mineralization site [27, 96]. This is clearly an excellent strategy for
maintaining control.

In vitro experiments by Iijima and her colleagues demonstrate the advantage
of separating the cation and the anion in terms of collagen mineralization [97].
The setup used involved diffusion of calcium and phosphate from opposite
sides of collagen membranes produced from the Achilles tendon – a tissue that
normally never mineralizes. Crystal formation, density, morphology and orientation were drastically influenced by the orientation of the fibers within the

organized medium, and the pathways of diffusion. In fact these are the first experiments that succeeded in inducing the nucleation and growth of a calcium phosphate mineral within the collagen fibril and with preferred orientation of their crystallographic axes relative to the fibril axis. The mineral formed *in vitro* was octacalcium phosphate, which is similar to the normally formed mineral in bone, carbonated apatite. In bone the crystals are mostly inside the fibril and their *c*-axes are aligned with the fibril axis.

Aizenberg *et al.* [98] showed that on a patterned surface comprising charged substrates separated by relatively hydrophobic areas, the patterning itself influences the extent of saturation of the medium in the proximity of the differently charged regions. Furthermore, once calcite nucleation occurs, the liquid immediately surrounding the growing crystal is depleted in ions and no additional crystals nucleate close to the growing crystal. This is in part a result of the diffusion gradients set up by the ions moving towards the growing crystal.

The above observations clearly point towards diffusion control as potentially being an important component of the control process operating in biological mineralization.

1.3.5 The microenvironment of nucleation

The microenvironment in which crystals form is important for understanding the control over mineral formation in biology. In each of the three systems described above, the microenvironments are very different in terms of shape and scale, confining components and the manner in which the space is further sub-divided. Clearly however a key element of the microenvironment is the nucleation site itself. There are numerous indirect observations *in vivo* and *in vitro* demonstrating that much of the control over mineral formation is exercised through the design and structure of the nucleation site. We know very little, however, about actual nucleation sites *in vivo*. In fact to date no protein or other specific components of a nucleation site have been unequivocally identified, let alone characterized. This is certainly one of the major immediate goals of this field.

One of the best understood mineralization systems in terms of the microenvironment of nucleation is the mollusk shell, and in particular one of the seven structural types of mollusk shell, the nacre [99]. The nacre has a simple geometry and is therefore a most convenient system to study in terms of nucleation. Observations of growing aragonitic tablet-shaped crystals show that the matrix surface at the location where the first crystals form is different from the remaining surfaces. This location can be differentially stained using dyes that have affinities for calcium and sulfate, for example [100]. The surface topography is also different in this region. This is assumed to be the nucleation site. It has been shown, based on electron diffraction patterns, that a stack of several adjacent crystals all have well-aligned crystallographic axes [101]. This, combined with the observation that there are pores in the matrix and possibly mineral bridges that link adjacent layers, suggest that not every tablet is nucleated *de novo*, but one nucleation event may produce a series of tablets that grow through the pores [102]. Nucleation must still occur for individual crystals or for individual stacks, that are estimated to comprise only 3–10 tablets in bivalves, but at least ten-fold more in gastropod shells.

The observation that there is a clear-cut spatial relation between matrix components and aragonite crystal axes is a good indication that nucleation is controlled at the molecular level [103]. Furthermore, *in vitro* systems in which natural matrix components are assembled and then mineralized, show that the specific polymorph (aragonite or calcite) formed is dependent on the presence of one or more nucleating matrix proteins [104–106]. In the system used by Falini *et al.* [105], the components assembled are β-chitin, silk fibroin and then aspartic acid-rich proteins from either the calcitic or aragonitic layers of mollusk shells. Ultrastructural investigations of the chitin and the silk using microtomed specimens preserved in vitrified ice, show that the chitin forms the rigid structural framework with interfibrillar distances in the range of 10–20 nm [107]. Surprisingly, the silk is not easily visualized and it is not clear how it is distributed within the framework. No information to date is available on the specific locations of the asp-rich proteins. The crystals that form in this framework are in intimate contact with the matrix, and the material produced is a true composite of mineral and organic material. All three components are essential in order to obtain polymorph specificity [105], thus clearly demonstrating the necessity of forming the right microenvironment. At this stage, however, we can only speculate about the chemical environment within this framework.

The dimensions of the space available are in the nanometer range. At this length scale, water is supposed to behave as a constrained liquid, which follows rules of diffusion, flow and structuring more akin to those of gels than to those observed in free liquids [108]. Furthermore, if the silk is present as a gel phase, the structured nature of the water molecules is even more enhanced. This in turn affects the activities of the ions within this medium, especially where polyelectrolytes are also involved. This speculative scenario envisages that the chemical environment of nucleation is very different from a simple saturated solution, and that the thermodynamics and kinetics of nucleation are more akin to crystallization from hydrogels. The same situation exists also in collagen-mediated mineralization, where the tiny apatite crystals form inside the grooves of the collagen fibrils. The space is in the nanometer scale, and the medium is thus expected to behave very differently from a free liquid [109].

1.4 Concluding remarks

In all three mineralization processes discussed, the interplay between cellular involvement and self assembly is apparent in terms of setting up the mineralization compartment and microenvironment, as well as the mineralization process *per se*. The current state of knowledge even in these relatively well-studied biological systems is not sufficient to draw the line between cell and self assembly. Furthermore, there are other factors such as diffusion and the nature of the solution/medium from which the mineral forms, that need to be reevaluated in terms of their importance in controlling the mineralization processes. This subject challenges us to focus in on the interface between the highly structured biologically produced surfaces and the seemingly well controlled 'solution' chemistry, and all this in a unique microenvironment.

1.4.1 Acknowledgements

We acknowledge support from a United States-Israel Binational Foundation grant and from US Public Service grant DEO6954 from the National Institute of Dental and Craniofacial Research. L.A. holds the Dorothy and Patrick E. Gorman Professorial Chair, and S.W. holds the Dr Walter and Dr Trude Borchard Professorial Chair in Structural Biology.

1.5 References

1. Vincent, J. F. V. (1982). *Structural Biomaterials*. Macmillan Press, London.
2. Wainwright, S. A., Biggs, W. D., Currey, J. D. and Gosline, J. M. (1976). *Mechanical Design in Organisms*. Princeton University Press, Princeton, New Jersey.
3. Currey, J. D. (1984). *The Mechanical Adaptions of Bones*. Princeton University Press, Princeton, New Jersey.
4. Haeckel, E. (1872). *Die Kalkschwamme*. Berlin.
5. Schmidt, W. J (1924). *Die Bausteine des Tierkorpers in Polarisiertem Lichte*. F. Cohen Verlag, Bonn.
6. Lowenstam, H. A. and Weiner, S. (1989). *On Biomineralization*. Oxford University Press, New York.
7. Simkiss, K. and Wilbur, K. (1989). *Biomineralization. Cell Biology and Mineral Deposition*. Academic Press, Inc., San Diego.
8. Mann, S. (1989). In *Biomineralization: Chemical and Biochemical Perspectives* (S. Mann, J. Webb and R. J. P. Williams, eds.), pp. 35–62. VCH Publishers, New York.
9. Lowenstam, H. A. (1981). *Science* **211**, 1126–1131.

10. Okazaki, K., McDonald, K. and Inoue, S. (1980). In *Mechanisms of Biomineralization in Animals and Plants* (M. Omori and N. Watabe, eds.), pp. 159–168. Tokai University Press, Tokyo.
11. Lowenstam, H. A. (1962). *Geol. Soc. Am. Bull.* **73**, 435–438.
12. Lowenstam, H. A. (1972). *Chem. Geol.* **9**, 153–166.
13. Kirschvink, J. L, and Lowenstam, H. A. (1979). *Earth Planet. Sci. Lett.* **44**, 193–204.
14. Nesson, M. H. (1969). PhD thesis, pp. 250. California Institute of Technology, Los Angeles.
15. Nesson, M. H. and Lowenstam, H. A. (1985). In *Magnetite Biomineralization and Magnetoreception in Organisms* (J. L. Kirschvink, D. S. Jones and B. J. MacFadden, eds.), pp. 333–363. Plenum Press, New York.
16. Runham, N. W. (1963). *Ann. Histochim.* **8**, 433–442.
17. Lowenstam, H. A. and Weiner, S. (1985). *Science* **227**, 51–53.
18. Evans, L. A., Macey, D. J. and Webb, J. (1990). *Phil. Trans. R. Soc. Lond. B* **329**, 87–96.
19. Bouligand, Y. (1972). *Tiss. Cell* **4**, 189–217.
20. Towe, K. M. and Lowenstam, H. A. (1967). *J. Ultrastr. Res.* **17**, 1–13.
21. van der Wal, P., Videler, J. J., Havinga, P. and Pel, R. (1989). In *Origin, Evolution, and Modern Aspects of Biomineralization in Plants and Animals* (R. E. Crick, ed.), pp. 153–166. Plenum Press, New York.
22. Webb, J., Evans, L. A., Kim, K., St. Pierre, T. G. and Macey, D. J. (1991). In *Mechanisms and Phylogeny of Mineralization in Biological Systems* (S. Suga and H. Nakahara, eds.), pp. 283–290. Springer-Verlag, Tokyo.
23. Lee, A. P., Webb, J., Macey, D. J., van Bronswijk, W., Savarese, A. R. and der Witt, G. C. (1998). *J. Biol. Inorg. Chem.* **3**, 614–619.
24. Lowenstam, H. A. (1967). *Science* **156**, 1373–1375.
25. Goldberg, M., Septier, D., Lecolle, S., Chardin, H., Quintana, M. A., Acevedo, A. C., Gafni, G., Dillouya, D., Vermelin, L., Thonemann, B. *et al.* (1995). *Int. J. Dev. Biol.* **39**, 93–110.
26. Ten Cate, A. R. (1994). *Oral History: Development, Structure and Function.* Mosby-Year Book, Inc., St. Luis.
27. Veis, A. (1989). In *Biomineralization: Chemical and Biological Perspectives* (S. Mann, J. Webb and R. J. P. Williams, eds.), pp. 189–222. VCH, Weinheim, New York.
28. Weiner, S. and Wagner, H. D. (1998). *Ann. Rev. Mat. Sci.* **28**, 271–298.
29. Dassule, H. R. and McMahon, A. P. (1998). *Dev. Biol.* **202**, 215–227.
30. Thesleff, I. and Aberg, T. (1999). *Bone* **25**, 123–125.
31. Bernard, G. W. (1972). *J. Ultrastruct. Res.* **41**, 1–17.
32. Reith, E. J. (1967). *J. Ultrastruct. Res.* **17**, 503–526.
33. Wang, R. and Weiner, S. (1998). *J. Biomech.* **31**, 135–141.
34. Birk, D. E., Zycband, E. I., Winkelman, D. A. and Trelstad, R. L. (1989). *Proc. Natl. Acad. Sci. U.S.A.* **86**, 4549–4553.
35. Leblond, C. P. (1989). *Anat. Rec.* **224**, 123–138.
36. Weinstock, M. and Leblond, C. P. (1974). *J. Cell Biol.* **60**, 92–127.
37. George, A., Bannon, L., Sabsay, B., Dillon, J. W., Malone, J., Veis, A., Jenkins, N. A., Gilbert, D. J. and Copeland, N. G. (1996). *J. Biol. Chem.* **271**, 32 869–32 873.
38. Salih, E., Ashkar, S., Gerstenfeld, L. C. and Glimcher, M. J. (1997). *J. Biol. Chem.* **272**, 13 966–13 973.

39. Evans, J. S., Chiu, T. and Chan, S. I. (1994). *Biopolymers* **34**, 1359–1375.
40. Evans, J. S. and Chan, S. I. (1994). *Biopolymers* **34**, 507–527.
41. Corneiro, J. and Lebond, C. P. (1959). *Exp. Cell Res.* **18**, 291.
42. Septier, D., Hall, R. C., Lloyd, D., Embery, G. and Goldberg, M. (1998). *Histochem. J.* **30**, 275–284.
43. Boskey, A. L., Spevak, L., Doty, S. B. and Rosenberg, L. (1997). *Calcif. Tissue Int.* **61**, 298–305.
44. Chen, C. C. and Boskey, A. L. (1985). *Calcif. Tissue Int.* **37**, 395–400.
45. Paschalakis, P., Vynios, D. H., Tsiganos, C. P., Dalas, E., Maniatis, C. and Koutsoukos, P. G. (1993). *Biochim. Biophys. Acta* **1158**, 129–136.
46. Volpi, M., Skita, V., Chester, D. W. and Katz, E. P. (1999). In *6th International Conference on Chemistry and Biology of Mineralized Tissues*, OS-VI.
47. Hodge, A. J. and Petrushka, J. A. (1963). In *Aspects of Protein Structure* (G. N. Ramachandran, ed.), pp. 289–300. Academic Press, New York.
48. Robinson, R. (1952). *J. Bone Joint Surg.* **34A**, 389–434.
49. Weiner, S. and Price, P. (1986). *Calcif. Tissue Int.* **39**, 365–375.
50. Boyde, A. and Pawley, J. B. (1976). *Calcif. Tissue Res.*, 117–123.
51. Traub, W., Arad, T. and Weiner, S. (1989). *Proc. Natl. Acad. Sci. U.S.A.* **86**, 9822–9826.
52. Weiner, S. and Traub, W. (1986). *FEBS Lett.* **206**, 262–266.
53. Traub, W., Arad, T. and Weiner, S. (1992). *Conn. Tissue Res.* **28**, 99–111.
54. Fitton Jackson, S. (1956). *Proc. R. Soc. Lond.* B **146**, 270–280.
55. Schmidt, W. J. (1936). *Naturwissenschaften* **24**, 361.
56. Arnold, S., Plate, U., Wiesmann, H. P., Kohl, H. and Hohling, H. J. (1997). *Cell Tissue Res.* **288**, 185–190.
57. Veis, A. and Perry, A. (1967). *Biochemistry* **6**, 2409–2416.
58. Boskey, A. L. (1991). *Crit. Rev. Oral Biol. Med.* **2**, 369–387.
59. Weiner, S., Veis, A., Beniash, E., Arad, T., Dillon, J. W., Sabsay, B. and Siddiqui, F. (1999). *J. Struct. Biol.* **126**, 27–41.
60. Landis, W. J., Song, M. J., Leith A., McEwen, L. and McEwen, B. F. (1993). *J. Struct. Biol.* **110**, 39–54.
61. Wang, R. Z. and Weiner, S. (1998). *Conn. Tissue Res.* **39**, 269–279.
62. Wuthier, R. E. (1989). *Connect. Tissue Res.* **22**, 27–33; discussion 53–61.
63. Wagner, H. D. and Weiner, S. (1992). *J. Biomech.* **25**, 1311–1320.
64. Ziv, V., Wagner, H. D. and Weiner, S. (1996). *Bone* **18**, 417–428.
65. Takuma, S. (1960). *J. Dent. Res.* **39**, 873–981.
66. Beniash, E., Aizenberg, J., Addadi, L. and Weiner, S. (1997). *Proc. R. Soc. Lond.* B **264**, 461–465.
67. Ameye, L. (1999). PhD thesis, pp. 111. Université Libre de Bruxelles, Bruxelles.
68. Okazaki, K. (1975). *Am. Zool.* **15**, 567–581.
69. Wilt, F. H. (1999). *J. Struct. Biol.*, in press.
70. Gibbins, J. R., Tilney, L. G. and Porter, K. R. (1969). *J. Cell Biol.* **41**, 201–226.
71. Millonig, G. (1970). *J. Submicrosc. Cytol.* **2**, 157–165.
72. Okazaki, K. (1960). *Embryologia* **5**, 283–329.
73. Wolpert, L. and Gustafson, T. (1961). *Exp. Cell Res.* **25**, 311–325.
74. Beniash, E., Addadi, L. and Weiner, S. (1999). *J. Struct. Biol.* **125**, 50–62.
75. Benson, S., Jones, E. M., Crise, B. N. and Wilt, F. (1983). *Exp. Cell Res.* **148**, 249–253.
76. Killian, C. E. and Wilt, F. H. (1996). *J. Biol. Chem.* **271**, 9150–9159.

77. Berman, A., Hanson, J., Leiserowitz, L., Koetzle, T. F., Weiner, S. and Addadi, L. (1993). *Science* **259**, 776–779.
78. Okazaki, K. and Inoue, S. (1976). *Dev. Growth Diff.* **18**, 413–434.
79. Decker, G. L., Morrill, J. B. and Lennarz, W. J. (1987). *Development* **101**, 297–312.
80. Thompson, D. (1942). *On Growth and Form.* Cambridge University Press, Cambridge.
81. Emlet, R. B. (1982). *Biol. Bull.* **163**, 264–275.
82. Addadi, L. and Weiner, S. (1992). *Angew. Chem. Int. Ed. Engl.* **31**, 153–169.
83. Ameye, L., Compère, P., Dille, J. and Dubois, P. (1998). *Histochem. Cell Biol.* **110**, 285–294.
84. Webb, J., Macey, D. J. and Mann, S. (1989). In *Biomineralization: Chemical and Biochemical Perspectives* (S. Mann, J. Webb and R. J. P. Williams, eds.), pp. 345–387. VCH Publishers, New York.
85. Weismann, H. P., Plate, U., Zierold, K. and Hohling, H. J. (1998). *J. Dent. Res.* **77**, 1654–1657.
86. Crenshaw, M. A. (1972). *Biol. Bull.* **143**, 506–512.
87. Lowenstam, H. A. (1963). In *The Earth Sciences: Problems and Progress in Current Research* (T. W. Donnely, ed.), pp. 137–195. University of Chicago Press, Chicago.
88. Taylor, M. G., Simkiss, K., Simmons, J., Wu, L. N. and Wuthier, R. E. (1998). *Cell Mol. Life Sci.* **54**, 196–202.
89. Wu, L. N., Genge, B. R., Dunkelberger, D. G., LeGeros, R. Z., Concannon, B. and Wuthier, R. E. (1997). *J. Biol. Chem.* **272**, 4404–4411.
90. Sawada, K. (1997). *Pure and Appl. Chem.* **69**, 921–928.
91. Aizenberg, J., Lambert, G., Addadi, L. and Weiner, S. (1996). *Adv. Mater.* **8**, 222–226.
92. Xu, G. F., Yao, N., Aksay, I. A. and Groves, J. T. (1998). *J. Am. Chem. Soc.* **120**, 11 977–11 985.
93. Raz, S., Weiner, S. and Addadi, L. (1999). *Adv. Mater.* **12**, 38–42.
94. Wang, R. Z., Addadi, L. and Weiner, S. (1997). *Phil. Trans. R. Soc. Lond.* B **352**, 469–480.
95. Worms, D. and Weiner, S. (1986). *J. Exp. Zool.* **237**, 11–20.
96. Miyamoto, H., Miyashita, T., Okushima, M., Nakano, S., Morita, T. and Matsushiro, A. (1996). *Proc. Natl. Acad. Sci. U.S.A.* **93**, 9657–9660.
97. Iijima, M., Moriwaki, Y. and Kuboki, Y. (1997). *Conn. Tissue Res.* **36**, 51–61.
98. Aizenberg, J., Black, A. J. and Whitesides, G. M. (1999). *Nature* **397**, 4500–4509.
99. Boggild, O. (1930). *K. Dan. Vidensk, Selsk. Skr. Naturvidensk. Math. Afd.* **9**, 233–326.
100. Crenshaw, M. A. and Ristedt, H. (1976). In *The Mechanisms of Mineralization in the Invertebrates and Plants* (N. Watabe and K. M. Wilbur, eds.), pp. 355–367. University of South Carolina Press, Colombia.
101. Feng, Q. L. W., Su, X., Cui, F. Z. and Li, H. D. (1995). *Biomimetics* **3**, 159–169.
102. Schaffer, T. E., Ionescu-Zanetti, C., Proksch, R., Fritz, M., Walters, D. A., Almqvist, N., Zaremba, C. M., Belcher, A. M., Smith, B. L., Stucky, G. D., Morse, D. E. and Hansma, P. K. (1997). *Chem. Mat.* **9**, 1731–1740.
103. Weiner, S. and Traub, W. (1984). *Phil. Trans. R. Soc. Lond.* B **304**, 421–438.
104. Belcher, A. M., Wu, X. H., Christensen, R. J., Hansma, P. K., Stucky, G. D. and Morse, D. E. (1996). *Nature* **381**, 56–58.

105. Falini, G., Albeck, S., Weiner, S. and Addadi, L. (1996). *Science* **271**, 67–69.
106. Levi, Y., Albeck, S., Brack, A., Weiner, S. and Addadi, L. (1998). *Chem. Eur. J.* **4**, 389–395.
107. Levi-Kalisman, Y., Addadi, L. and Weiner, S. (2001) *J. Struct. Biol.* (in press).
108. Israelachvili, J. and Wennerstrom, H. (1996). *Nature* **379**, 219–225.
109. Katsura, N. (1991). In *Mechanism and Phylogeny of Mineralization in Biological Systems* (S. Suga and H. Nakahara, eds.), pp. 193–197. Springer-Verlag, Tokyo.

2

Mesoscale materials synthesis and beyond

IVANA SOTEN AND GEOFFREY A. OZIN

2.1 Opening-up materials chemistry

For the latter half of the 'Solid-State Twentieth Century' materials science has been the engine that propelled technology. As we enter the 'Materials Twenty-first Century' it is abundantly clear that the insatiable demand for new materials for emerging technologies is driving materials science. Materials chemistry will play a central role in this endeavor through the creation of new materials with structures and properties able to meet the demands required by up-and-coming technologies. Recently, solid-state chemistry has moved beyond its first 50 years of thermodynamic phases and microscale structures to a new era of self-assembly materials chemistry and larger length scales, hierarchical structures and systems of integrated components. This phase in the development of solid-state chemistry focuses on metastable phases and on mesoscale and macroscale structures with accessible surfaces and well-defined interfaces between components that determine the overall function and utility of the material. The approach is interdisciplinary – it combines synthesis, solid-state architecture and functional hierarchy to create an innovative strategy for materials chemistry in the new millennium. The attractive feature of this direction is the ability to assemble complex structures rationally from inorganic, polymeric and organic modular components and to integrate them into self-assembling constructions for a range of perceived applications. By creating a series of purposeful materials design strategies it is believed that truly revolutionary advances in materials science and technology can result from this approach.

Particularly noteworthy is research in recent years in the burgeoning field of 'self-assembling supramolecular materials'. This is spectacular and imaginative work – it has shown the way to a world of materials chemistry that had not previously existed. The scientific significance of this outgrowth of solid-state chemistry is to bring control over the organization of inorganic and organic building-units to create novel materials with unprecedented structures that span ångström to millimeter length scales. It introduces the paradigm of 'panoscopic materials' – the synthesis of materials over 'all' length scales. This is definitely a breakthrough, a discontinuity in the field of materials research – it introduces complexity into inorganic materials chemistry and the possibility of creating integrated chemical systems purely through synthesis, which, until this work, was reserved for Nature's inspiring biominerals. In essence, it is becoming apparent that supramolecular templating over micro-, meso- and macroscopic dimensions can provide synthetic pathways to materials whose architecture at every level of construction and overall form is under control and determines their properties, function and utility. What is so impressive about this work is that it is deeply unusual. It combines synthesis, structure, topology and visual perception of a type that has not been seen before in inorganic materials chemistry – it is exceptionally original, extremely interesting and potentially very important technologically.

2.2 Materials filled with holes

Material chemists have always been interested in making porous materials because the air space plays an active role in determining the material properties. For example, high thermal and acoustic insulation properties and low dielectric constant behavior are observed in highly porous organic, polymeric and inorganic structures. Silicon, the archetype non-luminescent semiconductor, when made porous displays bright visible photo-luminescence and electroluminescence that can be exploited in light-emitting diodes and displays. Crystalline porosity imparts special properties to a material, which may not be found with a random distribution of poly-dispersed void spaces, and is therefore expected to expand the potential utility of the material. In many fields of application that exploit the special properties of solids filled with holes, such as catalysis, membrane separation, size exclusion chromatography, enzyme immobilization and fuel cell electrodes, periodicity in the arrangement of air spaces is often fundamental to the intentional design of structure–property–function relations. Catalysts based on porous structures usually work best with materials in which there exists a uniform distribution of structurally precise active sites organized within a periodic array of crystallographically defined void spaces. Crystalline porosity enables the electronic band structure of a material to be defined by the unit cell symmetry and elemental content of the structure. This would affect the description of electrical transport and optical properties for crystalline porous solids of interest in electronics and photonics. Encapsulation of single size and shape guests, such as polymer, ceramic, magnetic, semiconducting or metallic nanowires or nanoclusters within periodic and strictly uniform diameter pores of a host material, allows one to design specific properties and a particular function for the hybrid structure. An example is a regular arrangement of monodispersed semiconductor clusters encapsulated within a periodic porous host lattice to create a semiconductor cluster superlattice whose properties depend upon the extent of cluster spatial confinement and the strength of coupling between the clusters themselves and the clusters with the host. This molecular-scale integration and control over the interaction or interconnection of components in a host matrix is advantageous if one desires a nanocomposite material with well-defined and tailorable properties amenable to quantification. The existence of a regular distribution of guests in the host that interact cooperatively can lead to new materials with novel properties that are greater than the sum of the component parts.

2.2.1 Crystalline porosity by design

The discovery of synthetic crystalline porosity came in 1930 with the first successful preparation of microporous zeolites, aluminosilicate open framework materials with structures analogous to those formed geologically. This breakthrough opened up great opportunities in many areas of materials science, with catalysis being the most immediate and prominent one. Zeolites found widespread utility as catalytic and sorption materials because of their large surface area and high acidity. During the 1950s, industrial scientists realized that zeolites could not only separate hydrocarbons from crude oil based on their size and shape, but they were able selectively to catalyze important chemical transformations. Among many other applications, they were used in detergency as ion-exchangers to replace toxic phosphates, as storage matrices for nuclear waste disposal, in cosmetic and hygiene products, and as molecular-size reaction vessels for host–guest chemistry. In the 1960s, organic molecule templating was discovered as a way to control the structures, cavity and channel sizes of zeolite frameworks. This seminal discovery led to an explosion of activity in the field of zeolite science. Since that time there has been an intense desire to expand the range of structure types, void sizes and elemental compositions across a periodic table of crystalline porous materials to explore a range of fundamental scientific issues and address diverse and increasing demands from industry. To name a couple of areas of interest, separating and chemically transforming large molecules would benefit from materials with crystalline mesoporosity. As well, organizing nanoclusters of metals and semiconductors inside materials with regular arrangements of mesopores would provide new nanocomposites with electrical, optical and magnetic properties distinct from those of the parent materials.

Before the discovery in 1988 of a 12 Å channel diameter VPI-5 aluminophosphate molecular sieve, the consensus was that a cavity or channel diameter of 7–8 Å in a crystalline porous solid was somehow a 'natural limit' [1]. In 1991, the synthesis of Cloverite, an extra-large-pore gallophosphate molecular sieve, showed that this was not true and took the scale bar even higher in an appealing cubic symmetry open framework structure with clover-shaped window entrances into a 30 Å diameter cavity. By increasing crystalline pore sizes to the mesoscale, 20–100 Å, materials properties enter the 'fuzzy' region between the microscale (<20 Å) and the macroscale (>500 Å) where electrical, magnetic, optical, mechanical, photochemical and electrochemical properties become dependent on physical dimension. This is where the story of 'materials filled with periodic arrangements of holes' becomes especially interesting.

2.3 Escape from the microporosity prison

In the early 1990s, the synthesis of a siliceous mesoporous material called FSM-16 was described. The method involved the intercalation of long-chain alkyltrimethylammonium cations into the gallery space of a layered silicate, known as kanemite [2]. Around the same time, researchers at Mobil Corporation announced the surfactant-template-based synthesis of periodic mesoporous silica structures denoted MCM-41 [3]. The latter self-assembly process is reminiscent of that utilized for making zeolites in that an organic molecule directs the size and shape of void spaces in the resulting inorganic material, the major difference being that a single organic molecule templates zeolites whereas a supramolecular assembly of organics templates mesoporous materials. The discovery that self-assembling micellar systems are able to function as structure-directing agents for the synthesis of mesostructured materials has attracted the attention of the materials chemistry community for several years. It was not only the new mesoscale cavity, channel and layer spaces of the materials, which could be synthesized for the first time that was so appealing, but the introduction of a new way of thinking about the synthesis of materials over different length scales was quite inspiring. In essence, molecular-scale control over the synthesis of solid-state materials at different levels of construction was achieved from the 'atom up' rather than from the 'bulk down' [4]. The complex structures and unusual growth and form of materials found in biology was emerging for the first time in materials chemistry through intentional synthetic design [5]. This way of chemically constructing inorganic materials with at least one dimension in the 20–100 Å 'mesoscopic' regime provided a new opportunity for synthetic chemists to tailor the structure-morphology and property–function relations of solids for a range of perceived applications [6–8].

This was only the 'tip of the iceberg'. Soon, it became obvious that this approach could be extended to the synthesis of materials over 'all' length scales and to the realization through chemistry of purely synthetic integrated systems [9]. Once the paradigm was established, the new 'synthesis tool box' could be expanded. This emerged with the discovery of a wider range of templating structures as well as the combined use of microfabrication techniques with self-assembly and templating. New templates included polymers and colloidal crystals, while microfabrication techniques included patterning and molding. Methods for synthesizing hierarchical materials with structural units integrated over multiple length scales encompassed self-assembly in conjunction with templating and/or microfabrication (Fig. 2.1). Synthetic integrated chemical, physical and biological systems had at last become a reality.

Chemists are now able intentionally to plan the synthesis of materials with a particular structure and a predetermined function by exerting unprecedented control over the assembly of individual construction units to create structures that span microscopic to macroscopic dimensions. Therefore, although our presentation will be on new directions in the field of self-assembling materials, a 'panoscopic' view of materials chemistry over 'all' dimensions will be the common thread that ties the subject matter together. The 'panoscopic' integration of components over disparate length scales to yield materials with properties and functions that do not exist in the individual components is an important and timely concept that has only just become possible because of recent developments in the field of self-assembly. Panoscopic materials may find utility in next-generation solid-state batteries and fuel cells, chemical sensors and photovoltaics, catalysts and membranes, bone implants and chemical delivery systems. It is only in the past year or so that an abundance of articles has emerged, which convincingly demonstrated that hierarchy, familiar in biology, has established itself in materials chemistry.

2.3.1 Formation mechanisms

The general consensus is that mesoporous silica materials synthesized using surfactant templates can form by two distinct yet related reaction pathways. The first involves the hydrolytic poly-condensation of a silicatropic mesophase while the second works through co-assembly of silicate and surfactant micellar building blocks. For ionic surfactants, cooperative-assembly with silicate precursors is the result of charge density and geometry matching and multi-dentate

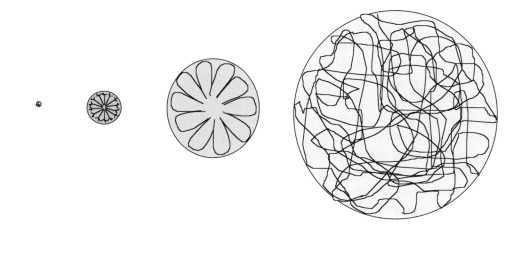

0.3–2 nm   **2–10 nm**   **10–50nm**   **50–1000 nm**

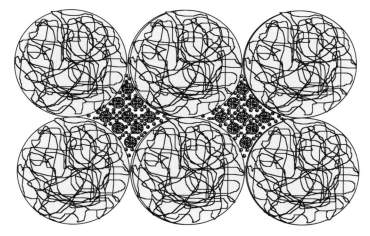

Figure 2.1. Schematic representation of 'panoscopic' materials formation over multiple length-scales *via* self-assembly. Reproduced with permission from [9].

binding [10] whereas nonionic surfactants interact with the silicates through hydrogen bonding [11]. Surfactant-based assembly of a range of metal oxide precursors can be assisted by the formation of metal–ligand bonds between, say, a metal alkoxide and an amphiphilic amine [12]. Whatever the building blocks and formation mechanism, it is generally agreed that surfactants form effective structure-directing agents for organizing diverse kinds of inorganics into a variety of mesostructured materials. However, the rational design of periodic mesoporous materials, with extensions to macroscopic analogues, necessitated a greater understanding of the supramolecular templating mechanism. To this end, an *in situ* X-ray diffraction study of the early stages in the formation of hexagonal mesoporous silica MCM-41 was recently described [11, 12]. At surfactant concentrations less than those required for the formation of lyotropic liquid crystal mesophases, the proposed assembly mechanism for mesoporous silica invoked strong cooperative interactions between silicate precursors and the organic template. Depending upon the choice of reaction conditions, such as pH and silicate precursors, different pathways to hexagonal phases were postulated, some of which involved layered intermediates.

In a separate study, the mode of formation of hexagonal mesoporous silica materials MCM-41 and FSM-16 has been followed by utilizing time-resolved energy dispersive *in situ* X-ray diffraction [13]. In addition to the observation of a mesostructured layered phase resulting from CTA^+ intercalation of kanemite, a broad diffraction peak was observed and assigned to a partially ordered silica–surfactant mesophase. This was explained in terms of kanemite dissolution products interacting with surfactant micelles, which contributed to the formation of the hexagonal mesoporous silica product phase after pH adjustment that favored the condensation–polymerization reaction. It was concluded that the silicate–surfactant precursor to FSM-16 forms from a reaction medium containing a number of intercalated silicate phases, while the hexagonal mesophase precursor to MCM-41 forms directly without any other phases detected. In addition, the silicate–surfactant mesophases that formed were highly dependent on the reaction medium, the silicate precursor being the most important. In similar studies the direct formation of the MCM-41 hexagonal phase was observed without any intermediates and a mechanism that depended on the reaction conditions. Rapid formation of the hexagonal mesophase was attributed to co-operative assembly of silicates and surfactants.

2.3.2 Silica mesostructures

As mentioned previously, mesoporous inorganic materials can be formed by one of two reaction pathways, liquid-crystal templating and cooperative self-assembly. Pre-assembled liquid crystals were first proposed as the species responsible for the surfactant-templated synthesis of MCM-41. Eventually it was realized that at the surfactant concentrations employed for the synthesis, often below the critical micelle concentration (cmc), it was unlikely that the hexagonal lyotropic mesophase was responsible for the assembly process that led to the formation of hexagonal mesoporous silica. Subsequent research showed that cooperative self-assembly of silicates with surfactant micelles was taking place at the low concentrations of surfactant used in these syntheses [14]. Three formation pathways to mesoporous silica have been envisaged – a layer puckering model [15], a silicate rod assembly model [10] and a cooperative charge density matching model [16]. It was recognized that the formation mechanism depends on a subtle contribution of many factors and each system has to be individually examined, to establish which effects are dominant [14, 17].

A generalized model for cooperative templating of inorganic mesostructured materials has been put forth based on the specific type of electrostatic interactions between an inorganic precursor and surfactant head group [9]. This model embraces three kinds of synergistic interaction between inorganics and surfactants, namely electrostatic, hydrogen and covalent bonding. During the ground-breaking years soon after the discovery of MCM-41, various strategies for synthesizing mesoporous silica materials were developed, which provided an enhanced understanding of surfactant-based templating mechanisms. This work led to greater control over the mesostructure and the quality of the materials so obtained. It was established that the molar ratio of surfactant to silicate in the S^+I^- synthetic route was the key factor that determined which particular mesophase formed [18]. Small ratios resulted in the hexagonal phase while larger ratios produced the cubic and lamellar phases. In general, synthetic routes to mesoporous silica materials involved acid- or base-catalyzed hydrolytic poly-condensation reactions. Acid-derived silica mesostructures were found to have thicker walls and frameworks, which allowed the extraction of surfactant template by an ethanol reflux. Base-catalyzed reactions produced frameworks, which allowed the surfactant to be removed by an acid–ethanol reflux.

Hydrogen bonding interactions between neutral or nonionic surfactants and silicate precursors were reported also to yield mesoporous structures [17–19]. The materials lacked long-range ordering of the mesopores and had a high degree of textural porosity, likely to be caused by the lack of electrostatic interaction that would have controlled well-ordered packing of rod shaped micelles. The pores exhibited a 'wormlike' architecture and the silicate framework was neutral, which allowed for surfactant removal under mild conditions. Recent reports have described the catalytic properties of silica materials with a wormhole mesostructure and textural mesoporosity, denoted HMS-HTx [20]. The wormhole channel motif was considered to be an important structural feature for favorable catalytic activity because channel branching within the framework facilitates access to reactive framework sites. Textural mesoporosity, arising from the spaces between particles (<200 nm), was found to be a benefit for the efficient mass transport of reactants within the material. In fact, aluminum-substituted Al-HMS-HTx containing wormhole and textural mesoporosity proved to be a more effective catalyst than the same wormhole material without textural mesoporosity, specifically towards the alkylation of 2,4-ditert-butylphenol with cinnamyl alcohol.

2.3.3 *Metal oxide and metal sulfide mesostructures*

Surfactant-based synthesis of mesoporous metal oxides and metal sulfides emerged about four years after the initial report of MCM-41 [21–36]. High surface area and thermally robust mesoporous metal oxides and sulfides represent a new class of materials with diverse opportunities for the development of improved fuel and solar cells, batteries, membranes, chemical delivery vehicles, heavy metal sponges, sensors, magnetic devices and new catalysts. All of these applications could benefit from tailorable Brønsted and Lewis acidity and basicity, flexible oxidation states, and tunable electronic, optical and magnetic properties.

First reported syntheses of surfactant-templated metal oxides led to the formation of mainly layered Sb, Fe, Zn, Pb, W and Mo oxide materials. They were unstable and collapsed upon removal of the template. From these early attempts to prepare analogues of mesoporous silica it was realized that it was not a straightforward matter to extend surfactant templating to the synthesis of mesoporous metal oxides and metal sulfides. One of the problems was the difficulty in finding suitable precursors that would undergo a sufficiently slow and controlled hydrolysis, nucleation and growth process in the microphase separated domains of a surfactant micellar or liquid crystalline template without disrupting the structure of the templating mesophase. A solution to this problem involved the use of a sol-gel mesophase composed of a surfactant with a coordinating head group, like acetylacetonate, amine, phosphate or sulfate, that chelated to the transition metal alkoxide precursor [21]. For example, phosphate surfactants were used to obtain hexagonal mesoporous titania displaying a narrow pore-size distribution and a BET surface area of 200 m^2/g. Phosphate surfactants were able to template mesoporous zirconia in a similar way [26]. In both cases phosphate groups were retained in the structure after calcination and are believed to be responsible for maintaining the structural integrity of the material after template removal. The ligand-assisted templating strategy using amine surfactants also worked well for the synthesis of mesoporous niobium and tantalum oxides [22, 25].

Catalysts and catalytic supports are often formulated from aluminas owing to their low cost, and their thermal, chemical and mechanical stability [27]. To this end, nonionic surfactant templating with polyethylene oxide surfactants was used to synthesize mesoporous alumina [28]. Partially ordered mesoporous materials with 'wormhole' channels were obtained – this seems to be a characteristic of templating with nonionic surfactants. Cationic surfactants have recently been found to template mesoporous aluminum oxide with pore sizes that could be fine-tuned by varying the molar ratio of the reactants [29]. The material obtained was thermally stable to 900 °C and exhibited a narrow pore size distribution.

Both aluminum oxide and zirconium oxide are catalytically interesting materials. Pure zirconium oxide is a weak acid catalyst and to increase its acid strength and thermal stability it is usually modified with anions such as phosphates. In the context of mesoporous zirconia prepared from zirconium sulfate using the $S^+X^-I^+$ synthesis route it was found that by ion exchanging sulfate counter-anions in the product with phosphates, thermally stable microporous zirconium oxo-phosphates could be obtained [30–32]. Thermally stable mesoporous zirconium phosphate, zirconium oxo-phosphate and sulfate were synthesized in a similar way [33, 34]. The often-encountered thermal instability of transition metal oxide mesoporous materials was circumvented in these studies by delayed crystallization caused by the presence of phosphate or sulfate anions.

There have been only a few reports of mesostructured metal sulfides. Mesoporous cadmium sulfide was prepared from polyethylene oxide surfactants and cadmium salts exposed to hydrogen sulfide [35]. A study of the effects of the counter-anion on the formation of CdS mesostructures led to the conclusion that the use of cadmium nitrate and perchlorate salts improved the degree of order of the mesostructure over the chloride, sulfate and acetate salts. This effect was attributed to the stronger acidity of conjugate acid by-products of the reaction in the case of nitrates that leads to the dissolution of high-energy defects and enhances structural order.

Layered tin sulfide mesostructures were synthesized using a cationic surfactant as template, and tin chloride and sodium sulfide as sources of tin and sulfide [36]. The structure was composed of $Sn_2S_6^{4-}$ dimers charge-balanced by dodecylammonium cations. A mesostructured tin sulfide mesh phase was synthesized by reacting $SnCl_4$, $(NH_4)_2S$ and hexadecylamine (HDA) under aqueous basic conditions at 150 °C [37]. The structure was found to be composed of mesoporous $(SnS_{2.07})^{-0.14}$ anionic layers sandwiched between well-ordered partially protonated HDA charge-balancing bilayers. It was discovered that the material displayed thermotropic liquid-crystal behavior displaying first an endothermic 'crystalline to semi-liquid crystalline' transition at 45°C in which the HDA bilayers became disordered and the mesoporous $(SnS_{2.07})^{-0.14}$ layers remain registered. This was followed at 85°C by an endothermic 'semi-liquid crystalline to liquid-crystalline' transition at which point both the HDA layers and mesoporous $(SnS_{2.07})^{-0.14}$ layers became disordered. In the fully liquid-crystalline state, electrical conductivity measurements showed a 10^3 increase in conductivity compared with the crystalline form, from a value of $10^{-8} \ \Omega^{-1}cm^{-1}$ to $10^{-5} \ \Omega^{-1}cm^{-1}$ at which point the material is best described as a semiconducting metallomesogen. It is worth noting that inorganic thermotropic liquid crystals of this type are rare. They are anticipated to combine the fluid properties of liquid crystals with, for instance, the semiconducting and optical properties of inorganic materials, making them potentially attractive for a range of display, optical and chemical sensing technologies.

2.3.4 Non-aqueous self-assembly of mesostructured materials

Non-aqueous synthetic methods have recently been used to assemble mesoporous transition metal oxides and sulfides. This approach may afford greater control over the condensation–polymerization chemistry of precursor species and lead to enhanced surface area materials and well ordered structures [38, 39]. For the first time, a rational synthesis of mesostructured metal germanium sulfides from the co-assembly of adamantanoid $[Ge_4S_{10}]^{4-}$ cluster precursors was reported [38]. Formamide was used as a solvent to co-assemble surfactant and adamantanoid clusters, while $M^{2+/1+}$ transition metal ions were used to link the clusters (see Fig. 2.2). This produced exceptionally well-ordered mesostructured metal germanium sulfide materials, which could find application in detoxification of heavy metals, sensing of sulfurous vapors and the formation of semiconductor quantum 'anti-dot' devices.

In a separate study, a generalized non-aqueous synthesis to the preparation of transition metal oxides via glycometallate precursors was described [39]. Again, as in the germanium sulfide work, electrostatic interactions and cooperative self-assembly processes were responsible for the formation of well-ordered mesostructured titania, niobia and zirconia. Following treatment with a disilane-coupling reagent, thermally stable mesoporous metallosilicates that displayed a high surface area and narrow pore size distribution were obtained. This synthetic strategy enables a variety of glycometallates to be incorporated into the precursor mesophase, thereby creating electroactive binary or ternary

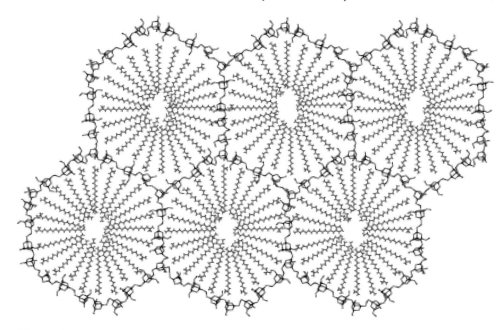

Figure 2.2. Proposed architecture for mesostructured metal germanium sulphides. Adamantanoid clusters $[Ge_4S_{10}]^{4-}$ link upon addition of a metal salt to form a supramolecular mesostructured material where surfactant molecules charge balance the negative charge on the framework. Reproduced with permission from [38].

mesostructured composites, for possible use in batteries, fuel cells, photovoltaics and chemical sensors.

2.4 Morphogenesis: mesostructure form and function

So far, we have discussed how self-organized organic amphiphiles have been employed to expand the pore size of inorganic materials from the microscopic to the mesoscopic length scale and how this provided control over the chemical structure and elemental composition of the framework. It was only 'natural' that 'morphogenesis', the origin and control of shape, became an issue for this class of synthetic materials formed by the organization of inorganics by organic templates over length scales commonly found for Nature's inorganic–organic composites. Nature, through the process of biomineralization, uses organics to nucleate and grow inorganics in the form of hierarchical composite materials that exhibit a diversity of shapes and functions. This has been a dream of materials chemistry and the inspiration to aspire to new heights came from liquid-crystal templating of inorganic mesostructures.

(a)   (b)

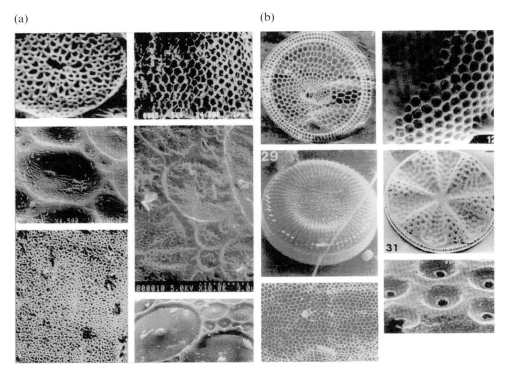

Figure 2.3. Scanning electron microscopy images of (a) representative mesolamellar aluminophosphate synthetic morphologies and (b) radiolaria and diatom microskeleton morphologies. Reproduced with permission from [40].

This phase in the development of the field of self-assembling supramolecular materials can be traced to the synthesis of lamellar aluminophosphate mesostructures with morphologies that displayed natural form (Fig. 2.3) [40]. Millimeter-sized spheroids and hollow shells of mesolamellar aluminophosphates were formed with surface patterns composed of micrometer-dimension pores, bowls and star dodecahedra resembling radiolaria and diatom microskeleton. The morphologies were found to originate in the liquid crystal templating behavior of the organic decylammonium-dihydrogenphosphate coupled with the multifunctional role of the solvent tetraethyleneglycol solvent. The latter served as a co-solvent with water, a poly-dentate ligand for Al^{III}, and a co-surfactant for inducing bilayer curvature.

In addition to mesostructured aluminophosphates, mesoporous silica was also discovered to form natural shapes [41–44]. It was found that quiescent, acidic, dilute and homogeneous solution phase conditions were the key ingredients for morphogenesis of curved shapes in the mesoporous silica system. Acidity, concentration and temperature of a synthesis mixture caused

(a)

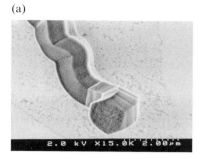

(b)

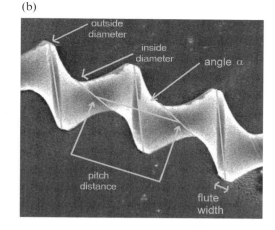

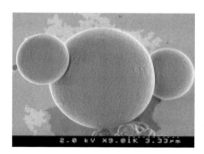

Figure 2.4. Scanning electron microscopy images of mesoporous silica synthetic morphologies: (a) low- and high-curvature morphologies, (b) hollow helicoidal morphologies. Reproduced with permission from [41] and [47].

morphologies to change from low-curvature fibers to high-curvature discoids, gyroids, spheres and spirals (Fig. 2.4). Liquid-crystalline defects were shown to be important in determining which particular mesoporous silica shape emerged in a synthesis. Colloidal electrical double-layer forces, solvent dielectric constant and ionic strength were also shown to play a key role in the creation of a particular form of mesoporous silica.

The possibility of synthesizing mesoporous materials with interesting shapes suggested avenues of exploration where the form of the material determined its function and potential utility. In addition to above-mentioned silica shapes whose growth and form were driven by topological defects and colloidal interactions [44, 45], intriguing radial patterns were also seen on the surface of observed morphologies [46]. It was found that subtle elastic and compressive forces had to be taken into account to explain these patterns [46]. A theoretical model based on the theory of elasticity was able to simulate the observed radial patterns. It involved minimizing the free energy density of a discoid-shaped hexagonal silicate liquid crystal including the surface tension term. To explain the origin of radial patterns observed on sunken discoid shapes, longitudinal and radial differential contraction of micelle rods were taken into account. Contraction-induced stresses were believed to originate from the polymerization of the silicates where the more contracted parts were located in the center of the discoid where the polymerization starts. Hence, sunken radially patterned discoid shapes usually emerge.

Differential contraction was again found to be responsible for the morphogenesis of hollow helicoid shapes of hexagonal mesoporous silica [47]. Under the synthesis conditions used in this study, the polymerization of silica in a 'patch' of hexagonal mesoporous film that formed at the air–water interface resulted in radial and longitudinal differential contraction effects and the formation of helical structures with the form of an Archimedian screw. TEM images revealed that the helicoids were hollow with shells of 1 μm thickness composed of hexagonal mesoporous silica. The mesoscale channels were found to follow the helical generating curve of the screw.

Tapping-mode AFM and TEM imaging techniques were used to probe the early stages of hexagonal mesoporous silica shape formation [48]. In addition, the growth kinetics, responsible for the formation of various hexagonal mesoporous silica shapes, were analyzed. The nucleation and growth of fiber and gyroid shapes was found to adhere well to a three-parameter Avrami equation that took into account an incubation period during which viable growth nuclei were assembled. The Avrami indices obtained from the analysis suggested that one- and two-dimensional growth processes were responsible for the emergence of fiber and discoid shapes, respectively. The proposed growth model based on AFM/TEM imaging and kinetic results involves formation of a silicate liquid-crystal seed, approximately 50–70 nm in size. Depending on the pH, ionic and dielectric strength of the solution phase environment experienced by the seed, particular shapes are formed as a consequence of selective anchoring of silicate micelles to the seed. This growth process is controlled by a compromise between electrical double-layer and van der Waals repulsive and attractive forces between reacting colloidal particles.

Mesoporous metal oxides in the form of vesicles are of great interest as catalysts and sorbents because textural porosity enables the efficient transport of guest species to the pore structure. A method has been described that produces vesicle shaped mesoporous silicas with very thin shells and a high degree of silica cross-linking [49]. An electrically neutral hydrogen-bonding formation pathway based on a Gemini surfactant assembly was used to provide vesicle-based hierarchical structures. PXRD and TEM results suggested a growth process analogous to a L_α–L_3 phase transformation of the surfactant via an undulated lamellar intermediate. All of the structures obtained were found to have the form of vesicles, ruptured vesicles or bowls with sizes ranging from 20 nm to 1400 nm. Shell thickness varied in the range 3–70 nm. A three-dimensional highly cross-linked silica pore network was thought to be responsible for the unprecedented thermal and hydrothermal stability of this system.

A completely different approach involves an aerosol-based process for synthesizing spherical-shaped nanoparticles with hexagonal, cubic and lamellar silica mesostructures [50]. This method is founded upon evaporation-induced self-assembly (EISA) of a silicate–surfactant solution confined to a spherical aerosol droplet. The heating and collecting of aerosol particles is fast and suitable for the formation of a range of nanoparticle composites including organic/silica, metal/silica, enzyme/silica and polymer/silica. Crucial for the formation of highly ordered mesostructures is maintenance of the liquid or liquid-crystalline state throughout the condensation–polymerization process that starts at the surface of the droplet and proceeds inwards in a radial fashion. The same method affords a convenient route to the formation of continuous mesoporous silica films through dip coating.

2.5 Only natural

Nature is known to produce an exquisite variety of biological minerals by controlling four main processes, pre-organization of organic components, interfacial molecular-scale recognition, vectorial regulation of crystal growth and cellular processing. These processes enable control over structure–property–function relations, something that material chemists would like to emulate. As a consequence of this desire, the field of biomimicry emerged and spawned an impressive array of synthetic, topologically complex, three-dimensional structures that were perceived as being useful as biomedical implants, light-weight ceramics or semiconductor devices [51]. In all of these studies, organic molecules of varying degrees of complexity were used to template, modify and manipulate the structure and physical properties of materials at different length scales in ways that were analogous to those employed by living organisms.

Calcium carbonate is an important skeletal mineral found in many biological organisms. Mimicry of biological calcification has been achieved using oil–water–surfactant microemulsions as a templating medium to synthesize macroporous aragonite, one of the calcium carbonate polymorphs [52]. By increasing the oil content of the microemulsion the pore size was found to increase from mesoscopic (< 50 nm) to macroscopic (average size ~ 130 nm) dimensions. The proposed mechanism for the formation of these macroporous structures involved rapid mineralization of aragonite within a water layer surrounding phase-separated oil droplets. Complete removal of the oil and surfactant by chloroform and subsequent heat treatment resulted in a mineralized replica of the foam-like templating medium. The structure produced depended sensitively upon the rate of inorganic nucleation and foam assembly and the success of creating a porous calcium carbonate structure requires compatibility between the rate of calcite nucleation and the rate of the oil-droplet removal.

Synthetic analogues of bone are being actively pursued for biomedical applications in the field of bone replacement, augmentation and repair. Numerous criteria have to be satisfied for a biomaterial to be acceptable as a bone implant, including the ability to integrate into bone and not cause any side effects. Recently a biomimetic self-assembly synthetic strategy was described for a new type of bone analogue material [53]. The material was composed of a porous hydroxyapatite–octacalcium phosphate–calcium dodecylphosphate (CaDDP) composite film grown on a titania–titanium substrate. The inorganic phase was composed of a macroporous and oriented OHAp-OCP mineral. This phase nucleated and grew, under physiological conditions, from TiOH Brønsted acid and base anchoring sites on the surface of a sputter-deposited titania film. The mesolamellar CaDDP phase was co-assembled with the OHAp phase in a 'mesoepitaxial' fashion. It was envisioned that synthetic hierarchical materials of this genre might be able to integrate into bone and simultaneously deliver bioeffectors and drugs, from the hydrophobic region of the CaDDP mesolamellar phase, to stimulate bone growth and combat disease. Notably, the macroporous structure of the material would facilitate the influx of cells and blood vessels to the site of the bone implant.

Using a surfactant-based self-assembly synthetic method, semiconductor quantum dot and anti-dot materials have been realized [54–56]. Templating and connecting quantum dots into a periodic array or 'inverting' the structure by templating cavities in the structure is expected to produce materials with novel electronic properties and new uses. The main feature of these experiments is that surfactants are able to control the size and shape of nanocrystals that form and the process by which they are connected into highly ordered microstructures analogous to colloidal crystals. Monodispersed organically coated CdSe nanocrystallites have been formed in this way that are able to assemble into an ordered assembly on a substrate or into a micrometer-size crystal, leading to a superlattice of semiconductor quantum dots [55, 57]. In one study, semiconductor quantum dots were joined with chemical linkers forming dimers of CdSe exhibiting predefined inter-particle spacing [58].

Conversely, a periodic array of anti-dots could modify the electronic properties of a semiconductor. In addition, periodic cavities could be used to selectively adsorb or transport molecules or to perform chemical modification of guest molecules by utilizing electronic or photonic properties of semiconductors. Several studies showed that it was possible irreversibly to bind templating molecules within the semiconductor lattice and through inorganic–organic interactions to modify electronic and photonic properties [59a, b].

Recently, surfactant molecules were used to assemble inorganic nanoparticles into well-defined superstructures [60]. Three different barium chromate nanostructures were obtained – linear chains, rectangular superlattices and long filaments, resulting from the interdigitation of surfactant molecules attached to a specific nanoparticle crystal face. To expand, when barium bis(2-ethylhexyl)sulphosuccinate $(Ba(AOT)_2)$ reverse micelles were mixed with a sodium chromate-containing NaAOT microemulsion, transfer of water molecules from the chromate-containing microemulsion to the 'dry' $Ba(AOT)_2$ reverse micelles provides a strong driving force for the exchange of ions. This, in turn, increases the intramicellar supersaturation, which results in nucleation and growth of surfactant-encapsulated $BaCrO_4$ crystals. Nanocrystal self-assembly into ordered superstructures was observed only at molar equivalence between barium and chromate. There is no net charge at the crystal surfaces at these conditions, so the particles develop regular faces and a prismatic morphology, coated with surfactant molecules. Interdigitation of the alkane chains of surfactant monolayers is induced as crystal faces develop in shape and size, resulting in one- and two-dimensional aggregation of the nanocrystals. Superlattice and chain architectures formed from rectangular and prismatic $BaCrO_4$ nanoparticles are shown in Fig. 2.5.

(a)   (b)

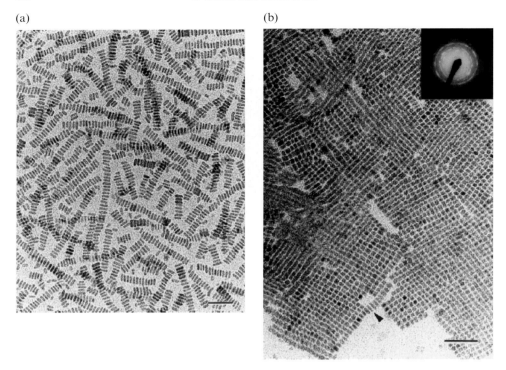

Figure 2.5. Transmission electron micrographs of prismatic $BaCrO_4$ nanoparticles ordered into chains (a) and rectangular superlattice of $BaCrO_4$ nanoparticles (b). Reproduced with permission from [60].

2.6 Expanding the synthesis 'tool box': beyond surfactant self-assembly

2.6.1 Block co-polymers templates

In addition to surfactants, which represent the most studied self-assembling system, block co-polymers were found to exhibit similar self-association properties and several reports of block co-polymer templating of mesoporous materials have since emerged [61–63]. A combination of electrostatic and hydrogen-bonding interactions was presumed to be responsible for mesoscale silica-triblock co-polymer assembly. Silica interacts with the more hydrophilic portion of the polymer tri-block and thus promotes cooperative self-assembly [61, 62]. In principle, this assembly process is applicable to any system in which network-forming species are selectively partitioned between components. Therefore, by slight modification of the procedure, a range of thermally stable, large pore mesoporous metal oxides and mixed-metal oxides were obtained [63] (Fig. 2.6). Tri-block copolymers have been co-assembled with inorganic halides

(a)   (b)

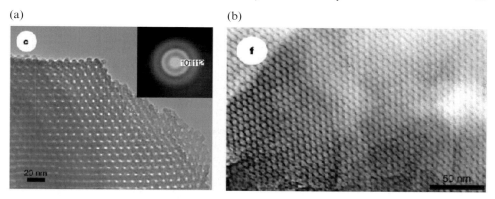

Figure 2.6. Transmission electron micrographs of mesoporous metal oxides (a) ZrO_2 and (b) $SiAlO_{3.5}$. Reproduced with permission from [63].

in a non-aqueous solvent in order to slow the rate of hydrolysis/condensation and inhibit crystallization of bulk metal oxide. Water slowly diffused from the atmosphere into the mesophase and promoted ordering of the composite mesostructure directed by microphase separation of the block co-polymer and polymerization of the inorganic.

Block co-polymers afforded inorganic frameworks with pore sizes up to 300 Å, which is considerably larger than pores obtainable with regular surfactants. Consequently, thicker walls were formed that could accommodate domains of nanocrystals resulting in improved thermal and hydrolytic stability of the materials. Surface areas were generally smaller then in the analogous surfactant-templated mesostructures owing to the increased wall thickness. Pore size and wall thickness were adjustable by varying the temperature and duration of the reactions or by addition of organic swelling agents [61, 62]. Processability of nonionic templated mesostructures proved possible with tri-block co-polymers, for example fibers could be easily drawn from reaction mixtures [62] or films deposited by dip or spin coating [62, 63].

2.6.2 Colloidal crystal templating

Application-driven research to easy, cheap and fast methods of producing highly ordered three-dimensional porous materials with adjustable pore sizes and compositions led to the discovery of latex or silica sphere colloidal crystal templating. Inorganic oxides, sulfides, metals and polymers have been templated by this method simply by varying the size of the latex or silica spheres that constitute the colloidal crystal [64–71]. By removing the templating colloidal crystal by solvent extraction or fluoride-based etching it has proven

Figure 2.7. SEM image of calcined periodic macroporous titania crystal. Reproduced
with permission from [66].

possible to obtain periodic macroporous materials, which are proving to be of
great interest for applications ranging from photonics to sensing, and from
quantum electronics to optical communications.

The method relies on the properties of monodispersed latex/silica spheres to
assemble, through colloidal interactions, into a well-ordered, face-centered-cubic
colloidal crystal upon centrifugation, sedimentation, electrophoresis, oscillatory
shear or pressing in the form of pellets. Following pre-assembly of the colloidal
crystal template, the precursor is infiltrated into the empty octahedral and tetrahedral interstitial sites that exist between the spheres. After conversion of the precursor to the desired material inside the voids, the template is removed leaving

behind a replica of the colloidal crystal template, namely a periodic macroporous crystal (inverse opal) of the desired material.

Using this strategy, periodic macroporous zirconia, titania and alumina materials were synthesized using a latex colloidal crystal as the template and a metal alkoxide as the precursor for the inorganic oxide replica material [66]. The procedure was expanded to include a range of inorganic oxides, phosphates and inorganic–organic hybrid materials [67]. Depending on the template removal procedure, periodic macroporous materials with either crystalline or amorphous walls could be obtained. SEM images of the materials revealed areas exhibiting close packing of macropores extending over tens to hundreds of micrometers (Fig. 2.7). Some control over the wall thickness and pore sizes was reported.

Syntheses of colloidal crystal-templated periodic macroporous metal films were also described [69]. Single-crystal colloidal silica multi-layers deposited onto glass substrates were used as templates. The colloidal silica crystal was coated with 3-mercaptopropylmethoxysilane producing a thiol-terminated sphere surface. Gold nanocrystals were subsequently affixed to the surface of the spheres, which catalyzed electrodeless deposition of nickel, copper, silver, gold and platinum within the interstitial voids of the colloidal crystal template. The silica colloidal crystal template was etched away in HF leaving behind three-dimensional, highly ordered and free-standing periodic macroporous metal films. Interconnected spherical voids were in the range of 200–400 nm. Macroporous metallic structures with long-range order of interconnected pores and high surface areas could lead to materials for novel photonic, sensing, optical and military-type 'stealth' applications.

In an analogous process, photonic crystals were synthesized from surfactant-capped, monodispersed cadmium selenide, semiconductor nanocrystals templated with a silica colloidal crystal (Fig. 2.8) [70]. A silica opal structure was pre-assembled into an ordered three-dimensional structure and surface functionalized with a long-chain aliphatic alcohol in order to facilitate, through hydrophobic interactions, incorporation of hydrophobically capped CdSe nanocrystals. The CdSe nanocrystals were diffused inside the structure from a decane solution and, after solvent evaporation, semiconductor quantum dots spontaneously assembled inside the silica opal structure. An optically clear solid was formed after the silica was etched from the CdSe material by aqueous HF. TEM images confirmed that it was composed of close-packed CdSe nanocrystals organized in the form of a periodic macroporous crystal. EDX revealed the presence of organic capping groups on each CdSe nanocrystal. Furthermore, a polycrystalline CdSe macroporous crystal was obtained when the as-synthesized structure was consolidated by sintering

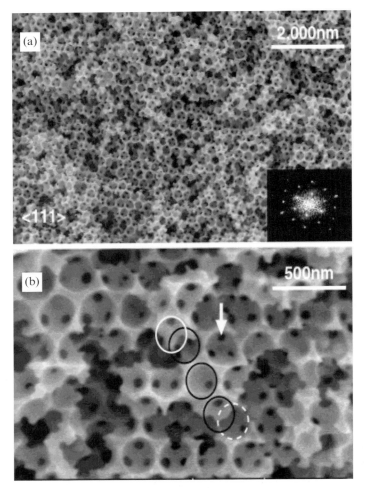

Figure 2.8. SEM images of a macroporous quantum dot solid. (a) Sample is fractured to expose the internal structure of the solid. The inset shows the Fourier transform of the image, illustrating long-range order in the <111> direction. (b) SEM image of the same sample at higher magnification. The position of the silica spheres reveals ABC sequence illustrated by different circles (dashed, black, white) and the arrow points to the place where the spheres are connected. Reproduced with permission from [70].

at 800 °C, which is below the softening point of the silica colloidal crystal but above the softening point of nanocrystalline CdSe, followed by removal of the silica by etching.

Monodispersed 35 nm colloidal silica spheres were used to template a periodic mesoporous polymer [71]. Divinylbenzene was used to make a cross-linked rigid polymer replica, ethylene glycol dimethacrylate ensured that it was flexible and shrinkable, while mixtures of the two were used to vary the degree of shrinkage. In addition, the mesoporous polymer replicas were themselves used as templates to produce monodispersed silica particles. Therefore, these materials can function as a 'nanoscale laboratory' for the production of quantum dot or quantum anti-dot lattices as well as membranes for various sorption and separation processes.

2.6.3 *Soft lithography and micromolding of mesostructured materials*

In addition to the aforementioned templating mechanisms, soft lithography and molding patterning strategies have been developed, mainly by Whitesides and co-workers. There are four main soft lithographic techniques capable of forming and transferring patterns at the ~ 30 nm scale: microcontact printing (mCP) [72,73], replica molding [74,75], micromolding in capillaries (MIMIC) [76], and microtransfer molding (μTM) [77]. Detailed reviews of this subject have been published [78] and so only highlights of methods that pertain to the topic described in this chapter will be described.

Microcontact printing utilizes self-assembled monolayers (SAMs) of functionalized alkanes to pattern structures with a smallest feature of *c.* 100 nm. It uses a silicon master with a relief structure on its surface made by lithography to prepare an elastomeric stamp. This stamp is inked with an alkane-thiol or alkane-alcoxysilane and transferred by contact to a suitable substrate. The non-covered part of the substrate is then etched leaving behind a SAM coated replica of the initial pattern. Replica molding, MIMIC and μTM are very similar in nature and are useful for creating nanostructures of polymeric and sol-gel materials. Replica molding against an elastomeric stamp master made from PDMS and patterned to *c.* 30 nm using X-ray lithography or electron beam writing have shown the feasibility of creating faithful replicas by this simple route. Using mechanical compression, bending or stretching of the elastomeric stamp, it was possible to control the size and shape of the desired replica. Both MIMIC and μTM use masters with a relief pattern of *c.* 1 μm with open ends. The technique relies on contact made between a substrate and the PDMS mold to create a network of microchannels that could be infiltrated with precursor solutions by capillary action. After solidification, removal of the mold and post-treatment, patterned polymer, inorganic salt, sol-gel or graphite replicas remained on the substrate. These methods could not be applied to features smaller than ~ 700 nm because the slow filling of very small capillaries limits its usefulness.

In addition to recent advances in block co-polymer templating of periodic mesoporous silica and silica colloidal crystal templating of periodic mesoporous polymers [79], SAMs have been used for polymer patterning [80]. Mixtures of two strongly incompatible polymers, a polystyrene (PS)–polyvinylpyridine (PVP) blend, were found to phase separate when placed on a

(a)   (b)

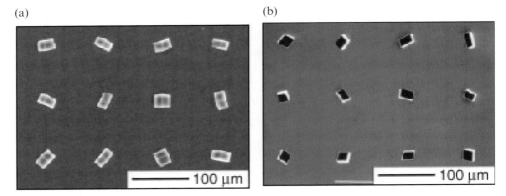

Figure 2.9. SEM images of ordered two-dimensional arrays of single calcite crystals. (a) Crystals with the density of nucleation sites $N \sim 100$ crystals/mm^2 grown selectively from the (015) plane on SAMs of $HS(CH_2)_{15}CO_2H$ supported on Au(111). (b) Crystals with the density of nucleation $N \sim 100$ crystals/mm^2 grown selectively from the (104) plane on SAMs of $HS(CH_2)_{15}OH$ supported on Au(111). Reproduced with permission from [81].

patterned hydrophobic SAM on a gold substrate – the more polar PVP adsorbed on the gold substrate while the less polar PS localized on the SAM. By contrast, a disordered, polymer-domain morphology formed on a SAM-modified glass because no preferential absorption took place on this patterned substrate. Optimization of several experimental parameters was required to achieve a particular pattern as two thermodynamic equilibria had to be balanced – phase separation and selective adsorption from the binary polymer mixture. Patterned polymer layers obtained by this strategy can be lifted off the substrate, which points the way towards multiple pattern replication by super-imposing several structured films on top of each other.

Microcontact printing of SAM patterns on gold resulted in spectacular advances for controlling crystal nucleation and growth process in various systems. A recent example deals with calcite $CaCO_3$ formation on SAMs having acid-terminated regions separated by methyl-terminated regions [81]. It is well known from the study of the growth and form of various biological crystals that nature controls the nucleation of specific crystallographic planes and subsequently the orientation of the mineral by charge, stereochemistry and geometry matching at the organic–inorganic interface. Therefore, having a patterned SAM with various terminal groups allowed for the first time precise control over the nucleation of specific crystallographic planes and growth in different crystallographic directions at a specific position on the substrate. Impressively ordered two-dimensional arrays of single size and shape calcite crystals with a preferred orientation were obtained (Fig. 2.9). Nucleation density was controlled by mass transport due to the fact that the rate of crystallization of calcite on SAMs terminated with polar groups, is much faster then on the methyl-terminated sites. Therefore mass transport to the growing crystals depletes ions from the slowly nucleating sites to the point of undersaturation.

2.7 Hierarchy from the bottom-up

So far, we have discussed various self-assembly and templating mechanisms geared towards the synthesis of porous, ordered materials at different length scales. As was mentioned previously, hierarchically ordered materials that simultaneously exhibit order over 'all' length scales are very attractive novel additions whose synthesis usually requires a combination of all of the techniques mentioned previously. Patterning of mesopores and macropores simultaneously achieves structures with order on several length scales.

One of the first examples of mesoscopic–macroscopic two-dimensional ordering within a structure involved a bacterial superstructure formed from the co-aligned multicellular filaments of *Bacillus subtilis* that was used to template macroporous fibers of either amorphous or ordered mesoporous silica [82]. The interfilament space was mineralized with mesoporous silica and, following removal of the organic, a macroporous framework with 0.5 μm wide channels remained. Mesoporous silica channel walls in this hierarchical structure were curved and approximately 100 nm in thickness. Dense, amorphous walls were obtained by replacing the surfactant-silicate synthesis mixture with a silica sol solution. The difference in the mode of formation between porous and non-porous wall structures was explained in terms of assembly from close-packed mesoporous silica coated bacterial filaments in the former compared to consolidation of silica nanoparticles within interfilament voids in the latter.

An elegant example of the successful integration of templated self-assembly and microfabrication to produce a hierarchical structure has been described [83]. The silica-based materials generated from this approach, exhibited structural ordering at length scales of 10, 100 and 1000 nm (Fig. 2.10). The method is sufficiently versatile that it can be generalized to other dimensions and compositions. To amplify, a PDMS micro-molded elastomeric stamp with a 1000 nm relief pattern was placed over a planar substrate and filled with 200 nm latex spheres by capillary action. Upon evaporation of the solvent, the latex spheres organized into a close-packed three-dimensional network within the confines of the micromold. A sol-gel block co-polymer solution was introduced into the channels of the micromold and infiltrated into the void spaces between the latex spheres by capillary action. Condensation–polymerization of the inorganic oxide species ensued to yield well-ordered hierarchical materials. This is an interesting example of the synthesis of a material in which control of structural features over micro-, meso- and macroscopic length scales has been achieved.

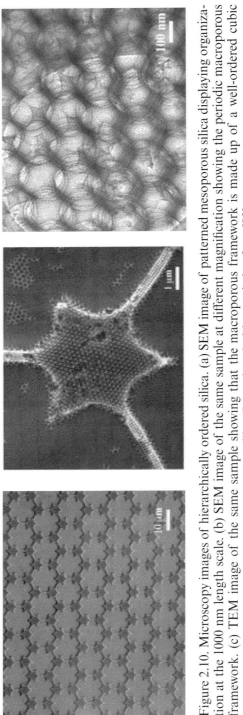

Figure 2.10. Microscopy images of hierarchically ordered silica. (a) SEM image of patterned mesoporous silica displaying organization at the 1000 nm length scale. (b) SEM image of the same sample at different magnification showing the periodic macroporous framework. (c) TEM image of the same sample showing that the macroporous framework is made up of a well-ordered cubic mesoporous silica. Reproduced with permission from [83].

Several approaches towards the synthesis of hierarchical meso- and macro-porous materials have been described. For instance, a mixture that comprised a block co-polymer and polymer latex spheres was utilized to obtain large pore silicas with a bimodal pore size distribution [84]. Rather than pre-organizing latex spheres into an ordered structure they were instead mixed with block-copolymer precursor sols and the resulting structures were disordered. A similar approach that utilized a latex colloidal crystal template was used to assemble a macroporous crystal with a mesoporous silica framework [67].

The influence of additives in the synthesis of mesoporous silicas has been investigated and the results show that hierarchically ordered mesoporous silica rods could be prepared by the addition of fluoride [85] to the synthesis mixture, whereas mesoporous–macroporous membranes were obtained in the presence of aqueous electrolytes [86]. This approach differs from those discussed previously in that no pre-organized templating structure was used to create the hierarchy, but instead the process relied on phase separation during self-assembly. It was proposed that droplets of an aqueous inorganic salt solution macroscopically phase separate from the inorganic oxide-block co-polymer mesophase during evaporation of co-solvent. As the solvent evaporates during the condensation–polymerization, water-rich and water-poor phases separate but remain in contact to allow silica-block co-polymer permeation into the interstitial regions between salt droplets. It was found that the silica membrane morphology depends on the electrolyte strength such that grapevine, toroid, dish or gyroid morphologies were obtained. The resulting three-dimensional structure has uniform but disordered macropores.

Phase separation is also believed to be responsible for the formation of hierarchically ordered mesoporous molybdenum oxide [87]. Mesoporous oxides of metals with partially filled d-shells could display novel magnetic and electronic properties. Toward that aim, toroidal mesoporous molybdenum oxides were prepared using a dimeric molybdenum ethoxide precursor that contained a bridging dodecylimido group. These materials displayed semiconducting properties with a band gap of 1.3 eV. The suggested formation pathway involves a two-fold templating role of the surfactant. It directs formation of individual channels and phase separates to organize the channels into ring structures. Control of the nucleation and growth processes led to a hierarchically ordered transition metal oxide with structure on both the meso- and macroscale. Upon hydrothermal treatment aimed towards improving the thermal stability of the material, a mixed-valency molybdenum oxide was observed that exhibited a lamellar structure and crystalline walls. Conductivity measurements revealed the metallic nature of the material.

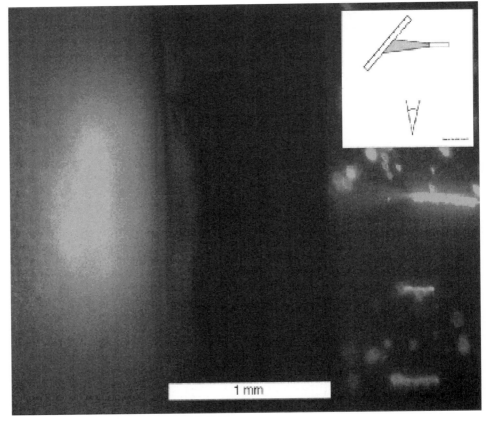

Figure 2.11. Photograph of the output beam and the emitting mesoporous silica fiber.
Inset shows the screen position with respect to the fiber. The picture was taken during a
single excitation pulse. Reproduced with permission from [89].

2.8 Host-guest assemblies

Self-assembled mesoporous materials have been used as hosts for a range of
inorganic and organic species and an extensive review of this field has recently
been published [88]. However, several recent reports have since emerged with
exciting results that will be mentioned here. 'Smart' materials have been
reported that combine the properties of host and guest as well as their mutual
interaction. One can imagine mesoporous materials that contain a dye molecule, which alters its lasing wavelength upon gas adsorption. Dye-sensitized
mesoporous semiconductor can also be envisioned, that functions as a photo-
cathode in a photoregenerative solar cell.

As a step in this direction, a 'one pot' self-assembly synthesis has been
described of hexagonal mesoporous silica fibers sensitized with a dye that functions as a laser (Fig. 2.11) [89]. In this example, large mesopores and process-

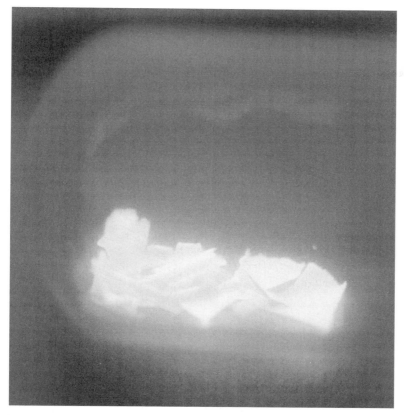

Figure 2.12. Photograph of silicon nanocluster–mesoporous silica composite film under UV irradiation at ~77 K, showing bright visible nanosecond lifetime lumin-escence. Reproduced with permission from [90].

ability of a material to give a shaped fiber with designed lasing and light-propa-gating properties was taken to its best advantage. A wide choice of dyes and tuning of the host–guest interactions are some of the options to obtain viable optical devices. Fibers were pumped by a frequency-doubled Nd:YAG laser and fluorescence spectra of the fiber with rhodamine-6G dye displayed spectra typical for amplified, gain-narrowed emission.

It has also been demonstrated that mesoporous materials are viable candi-dates for optical devices [90]. Silicon nanoclusters were formed inside optically transparent, free-standing, oriented mesoporous silica film by chemical vapor deposition (CVD) of disilane within the spatial confines of the channels. The resulting silicon–silica nanocomposite displayed bright visible photolumines-cence and nanosecond lifetimes (Fig. 2.12). The presence of partially polymer-ized silica channel walls and the retention of the surfactant template within the channels afforded very mild 100–140 °C CVD conditions for the formation of silicon nanoclusters in the aligned film, due mainly to facilitated adsorption, anchoring and reductive-elimination of dihydrogen from disilane. Raman, solid-state NMR and luminescence spectroscopic techniques indicated that the size of the silicon nanoclusters was c. 1–2 nm and that in this size range they exhibited distinct electronic properties compared to larger silicon nanocrystals and bulk silicon. This purely synthetic process to silicon–silica composite film with short luminescent lifetimes may enable the production of light emitting diodes, displays and chemical sensors.

In addition to creating semiconductor or metal replicas of the channels of mesoporous materials, which are expected to display electron and hole quantum confinement effects, forming fibers of polymers could lead to materials with novel electrical, magnetic, optical or mechanical properties. To this end, oxidative polymerization of aniline within the channels of mesoporous silica has been reported [91]. Convincing spectroscopic evidence for intrachannel polymerization of aniline to poly(aniline) was provided. Extracted polymer had a molecular weight considerably smaller than that of the bulk material under similar conditions indicative of a diffusion constraint imposed upon the polymerization and growth of monomer inside the channel space of mesoporous silica.

Confinement of poly(methylmethacrylate), poly(styrene) and poly (vinylacetate) inside mesoporous silica was also studied [92]. Interestingly, it was discovered that the average molecular mass of PMMA increased as the pore size of the host decreased. Poly(phenolformaldehyde) fibers with $\sim$ 2–3 nm diameters were imaged using a negative staining technique [93] in a study designed conclusively to identify formation of mesofibers inside the channels of mesoporous silica. The width of the fibers was controlled by the diameter of the host channels and the length by the dimensions of the host particles. Polymer mesofibers grown in this way displayed aspect ratios greater than 10^4.

The periodic and uniform diameter channels of mesoporous silica have opened an exciting and potentially technologically useful route to a completely new generation of polymerization catalyst [94]. In this system, the advantages of a well-defined and tunable metallocene catalyst that is anchored to the silica channel walls have been combined with the tailorable diameter and extended channels of fiber-shaped mesoporous silica. In the presence of ethylene, this nanochannel catalyst produces exceptionally high molecular weight (6 200 000) crystalline nanofibers of linear polyethylene with a diameter of 30–50 nm. It is envisioned that the crystalline mesoporosity of the silica framework can function as a 'nanoextruder' for the synthesis of oriented polymer nanofibers for a range of common polymers, thereby endowing them with new bulk properties. By contrast, a metallocene catalyst attached to a silica aerogel support can only lead to a structurally inhomogeneous polyethylene with a broad molecular weight distribution.

Metal-containing polymers grown and pyrolyzed within the spatial confines of the channels of mesoporous silica were shown to function as precursors to new-generation magnetic ceramic nanostructures [95]. To expand, multiple techniques were used to monitor *in situ* thermal ring-opening polymerization of strained ferrocenophanes trapped inside mesoporous silica. Pyrolysis of the polymer generated magnetic iron nanoparticles in a carbosilane matrix in greater than 90% ceramic yield and confined to the channels of mesoporous silica. Significantly, the iron nanoparticle–carbosilane–mesoporous silica composite was more thermally and electron-beam stable then mesoporous silica itself. In addition, cross-linked poly(ferrocenylsilane) materials synthesized by ring-opening polymerization of [1]ferrocenophanes have been found to be excellent precursors to shaped magnetic ceramics when sublimed inside mesoporous silica (Fig. 2.13) [96].

2.9 Anchored organics within the channel spaces of mesoporous silica

Attaching various organic species onto the walls of a mesoporous material by direct synthesis and post-grafting techniques is envisioned to offer a range of exciting opportunities. Anchored organics housed within channel and cavity spaces include magnetic materials, electroluminescent compounds, electrical or photoconductors, metal chelators, enzymes and bioeffectors. Several reports have described the functionalization of silica mesopores by co-condensing tetraethoxysilane with an organosilane, like vinyltriethoxysilane or phenyltriethoxysilane [97, 98]. This procedure yields a mesoporous material with organic groups nesting exclusively inside the host channels. Inorganic–organic composite materials of this genre portend applications such as chemical sensing, asymmetric catalysis, anion exchange, chromatographic separation and environmental remediation. Incorporation of reactive groups such as amines, vinyl and acids creates opportunities for coordinating metal species and retaining them inside the pores; this could be important for the removal of pollutants. Towards this goal, mercaptopropyltrimethoxysilane, when refluxed with mesoporous silica, forms a material that incorporates thiol functionality inside the channel spaces (Fig. 2.14) [99, 100]. These materials have been shown to be useful as heavy-metal sponges for selectively adsorbing mercury and lead from environmental water streams. Recent work, that takes advantage of this type of material, includes the size selective incorporation of reverse micelle synthesized CdS nanoparticles into the channels of alkanethiol-modified mesoporous silica [101]. Metal-chelated ligands immobilized on the channel

Figure 2.13. Photograph of the shaped magnetic ceramics obtained from crosslinked poly(ferrocenylsilane) precursor materials. Reproduced with permission from [96].

surface of mesoporous silica were reported as novel anion-exchange materials for the removal of arsenate and chromate from groundwater [102]. Imprint-coating was used to functionalize the surface of mesoporous silica with a metal ion coordinated to a specific ligand, which after the removal of the metal lead to increased performance of these materials towards removal of the same metal [103].

Coupling of functional groups onto the walls of mesoporous silica materials can also be achieved by surfactant-templated assembly of an organosilicon alkoxide and silicon alkoxide reagents. Recent reports that use this synthetic strategy include hybrids of methacrylate–mesoporous silica [104], phenyl-amino (thiol or allyl)–mesoporous silica [105], and organic chromophore–mesoporous silica [106].

Metal-doped mesoporous materials have also benefited from this approach.

Figure 2.14. Schematic drawing of functionalized monolayers on ordered mesoporous supports. One end group of the monolayers is covalently bonded to the silica surface and the other end group can be used to bind heavy metals or other functional molecules. Reproduced with permission from [99].

Complexation of metal ions and subsequent incorporation of the resulting metal complex into the oxide matrix during surfactant-templated synthesis prevents aggregation and leads to a homogeneous distribution of metal centers in the mesostructure. Copper- and vanadium-substituted mesoporous silicas were prepared in this way [107, 108]. Such materials have great potential in the field of catalysis.

2.10 Organic function 'inside' the channel walls of mesoporous silica

A new class of nanocomposite materials has recently emerged from a fusion of organic synthesis, sol-gel chemistry and supramolecular assembly, called periodic mesoporous organosilicas, PMOs [109–112]. What is so special about these materials is that organic functionality has for the first time been integrated 'inside' the channel walls of a silica-based material, which exhibits crystalline mesoporosity. The incorporation of 'bridge-bonded' organic groups within the walls of periodic mesoporous silica is considered a breakthrough for several reasons.

1. It allows the composition, hydrophobicity–hydrophilicity, and chemical properties of the mesoporous host to be tuned using chemistry.
2. There is potential to chemically change the mesoporous material to modify the physical and mechanical properties of the mesoporous host – no other material designed thus far can claim this possibility.
3. The functional groups do not hinder space inside of the channels, as do terminally grafted organic groups.

4. It permits a greater fraction (50%) of organic species to be placed in the framework of the material than other routes, which put organic groups inside the channels and is restricted to 25% before all order is lost.
5. It ensures a homogeneous distribution of organic groups inside the walls of the materials.

Having reactive organic groups 'within' the wall of a periodic mesoporous silica, rather than simply 'hanging' in the channel, may be advantageous for many applications. For example, chemical reaction with the organic groups that constitute the wall could result in changes in the dimensional, mechanical and physical properties of the materials. Chiral organics inside the walls could enable molecular recognition in a size-tunable channel space, which may permit new kinds of enantioselective separations and asymmetric syntheses. Redox-active organics within the walls may lead to electrically tunable molecular filters or vehicles for the controlled delivery of chemicals. Photoisomerizable organics within the walls could lead to smart channels that change their diameter reversibly when irradiated with light of the right wavelength. Metallocene polymerization catalysts housed in the channel walls may provide a route to oriented polymer fibers in what is effectively a nano-extrusion process. Modified crown ethers inside the wall may be very effective for removing alkali metals from solutions. Organic groups with acidic functionality may facilitate fast proton conduction along the channel walls and new proton conducting mesostructured membranes for hydrogen/methanol–oxygen fuel cells. Coordinating ligands, such as bipyridine, may be inserted into the framework to generate mesoporous materials that strongly bind transition metals.

There are additional advantages that may arise from the incorporation of organics into the 'backbone' of the material. Owing to the flexibility of organic spacers, monoliths may be less prone to cracking, a problem that has plagued research into mesoporous silica monoliths. This property may also be advantageous to film formation. Moreover, the hardness and density of the bulk mesoporous organosilica may be tuned by changing the organic groups. This may prove to be a route to lightweight periodic mesoporous materials. Clearly, many opportunities abound for periodic mesoporous organosilicas with appropriately designed functional organics 'inside' the framework.

As described above for periodic mesoporous silica, the morphology of the material often dictates its function and utility [113]. In this regard, the PMOs are expected to be no exception. Non-ionic surfactant-based, lyotropic liquid-crystal templating has been successfully utilized for synthesizing optically transparent oriented film and monolith morphologies of the PMOs [114]. This breakthrough enabled polarization micro-Raman spectroscopy to be effectively used as a probe of the spatial organization of organic groups inside the silica channel walls of the PMOs [115]. The results showed that while the organosilica channels were oriented parallel to the surface of the film the channel structure at the microscale was glassy. The ability to synthesize oriented PMO film is an important step forward in the utilization of these materials.

2.10.1 *Periodic mesoporous organosilicas compared with aperiodic organosilica xerogels*

To put the work on PMO materials into context it is useful to contrast them with prior reports of organosilica composites [116]. They are considered to be distinct from organosilica xerogels, which are amorphous materials containing a broad distribution of pore sizes. While xerogels have attracted attention for potential application in catalysis, their lack of size and shape selectivity has limited their use in this regard. Although xerogels have incorporated organic groups into the silica structure since their conception [117], *no* periodic mesoporous silicas were known with organic moieties as an integral part of the framework. By analogy with polymer materials, for example polysiloxanes, this work puts organic and organometallic species into the 'backbone' of the silica framework where they can have a more dramatic effect on the properties of the structure than do grafted or 'side chain' groups.

2.11 Self-assembling mesostructured film

As discussed, the possibility of synthesizing mesoporous materials with predetermined and interesting shapes suggested new avenues of exploration where the form of the material determined its function and potential utility. The processing of mesostructured materials into film morphologies is a subject of great current interest since they are envisioned to find utility as sensors, coatings or optical and electronic components for which powder samples may not have much value. They are also considered promising for applications such as adsorption, separation and hosts for macromolecules and bio-molecules. Controlling the morphology of mesostructured materials is definitely a key for particular applications. Moreover, morphological traits such as shape, surface topology and growth kinetics of mesostructured materials can provide useful information on their mode of formation. To the present time, numerous reports on the formation of mesoporous silica thin film have emerged; however, equivalent reports on mesostructured metal oxide film are rare. Two approaches for the synthesis of mesostructured silica film have been described – the first involves processing silica-surfactant solutions by spin, dip or spray coating methods, whereas the second is based on spontaneous nucleation and growth at the boundary between the silica surfactant solution and air, liquid or solid surfaces.

2.11.1 Coating substrates with mesostructured silica film

Lamellar [118] and hexagonal [119] mesostructured silica film have been spin-coated from homogenous acid-catalyzed alkoxysilane-surfactant solutions onto non-porous substrates. This method affords a simple procedure for obtaining optically transparent film by a rapid solvent evaporation process. In this approach, TMOS was partially hydrolyzed by a sub-stoichiometric amount of water under acidic conditions at room temperature. After C_nTAC or C_nTAB addition, the resulting solution was spin-coated on Pyrex or fused silica glass substrates. Samples were air dried to remove solvent and to promote complete condensation–polymerization of the silica. Surfactants could be removed from the hexagonal mesoporous silica film by calcination to produce film samples with BET surface area of c. 1000 m^2/g. Because the films obtained were transparent, their application as hosts for photo-active species such as organic dyes was investigated. It was imagined that such films would lead to new kinds of functional supramolecular material. By using poly(methylmethacrylate) as the substrate it proved possible to dissolve the substrate, leaving behind free-standing mesostructured silica film.

Mesostructured metal film is of considerable interest for applications in catalysis, fuel cells, capacitors and sensors. While granular forms of mesostructured metals are useful in catalysis they are best made as film for applications such as electrochemical or sensor devices. In this context, mesoporous platinum films have been synthesized by electrodepositing platinum within the microphase separated domains of a lyotropic liquid crystal template on an electrode surface [120]. Specifically, the nonionic surfactant $C_{16}EO_8$, hexachloroplatinic acid and water were used for 'mesophase metal plating'. An electrodeposited platinum replica of the hexagonal mesophase on a gold electrode was conducted at temperatures between 25 °C and 65 °C. The electrode was subsequently rinsed with copious amounts of water to remove the surfactant. A mesoporous platinum structure was obtained that consisted of cylindrical 25 Å channels arranged in a hexagonal close-packed lattice with $\sim$ 25 Å wall thickness (Fig. 2.15). An STM study showed that the film was uniform over the entire area of the electrode with an RMS surface roughness of 20 ($\pm$5) nm and an estimated surface area of 22 m^2/g. This work represents a first attempt at producing well-defined mesoporous metal films with high surface areas, electri-

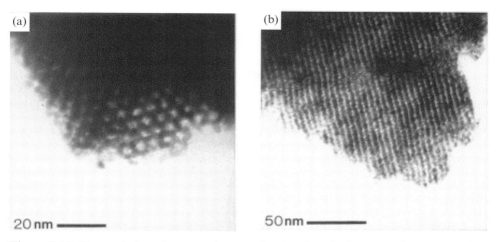

Figure 2.15. Transmission electron micrographs showing the hexagonal structure of Pt film electrodeposited onto a polished gold electrode. (a) End view of the pores; (b) side view of the pores. Reproduced with permission from [120].

cal connectivity, fast electrolyte diffusion and good mechanical and electro-chemical stability.

Sol-gel dip-coating methods have been used to prepare cubic and hexagonal mesoporous silica film [121]. Cubic mesoporous silica film is interesting because it has a three-dimensional network of channels that guarantees access-ibility of reagents into the film. Surface-acoustic-wave nitrogen-adsorption has been used to provide information on the mesopore surface area, volume and diameter of the film. Fluorescence depolarization allows *in situ* monitoring of the formation of the film (Fig. 2.16). The method aptly named evaporation-induced self-assembly (EISE) involves the use of an aqueous acidic ethanol precursor solution of surfactant and polymeric silica-sols, which during the dip-coating process experiences preferential evaporation of ethanol and conse-quent enrichment in acid, water and surfactant. This promotes micellization, organization of micelles into liquid-crystalline phases and the condensation of silica precursor. The method affords rapid synthesis of both hexagonal and cubic phases of mesoporous silica film on a solid substrate. Hydrophobic dyes incorporated into the interior of the micellar template allowed the formation of the film to be followed by means of fluorescence depolarization. The proposed assembly pathway involves co-assembly of silica-surfactant micellar species at substrate–liquid and liquid–air interfaces.

The dip coating method enabled the synthesis of nanocomposite silica film with the laminated structure of abalone nacre, which is known for its strength,

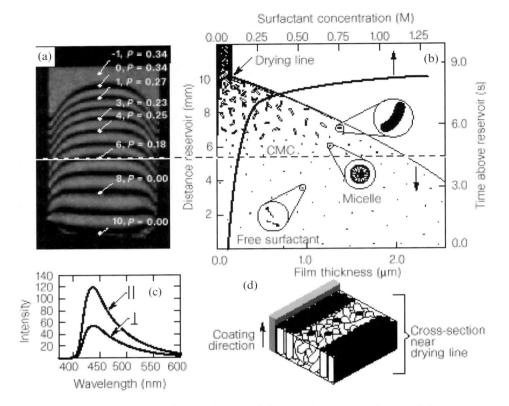

Figure 2.16. (a) Optical interference image of the steady state condition of the mesoporous silica film during dip-coating at 7.6 cm/min. Fringes represent changes in the optical thickness and are numbered from 0 at the drying line to 10 near the reservoir surface. (b) Steady-state film cross-section, corresponding to the interference image (a). (c) Fluorescence spectra of the probe molecule introduced at an initial concentration of 10^{-4} M and excited at 351 nm. (d) Cross-section of an ordered region of the film near the drying line, depicting formation of interfacial regions of liquid-crystalline order. Reproduced with permission from [121].

hardness and toughness [122]. In this case the precursor sol solution contained organic monomer, initiator, cross-linker and coupling agent in addition to the surfactant and silica source. Monomer and initiator were incorporated into the hydrophobic interior of the micellar template by evaporation-induced partitioning and continued evaporation during the dipping process. This promoted simultaneous co-assembly of silica-surfactant-micellar-organic additive species. Polymerization of the imbibed organics was induced by heat or light, together with condensation–polymerization of the silica to form the desired composite nanolaminate.

The ability to process nonionic surfactant/inorganic salt composites as film

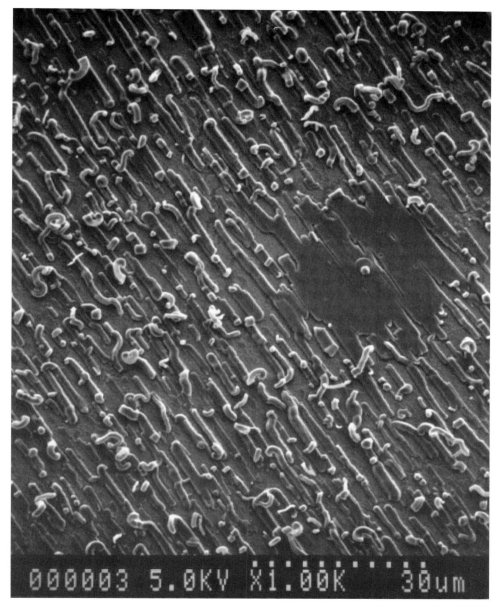

Figure 2.17. Scanning electron microscopy image of the nucleation and growth of a mesoporous silica film on mica. Reproduced with permission from [126].

and monolithic samples was demonstrated in a study of mesostructured lithium triflate-oligoethyleneoxide surfactant-silica [123]. NMR and Raman spectroscopy revealed that $LiCF_3SO_3$ was dissociated into free lithium and triflate ions, confined within the oligoethyleneoxide head group region. It was shown by ac impedance spectroscopy that such composites behave as a fast lithium ion conductor at room temperature, which may prove to be important in the field of polymer electrolytes and solid plasticizers for battery applications.

2.11.2 Nucleation and growth of mesoporous silica film

Templated growth of mesoporous silica film at solid–water and air–water interfaces, has been reported [124–129]. Oriented mesoporous silica film was grown on freshly cleaved mica under quiescent, aqueous acidic conditions [126]. The early stage of film formation involved the appearance of aligned stripes of mesoporous silica with a preferred growth direction that eventually coalesced into a continuous film structure (Fig. 2.17). The proposed mode of film growth involved matching of charge and structure at the interface between mica and surfactant-silica micellar precursors. Free-standing and oriented mesoporous silica films could also be grown by this method at the air–water interface [127]. Film formed in this way was continuous, well ordered and elastically deformable and exhibited a mean surface roughness of about 3 Å (Fig. 2.18). The proposed mode of formation involved cooperative interaction between silicate–surfactant micelles and a hemimicellar overlayer, known to exist at the air–water interface. It was also observed that these mesoporous silica films exhibited unusual surface designs and channel patterns such as concentric circles, herringbones, fingerprints, and hairpins [129]. It was suggested that these patterns were spatio-temporal silicified recordings of the two-dimensional polymerization and growth of a silicate liquid-crystal seed emerging at the air–water interface. Topological defects present in the precursor silicate liquid-crystal film were envisaged to cause distortions in the director field, which could lead to the observed liquid-crystalline texture in free-standing mesoporous silica film [130]. A hemimicellar cylindrical assembly was observed by AFM to be responsible for the templating of mesoporous silica film on hydrophobic substrates like graphite [128].

2.11.3 Patterning and molding mesoporous silica film

As was mentioned earlier, designing mesoporous materials with structural features that span several length scales could benefit perceived device applications. In this context a three-way-templating strategy that facilitates selected area

(a)

(b)

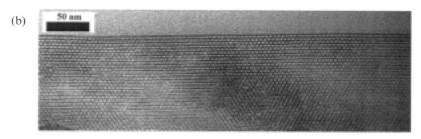

Figure 2.18. Scanning electron microscopy (a) and transmission electron microscopy (b) images of a free-standing mesoporous silica film grown at the air–water interface. Reproduced with permission from [127].

growth of patterned mesoporous silica films has been described [131, 132]. The process involved co-assembly of a soft lithographically patterned alkanethiol–surfactant heterobilayer on a gold surface with surfactant-silicate assemblies in solution. It was found that oriented mesoporous silica film grew preferentially on the patterned regions of the gold substrate thus creating the first example of a mesoporous silica film with a macroscale design (Fig. 2.19). EDX showed that more than 80% of the deposited mesoporous silica was located on the SAM-delineated regions implying a much greater rate of adhesion of silicate-micellar species to the SAM surface than on the bare gold surface. This was presumed to be caused by the hydrophobic interaction between the tails of an alkanethiolate SAM on gold with the tails of the surfactant to

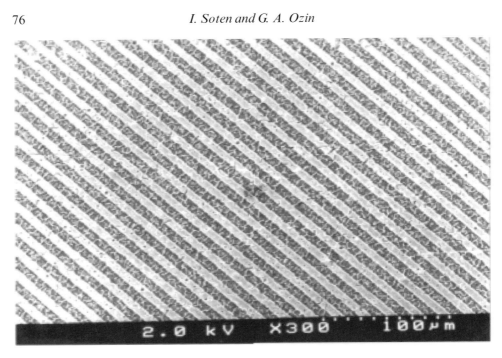

Figure 2.19. Scanning electron microscopy image of mesoporous silica selectively grown on a parallel line pattern of hexadecanethiolate SAM on a gold substrate. Reproduced with permission from [131].

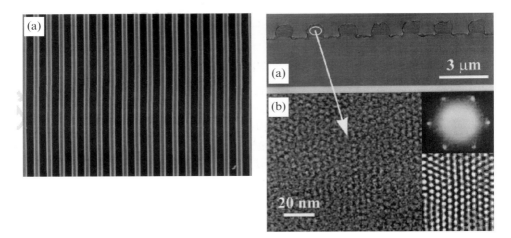

Figure 2.20. (a) Scanning electron microscopy image of 1 μm line mesoscopic patterns formed by guided growth within microcapillaries. (b) Transmission electron microscopy images of a patterned mesoscopic silica structure grown on a plastic substrate displaying hexagonally packed surfactant tubule structure within 1 μm size lines shown in (a). Reproduced with permission from [133].

create a surface confined heterobilayer that favored nucleation and growth of mesoporous silica from the capture, organization and polymerization of silicate-micellar solute species [130].

Micromolding in capillaries (MIMIC) was used to create microscopically patterned forms of mesoporous silica [133]. The MIMIC technique involves the use of a PDMS elastomeric stamp with a designed pattern of capillaries on its surface. When this stamp is placed on a substrate and a solution of surfactant-silicate precursor infiltrated inside the channels, a patterned mesoporous silica film is formed. Infiltration of reactants was facilitated by an applied electric field parallel to the substrate. The effect was threefold – inducing electro-osmotic fluid flow, aligning surfactant tubules, and causing localized Joule heating of the solution. The resulting structure exhibited micrometer-sized lines consisting of hexagonally packed silica channels oriented parallel to the substrate surface (see Fig. 2.20). Long-range alignment of channels was explained in terms of the action of the electric field and/or growth restricted to a confined space with specific surface interactions.

Hierarchically ordered silica, titania and niobia have also been synthesized using a creative modification of the above strategy [83]. The process involves the simultaneous use of micromolding, polystyrene sphere templating and cooperative assembly of inorganic sol-gel species with amphiphilic triblock co-polymers leading to 10, 100 and 1000 nm length scales of hierarchy. Such multiple scales of structural organization and variations of chemical composition, make it possible to tune the physical properties of the materials for a range of possible applications.

2.12 Room at the top and bottom

Not so long ago in materials chemistry it seemed that there was only 'room at the bottom'. The trend was to synthesize and organize nanoscopic materials. As we enter the new millennium it is becoming abundantly clear that there is also 'room at the top'. The paradigmatic shift comes from the realization that a creative combination of self-assembly and templating, patterning and molding, expand the materials chemist's tool box of opportunities to include synthesis and organization of materials at 'all' length scales. Hierarchy has been introduced into materials chemistry, and purely synthetic integrated chemical, physical and biological systems, designed to achieve a particular function, are becoming a reality. This panoscopic approach to the synthesis of micro-, meso- and macroscale structures and hybrid architectures offers ways to introduce novel function and form into materials. The unusual physical, chemical, electrical, optical, magnetic and mechanical properties envisaged for this new breed of materials might make them attractive as electrically tunable membranes and chemical delivery systems, chemical sensors and solar cells, batteries and fuel cells, bone implants and photonic structures, nanocomposites and heavy-metal sponges.

A common theme running throughout all of the work in this emerging subdiscipline of chemistry is the organization of inorganics by organic, polymeric and inorganic templates over multiple length scales and spatial dimensions. In all cases, the boundary between active components and their relative disposition in the structure are expected profoundly to influence the properties of the composite. Because of this, it is anticipated that an amalgamation of research methods in organic, polymer and inorganic chemistry will lead to the development of new solid-state materials with structures and compositions of a type not seen before in materials science. Materials of this genre may prove to be scientifically interesting and technologically relevant in the twenty-first century.

2.13 References

1. Breck, D.W. (1984). *Zeolite Molecular Sieves*. Kieger, Malabar.
2. Yanigisawa, T., Schimizu, T., Kiroda, K. and Kato, C. (1990). *Bull. Chem. Soc. Jpn* **63**, 988.
3. Kresge, C. T., Leonowicz, M. E., Roth, W. J., Vartuli, J. C. and Beck, J. S. (1992). *Nature* **359**, 710.
4. Ozin, G. A. (1992). *Adv. Mater.* **4**, 612.
5. Mann, S. and Ozin, G. A. (1996). *Nature* **382**, 313.
6. Ozin, G. A. (1998). *Sci. Am.* January 22, 7.
7. *Nanostructures and Mesoscopic Systems* (1992). M. Reed and W. P. Kirk, eds. Academic Press, Boston.
8. Molecular assemblies and nanochemistry (1997). M. R. Bryce, ed. Special Issue *J. Mater. Chem.* **7**, 1069.
9. Ozin, G. A. (2000). *J. Chem. Soc. Chem. Commun.* **6**, 419.
10. Huo, Q., Margolese, D. I., Ciesla, U., Feng, P., Gier, T. E., Sieger, P., Leon, R., Petroff, P., Shuth, F. and Stucky, G. D. (1994). *Nature* **368**, 317.
11. Attard, G. S., Glude, J. C. and Göltner, C. G. (1995). *Nature* **378**, 366.
12. Antonelli, D. M. and Ying, J. (1996). *Angew. Chem. Int. Ed. Engl.* **35**, 426.
13. O'Brien, S., Francis, R. J., Fogg, A., O'Hare, D., Okazaki, N. and Kuroda, K. (1999). *Chem. Mater.* **11**, 1822.
14. Firouzi, A., Kumar, D., Bull, L. M., Besier, T., Sieger, P., Huo, Q., Walker, S. A., Zasadzinski, J. A., Glinka, C., Nicol, J., Margolese, D., Stucky, G. D. and Chmelka, B. F. (1995). *Science* **267**, 1138.
15. Steel, A., Carr, S. W. and Anderson, M. W. (1994). *J. Chem. Soc. Chem. Commun.*, 1571.
16. Monnier, A., Schuth, F., Huo, Q., Kumar, D., Margolese, D., Maxwell, R. S., Stucky, G. D., Krishnamurty, M., Petroff, P. and Firouzi, A. (1993). *Science* **261**, 1299.
17. Tanev, P. T. and Pinnavaia, T. J. (1995). *Science* **267**, 1138.

18. Vartuli, J. C., Schmitt, K. D., Kresge, C. T., Roth, W. J., Leonowitz, M. E., McCullen, S. B., Hellring, S. D., Beck, J. S., Schlenker, J. L., Olsen, D. H. and Sheppard, E. W. (1994). *Chem. Mater*. **6**, 2317.
19. Bagshaw, S. A., Prouzet, E. and Pinnavaia, T. J. (1995). *Science* **269**, 1242.
20. Pauly, T. R., Liu, Y., Pinnavaia, T. J., Billinge, S. J. L. and Rieker, T. P. (1999). *J. Am. Chem. Soc.* **121**, 8835.
21. Antonelli, D. M. and Ying, J. (1995). *Angew. Chem. Int. Ed. Engl*. **34**, 2014.
22. Antonelli, D. M. and Ying, J. (1996). *Angew. Chem. Int. Ed. Engl*. **35**, 426.
23. Ulagappan, N. and Rao, C. N. R. (1996). *J. Chem. Soc. Chem. Commun.*, 1685.
24. Severin, K. G., Abdel-Fattah, T. M. and Pinnavaia, T. J. (1998). *J. Chem. Soc. Chem. Commun.*, 1471.
25. Antonelli, D. M. and Ying, J. (1996). *Chem. Mater*. **8**, 874.
26. Wong, M. S., Antonelli, D. M. and Ying, J. Y. (1997). *Nanostr. Mater*. **9**, 165.
27. Chang, H. L., Chun, C. M., Aksay, I. A. and Shih, W. H. (1999). *Ind. Eng. Chem. Res*. **38**, 973.
28. Bagshaw, S. A. and Pinnavaia, T. J. (1996). *Angew. Chem. Int. Edn Engl*. **35**, 1102.
29. Cabrera, S., Haskouri, J. E., Alamo, J., Beltrán, A., Beltrán, D., Mendioroz, S., Marcos, M. D. and Amorós, P. (1999). *Adv. Mater*. **11**, 379.
30. Reddy, J. S. and Sayari, A. (1996). *Catal. Lett*. **38**, 219.
31. Ciesla, U., Schacht, S., Stucky, G. D., Unger, K. K. and Schüth, F. (1996). *Angew. Chem. Int. Ed. Engl*. **35**, 541.
32. Liu, P., Reddy, J. S., Adnot, A. and Sayari, A. (1996). *Mater. Res. Soc. Symp. Proc.* **431**, 101.
33. Jiménez-Jiménéz, J., Maireles-Torres, P., Olivera-Pastor, P., Rodríguez-Castellón, E., Jiménez-López, A., Jones, D. J. and Rozière, J. (1998). *Adv. Mater*. **10**, 812.
34. Ciesla, U., Fröba, M., Stucky, G. and Schüth, F. (1999). *Chem. Mater*. **11**, 227.
35. Braun, P. V., Osenar, P. and Stupp, S. I. (1996). *Nature* **380**, 325.
36. Li, J., Delmotte, L. and Kessler, H. (1996). *J. Chem. Soc. Chem. Commun.*, 1023.
37. Sokolov, I., Jiang, T. and Ozin, G. A. (1998). *Adv. Mater*. **10**, 942.
38. MacLachlan, M., Coombs, N. and Ozin, G. A. (1999). *Nature* **397**, 681.
39. Khushalani, D., Kuperman, A. and Ozin, G. A. (1999). *J. Mater. Chem*. **9**, 1491.
40. Oliver, S., Kuperman, A., Coombs, N., Lough, A. and Ozin, G. A. (1995). *Nature* **378**, 47.
41. Yang, H., Coombs, N. and Ozin, G. A. (1997). *Nature* **386**, 692.
42. Ozin, G. A., Yang, H., Sokolov, I. and Coombs, N. (1997). *Adv. Mater*. **9**, 662.
43. Coombs, N., Khushalani, D., Oliver, S., Ozin, G. A., Shen, G. C., Sokolov, I. and Yang, H. J. (1997). *J. Chem. Soc., Dalton Trans.*, 3941.
44. Yang, H., Kresge, C. T. and Ozin, G. A. (1998). *Adv. Mater*. **10**, 883.
45. Ozin, G. A. (1999). *Can. J. Chem*. **77**, 2001.
46. Sokolov, I., Yang, H., Ozin, G. A. and Kresge, C. T. (1999). *Adv. Mater*. **11**, 636.
47. Yang, S. M., Sokolov, I., Coombs, N., Kresge, C. T. and Ozin, G. A. (1999). *Adv. Mater*. **11**, 1427.
48. Yang, S. M., Yang, H., Coombs, N., Sokolov, I., Kresge, C. T. and Ozin, G. A. (1999). *Adv. Mater*. **11**, 52.
49. Kim, S. S., Zhang, W. and Pinnanaia, T. J. (1998). *Science* **282**, 1302.

50. Lu, Y., Fan, H., Stump, A., Ward, T. L., Rieker, T. and Brinker, C. J. (1999). *Nature* **398**, 223.
51. Stupp, S. I. and Braun, P. V. (1997). *Science* **277**, 1242.
52. Walsh, D. and Mann, S. (1995). *Nature* **377**, 320.
53. Soten, I. and Ozin, G. A. (1999). *J. Mater. Chem.* **9**, 703.
54. Braun, P. V., Osenar, P. and Stupp, S. I. (1996). *Nature* **380**, 325.
55. Murray, C. B., Kagan, C. R. and Bawendi, M. G. (1995). *Science* **270**, 1335.
56. Tohver, V., Braun, P. V., Pralle, M. U. and Stupp, S. I. (1997). *Chem. Mater.* **9**, 1495.
57. Murray, C. B., Norris, D. J. and Bawendi, M. G. (1993). *J. Am. Chem. Soc.* **115**, 8706.
58. Peng, X., Wilson, T. E., Alivisatos, A. P. and Schultz, P. G. (1997). *Angew. Chem. Int. Ed. Engl.* **36**, 145.
59. (a) Mitzi, D. B., Wang, S., Field, C. A., Chess, C. A. and Guloy, A. M. (1995). *Science* **267**, 1473; (b) Tian, Y., Wu, C., Kotov, N. and Fendler, J. H. (1994). *Adv. Mater.* **6**, 959.
60. Li, M., Schnablegger, H. and Mann, S. (1999). *Nature*, **402**, 393.
61. Zhao, D., Huo, Q., Feng, J., Chmelka, B. F. and Stucky, G. D. (1998). *J. Am. Chem. Soc.* **120**, 6024.
62. Yang, P., Zhao, D., Chmelka, F. and Stucky, G. D. (1998). *Chem. Mater.* **10**, 2033.
63. Yang, P., Zhao, D., Margolese, D. I., Chmelka, F. and Stucky, G. (1998). *Nature* **396**, 152.
64. Velev, O. D., Jede, T. A., Lobo, R. F. and Lenhoff, A. M. (1997). *Nature* **389**, 447.
65. Park, S. H. and Xia, Y. (1998). *Chem. Mater.* **10**, 1745.
66. Holland, B. T., Blanford, C. F. and Stein, A. (1998). *Science* **281**, 538.
67. Holland, B. T., Blanford, C. F., Do, T. and Stein, A. (1999). *Chem. Mater.* **11**, 795.
68. Zakhidov, A. A., Baughman, R. H., Iqbal, Z., Cui, C., Khayrullin, I., Dantas, S. O., Marti, J. and Ralchenko, V. G. (1998). *Science* **282**, 897.
69. Jiang, P., Cizeron, J., Bertone, J. F. and Colvin, V. L. (1999). *J. Am. Chem. Soc.* **121**, 7957.
70. Vlasov, Y. A., Yao, N. and Norris, D. J. (1999). *Adv. Mater.* **11**, 165.
71. Johnson, S. A., Ollivier, P. J. and Mallouk, T. E. (1999). *Science* **283**, 963.
72. Xia, Y. and Whitesides, G. M. (1995). *J. Am. Chem. Soc.* **117**, 3274.
73. Xia, Y., Qin, D. and Whitesides, G. (1996). *Adv. Mater.* **8**, 1015.
74. Xia, Y., McClelland, J. J., Gupta, R., Qin, D., Zhao, X. M., Sohn, L. L., Celotta, R. J. and Whitesides, G. M. (1997). *Adv. Mater.* **9**, 147.
75. Xia, Y., Kim, E., Zhao, X. M., Rogers, J. A., Prentiss, M. and Whitesides, G. M. (1996). *Science* **273**, 347.
76. Xia, Y., Kim, E. and Whitesides, G. M. (1996). *Chem. Mater.* **8**, 1558.
77. Zhao, X. M., Xia, Y. and Whitesides, G. M. (1996). *Adv. Mater.* **8**, 837.
78. Zhao, X. M., Xia, Y. and Whitesides, G. M. (1997). *J. Mater. Chem.* **7**, 1069.
79. Göltner, C. G. and Weibenberger, M. C. (1998). *Acta Polymer* **49**, 704.
80. Böltau, M., Walheim, S., Mlynek, J., Krausch, G. and Steiner, U. (1998). *Nature* **391**, 877.
81. Aizenberg, J., Black, A. J. and Whitesides, G. M. (1999). *Nature* **398**, 495.
82. Davis, S. A., Burkett, S. L., Mendelson, N. H. and Mann, S. (1997). *Nature* **385**, 420.

83. Yang, P., Deng, T., Zhao, D., Feng, P., Pine, D., Chmelka, B. F., Whitesides, G. M. and Stucky, G. D. (1998). *Science* **282**, 2244.

84. Antoinetti, M., Berton, B., Göltner, C. and Hentze, H. P. (1998). *Adv. Mater.* **10**, 154.

85. Schmidt-Winkel, P., Yang, P., Margolese, D. I., Chmelka, B. F. and Stucky, G. D. (1999). *Adv. Mater.* **11**, 303.

86. Zhao, D., Yang, P., Chmelka, B. F. and Stucky, G. D. (1999). *Chem. Mater.* **11**, 1174.

87. Antonelli, D. M. and Trudeau, M. (1999). *Angew. Chem. Int. Ed. Engl.* **38**, 1471.

88. Ozin, G. A., Chomski, E., Khushalani, D. and MacLachlan, M. J. (1998). *Curr. Opin. Coll. Int. Sci.* **3**, 181.

89. Marlow, F., McGehee, D., Zhao, D., Chmelka, B. F. and Stucky, G. D. (1999). *Adv. Mater.* **11**, 632.

90. Dag, Ö., Ozin, G. A., Yang, H., Reber, C. and Bussière, G. (1999). *Adv. Mater.* **11**, 474.

91. Wu, C.-G. and Bein, T. (1994). *Science* **264**, 1757.

92. Llewellyn, P. L., Ciesla, U., Decher, H., Stadler, R., Schuth, F. and Unger, K. K. in Zeolites and Related Microporous Materials: State of the Art 1994, *Stud. Surf. Sci. Catal.* 1994. Proceedings of the 10th International Zeolite Conference Garmisch-Partenkirchen Germany 1994, vol. 84, pp. 2013–2020. New York: Elsevier Science.

93. Johnson, S. A., Khushalani, D., Coombs, N., Mallouk, T. E., Ozin, G. A. (1997). *J. Mater. Chem.* **1**, 13.

94. Kageyama, K., Tamazawa, J. I. and Aida, T. (1999). *Science* **285**, 2113.

95. MacLachlan, M. J., Aroca, P., Coombs, N., Manners, I. and Ozin, G. A. (1998). *Adv. Mater.* **10**, 144.

96. MacLachlan, M. J., Ginzburg, M., Coombs, N., Raju, N. P., Greedan, J. E., Ozin, G. A. and Manners, I. (2000). *Science* **287**, 1460.

97. Lim, M. H., Blanford, C. F. and Stein, A. (1997). *J. Am. Chem. Soc.* **117**, 4090.

98. Burkett, S. L., Sims, S. D. and Mann, S. (1996). *J. Chem. Soc. Chem. Commun.*, 1367.

99. Feng, X., Fryxell, G. E., Wang, L.-Q., Kim, A. Y., Liu, J. and Kemner, K. M. (1997). *Science* **276**, 923.

100. Mercier, L. and Pinnavaia, T. J. (1997). *Adv. Mater.* **9**, 500.

101. Hirai, T., Okubo, H. and Komasawa, I. (1999). *J. Phys. Chem. B.* **103**, 4228.

102. Fryxell, G. E., Liu, J., Hauser, T. A., Nie, Z., Ferris, K. F., Mattigod, S., Gong, M. and Hallen, R. T. (1999). *Chem. Mater.* **11**, 2148.

103. Dai, S., Burleigh, M. C., Shin, Y., Morrow, C. C., Barnes, C. E. and Xue, Z. (1999). *Angew. Chem. Int. Ed. Engl.* **38**, 1235.

104. Moller, K., Bein, T. and Fischer, R. X. (1999). *Chem. Mater.* **11**, 665.

105. Hall, S. R., Fowler, C. E., Lebeau, B. and Mann, S. (1999). *J. Chem. Soc. Chem. Commun.*, 201.

106. Fowler, C. E., Lebeau, B. and Mann, S. (1998). *J. Chem. Soc. Chem. Commun.*, 1825.

107. Karakassides, M. A., Fournaris, K. G., Travlos, A. and Petridis, D. (1998). *Adv. Mater.* **10**, 483.

108. Wei, D., Xiaobing, F., Chueh, W. T., Ravikivotch, P., Lyubovsky, M., Li, C., Takeguchi, T. and Haller, G. L. (1999). *J. Phys. Chem. B* **103**, 2113.

109. Asefa, T., MacLachlan, M. J., Coombs, N. and Ozin, G. A. (1999). *Nature* **402**, 867.
110. Inagaki, S., Guan, S., Fukushima, Y., Oshuna, T. and Terasaki, O. (1999). *J. Am. Chem. Soc.* **121**, 9611.
111. Melde, B. J., Holland, B. T., Blanford, C. F. and Stein, A. (1999). *Chem. Mater.* **11**, 3302.
112. Yoshina-Ishii, C., Asefa, T., Coombs, N., MacLachlan, M. J. and Ozin, G. A. (1999). *J. Chem. Soc. Chem. Commun.* **24**, 2539.
113. Ozin, G. A. (1999). *Can. J. Chem.* **77**, 2001.
114. Asefa, T., Yoshina-Ishii, C., MacLachlan, M. J. and Ozin, G. A. (2000). *J. Mater. Chem.* **10**, 1751.
115. Asefa, T., Coombs, N., Dag, Ö., Yoshina-Ishii, C., MacLachlan, M. J. and Ozin, G. A. (2000). *Abstr. Pap. Am. Chem. Soc.* **219**: U883 Part 1.
116. Lim, M. H. and Stein, A. (1999). *Chem. Mater.* **11**, 3285.
117. Corriu, R. J. P. and Leclercq, D. (1996). *Angew. Chem. Int. Ed. Engl.* **35**, 1420.
118. Ogawa, M. (1994). *J. Am. Chem. Soc.* **116**, 7941.
119. Ogawa, M. (1996). *J. Chem. Soc. Chem. Comm.*, 1149.
120. Attard, G. S. A., Bartlett, P. N., Coleman, N. R. B., Elliott, J. M., Owen, J. R. and Wang, J. H. (1997). *Science* **278**, 838.
121. Lu, Y., Ganguli, R., Drewien, C. A., Anderson, M., Brinker, C. J., Gong, W., Guo, Y., Soyez, H., Dunn, B., Huang, M. H. and Zink, J. I. (1997). *Nature* **389**, 364.
122. Sellinger, A., Weiss, P. M., Nguyen, A., Lu, Y., Assink, R. A., Gong, W. and Brinker, C. J. (1998). *Nature* **394**, 256.
123. Dag, Ö., Verma, A. and Ozin, G. A. (1999). *J. Mater. Chem.* **9**, 1475.
124. Aksay, I. A., Trau, M., Manne, S., Honma, I., Yao, N., Zhou, L., Fenter, P., Eisenberger, P. M. and Gruner, S. M. (1996). *Science* **273**, 892.
125. Schacht, S., Huo, Q., Voigt-Martin, I. G., Stucky, G. D. and Schüth, F. (1996). *Science* **273**, 768.
126. Yang, H., Kuperman, A., Coombs, N., Mamiche-Afara, S. and Ozin, G. A. (1996). *Nature* **379**, 703.
127. Yang, H., Coombs, N., Sokolov, I. and Ozin, G. A. (1996). *Nature* **381**, 589.
128. Yang, H., Coombs, N., Sokolov, I. and Ozin, G. A. (1997). *J. Mater. Chem.* **7**, 1285.
129. Yang, H., Coombs, N., Dag, Ö., Sokolov, I. and Ozin, G. A. (1997). *J. Mater. Chem.* **7**, 1755.
130. Yang, H., Coombs, N. and Ozin, G. A. (1998). *J. Mater. Chem.* **8**, 1205.
131. Yang, H., Coombs, N. and Ozin, G. A. (1997). *Adv. Mater.* **9**, 811.
132. Sigal, G. B., Mrksich, M. and Whitesides, G. M. (1997). *Langmuir* **13**, 2749.
133. Trau, M., Yao, N., Kim, E., Xia, Y., Whitesides, G. M. and Aksay, I. A. (1997). *Nature* **390**, 674.

3

Towards the rational design of zeolite frameworks

PAUL WAGNER AND MARK E. DAVIS

3.1 Introduction

Zeolites are crystalline, microporous materials composed of oxide tetrahedra of silicon and aluminum that are three-dimensionally connected to give an $O/[Si + Al] = 2$. Each Al^{+3} in the oxide framework requires an extra-framework charge-balancing cation to maintain electroneutrality of the material. The charge-balancing cations for as-synthesized zeolites can be inorganic, organic or organometallic and can be post-synthetically changed to industrially useful cations such as protons and transition metals.

In addition to aluminosilicates, crystalline microporous materials can be phosphate-based. The aluminophosphate ($AlPO_4$) framework is electroneutral (analogue of SiO_2), and the aluminum and/or phosphorus tetrahedral atoms can be substituted by a number of metal and non-metal atoms that result in producing charged frameworks [1–3], e.g. Si^{+4} substitution for P^{+5}. In addition, numerous other metal oxide, and nitride based microporous materials have been reported recently [4, 5].

To date, approximately 125 different microporous structures have been reported to the International Zeolite Commission (IZC) for structure codes [6] and the interested reader can access this information via the following website: http://www.iza-structure.org. Each microporous material is assigned a three-letter code by the IZC (the IZC code will be used throughout the remainder of this chapter). These microporous materials are distinguished by the connectivity of the framework atoms that determine the size and dimensionality of the pores and/or cages that they possess.

The cage or pore size is denoted by the number of oxygen atoms (likewise the number of tetrahedral atoms) circumscribing the pore or cavity, e.g. a pore circumscribed by n oxygen atoms is referred to as an n membered-ring pore, or more simply, n MR. If the framework is composed of cages that can only be accessed through windows that are 6 MR or smaller the material is referred to as a clathrate [7]. Microporous materials containing pores and/or cages with molecular-sized windows (containing 8 MRs or larger) can have industrial utility for separations, ion exchange, and catalysis [8–11].

The dimensionality of the pore system is the degree of interconnections between the pores within the structure. For example, a one-dimensional (1-D) pore system contains channels that do not interconnect with each other, a two-dimensional (2-D) pore system allows interconnection of pores in only one crystallographic direction, and a three-dimensionally connected (3-D) pore system contains interconnections of the pores in two crystallographic directions to allow molecules to travel through the material in three dimensions (in the direction of the pores or through the interconnections in either of the remaining two crystallographic directions). In any 2-D or 3-D material, it is best not to consider the void space as connected channels but rather cages (intersections) that are connected by n MRs.

3.2 Synthesis

In Nature, zeolites commonly form as geothermally heated ground water passes through silicate, volcanic ash. Typically, in silica-rich volcanic deposits, heulandite (HEU) and analcime (ANA) tend to form, whereas in low-silica deposits chabasite (CHA), phillipsite (PHI) and natrolite (NAT) are favored [12]. Early attempts to synthesize zeolites centered around recreating the high-pressure, high-temperature conditions found in Nature. Barrer demonstrated the first successful zeolite synthesis (mordenite (MOR)) [13], while Milton developed large-scale zeolite syntheses at low temperatures and pressures that allowed zeolites to gain industrial importance [14]. These zeolite syntheses relied on the presence of alkali-metal hydroxides in the synthesis mixture to serve as a mineralizing agent. The alkali-metal cations also play a role in the structure direction of the particular zeolite that forms. Building on the concept of cationic structure direction, Barrer and Denny [15], and Kerr and Kokotailo [16] subsequently expanded the range of cations from the inorganic metal cations to organic cations. Using a quaternized amine (tetramethylammonium, TMA), Barrer successfully synthesized sodalite (SOD) and a high-silica version of zeolite A (LTA). The TMA cations were found to be occluded within the sodalite cages [17]. Because the TMA molecules are too large to diffuse into the sodalite cages, they must have been occluded during the crystal growth. Following this initial discovery of organic structure direction, tetraethylammonium (TEA), tetrapropylammonium (TPA) and tetrabutylammonium (TBA) were employed to synthesize the zeolites beta (*BEA) [18], ZSM-5 (MFI) [19] and ZSM-11/ZSM-5 (MEL/MFI) [20], respectively. The syntheses are all carried out under basic, hydrothermal conditions (pH > 10), that enhance the silicate dissolution and condensation reactions necessary for zeolite crystallization from the silica-containing reagents, e.g. amorphous silica, fumed silica, etc., present in the reaction media.

3.3 Organic molecule properties favoring zeolite formation

The use of organic molecules to direct the formation of high-silica zeolitic phases has become commonplace and typically involves the use of a quaternized amine. Kubota and coworkers have studied the relationship between the ability of organic molecules to direct the formation of high-silica zeolites and their C/N^+ ratio, hydrophobicity, and rigidity [21]. Rigid organic molecules are found to favor zeolite formation, presumably because molecules that undergo fewer conformational changes are better able to stabilize a smaller number of oxide frameworks through non-bonded organic/inorganic interactions. An optimal C/N^+ ratio is found to be between 11–14, and is interpreted in terms of the hydrophobicity of the organic molecule. Partitioning studies of quaternized amines between an aqueous solution and an organic chloroform layer reveal that organic molecules possessing intermediate hydrophobicity/hydrophilicity (as indicated by the similar preference of the molecule for the organic layer and the aqueous solution) tend to direct the formation of zeolites (Fig. 3.1) [21, 22]. These findings have been interpreted as follows: if the organic is too hydrophilic then interactions with the aqueous solvent are favored over the more hydrophobic silica. If the organic is too hydrophobic, it phase separates from the aqueous medium.

3.4 Organic–inorganic interactions leading to zeolite formation

In examining the organic–inorganic interactions that lead to zeolite formation we have focused our attention on the high-silica zeolites. The electrostatic interaction energy between the disordered framework defect sites and the charge-balancing extra-framework cations in these high-silica zeolite systems is neglected as it is likely to be constant from framework to framework [23]. Therefore, the additional, non-bonded van der Waals energy dominates the interaction between the organic molecules and the inorganic-oxide framework. The effects of the organic–inorganic van der Waals interactions have been

P. Wagner and M. E. Davies

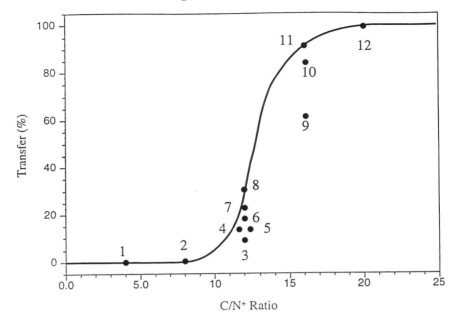

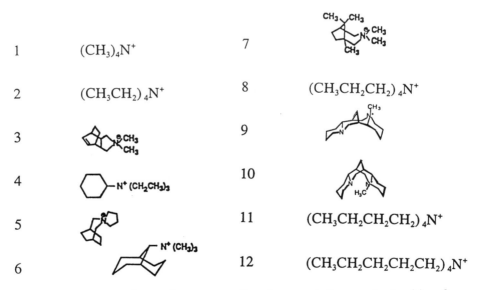

1	$(CH_3)_4N^+$	7	
2	$(CH_3CH_2)_4N^+$	8	$(CH_3CH_2CH_2)_4N^+$
3		9	
4		10	
5		11	$(CH_3CH_2CH_2CH_2)_4N^+$
6		12	$(CH_3CH_2CH_2CH_2CH_2)_4N^+$

Figure 3.1. Transport of organic structure directing agents from water to chloroform.

divided into three categories depending on the degree of interaction: pore filling, structure directing, and templating.

Pore filling describes a weak interaction of the organic molecule with the oxide framework in which there is not a unique specificity between the organic molecule and the inorganic framework. For example, there are at least 13 known organic molecules that can be used in the synthesis of ZSM-48, and many of these organic molecules can be employed to crystallize different zeolite phases depending on the synthesis conditions [23]. It is hypothesized that in the case of pore filling the organic molecules serve only to protect the internal surface from attack and dissolution by the mineralizing agent.

The term structure directing is employed to designate strong van der Waals interactions between the organic molecule and the oxide framework. However, there is not a correlation between symmetry of the organic molecule and the framework. Molecules that can act as structure-directing agents may possess some degree of rotational or translational freedom within the oxide framework. Structure direction of a high-silica zeolite by an organic molecule is a more common occurrence than templating. Pure-silica ZSM-5 that is structure directed by TPA is found to contain the TPA molecules at the channel intersections with the propyl arms extending into both the sinusoidal channels and the linear channels [24]. Owing to the lack of rigidity of the propyl moieties of the TPA molecules, the symmetry of the ZSM-5 structure does not conform to the symmetry of the TPA molecules; however, the TPA molecules are held tightly within the channel of the ZSM-5 material and can only be removed through calcination. This entrapment of the organic molecules occurs during crystal growth. Direct experimental support for the strong organic–inorganic van der Waals interaction is obtained from ^{1}H–^{29}Si cross-polarization magic-angle spinning (CP MAS) NMR spectra from freeze dried TPA-ZSM-5 samples that indicate that the protons of the organic molecules are in van der Waals contact with the oxide framework (~ 3.3 Å) [25, 26]. This short-range van der Waals interaction is observed to occur before the onset of the long-range order detected by X-ray diffraction. No ^{1}H–^{29}Si cross polarization occurs if TMA molecules are substituted for the TPA molecules, indicating that no van der Waals interactions occur between the TMA and the silica, and consequently no zeolite formation is observed.

The influence of the non-bonded organic/inorganic interactions in zeolite syntheses is illustrated by the following example in which sparteine derivatives are investigated as structure directing agents. Sparteine is a naturally occurring alkaloid that possesses a rigid, fused-ring, chiral structure. Nakagawa [27] and Lobo and Davis [28] succeeded in utilizing the sparteine-related compound

N(16)-methylsparteinium as a structure-directing agent to synthesize the high-silica molecular sieve with the AFI topology. Wagner *et al.* followed this work with the synthesis of CIT-5 (CFI) that is a high-silica molecular sieve with a 14-membered ring pore [29, 30].

The following two quaternized sparteine derivatives were synthesized recently and their performances in high-silica molecular sieve syntheses were investigated and compared [31].

(–)- α- Isosparteine   **A**

(–)-Sparteine   **B**

N(1)-methyl-α-isosparteinium (compound **A**) is found to form the pure-silica molecular sieve with the CFI topology that was first synthesized with its diastereomer N(16)-methylsparteinium (**B**). The crystallization time for the synthesis of pure-silica CFI using **A** as an organic agent is shorter than that with **B** and the crystal size of the product obtained with **A** ($\sim 2 \times 5 \mu m$) is also much smaller than that recovered when using **B** ($5 \times 15 \mu m$). Because the crystal size can be related to the number of nuclei, these results indicate that **A** is better able to stabilize the forming zeolite nuclei and increase the nucleation kinetics that result in a greater degree of nucleation and a faster rate of crystallization.

Molecular modeling studies reveal that the van der Waals energies of inter-action between the CFI framework and the SDA molecules calculated on the basis of 1 SDA per unit cell are -7.7 kcal for the A/CFI interaction and -5.3 kcal for the B/CFI interaction. These calculated stabilization energies correlate well with the experimental observations of a decrease in the crystal-lization time and an increase in the nucleation (smaller crystallite size) of CFI in the presence of **A** relative to **B**. The increased stabilization energy between **A** and the CFI framework compared with **B** appears primarily to be a result of the interaction between the N-methyl group of the organic agent with the eight-membered-ring pockets along the channel system of CFI. The cisoid orientation of the N(1)-methyl group (relative to N(16) lone-pair electrons) in **A** allows increased van der Waals contacts between the methyl hydrogens of **A**

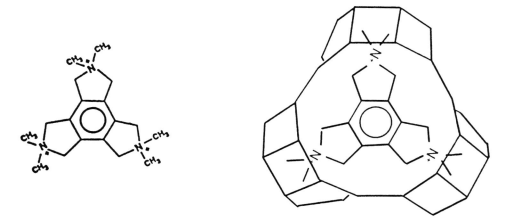

Figure 3.2. Triquat used to template the zeolite ZSM-18.

and the framework oxygens of the eight-membered-ring pocket relative to the transoid configuration of the N(16)-methyl group of **B**. These results indicate that although both **A** (N(1)-methyl-α-isosparteinium) and **B** (N(16)-methyl-sparteinium) serve specific structure-directing functions for the formation of pure-silica CFI, **A** is better able to stabilize the unique pore structure of CFI due to the increased van der Waals interactions of the cisoid N-methyl group with the side pockets along the CFI channels.

Finally, in the case of templating, there is a one-to-one correspondence between the organic molecule and the framework resulting in a silica structure that possesses similar symmetry to the organic molecule. ZSM-18 (MEI), synthesized using the triquaternium cation pictured in Fig. 3.2, is an exceptional example of organic templating of an inorganic zeolite structure [32]. The point-group symmetry of the triquat is D_{3h}, and is located within the ZSM-18 cages that possess the same symmetry as the organic. Furthermore, molecular modeling studies by Lobo and Davis reveal that the molecules are rigidly confined by the oxide framework and are not free to rotate or translate within the ZSM-18 cages [33]. As pointed out by Lobo and Davis, if the organic molecules are free to rotate within the zeolite pores then the framework needs to conform to the geometry of the volume generated by the rotation of the molecule rather than the volume of the molecule itself. This lack of free rotation is a necessary feature for templating, and implies that the synthesis of the highly sought-after chiral zeolite beta from optically active molecules must proceed through organic templating that inhibits free rotation of the molecules within the pores and imparts the symmetry of the chiral template molecule to the oxide framework.

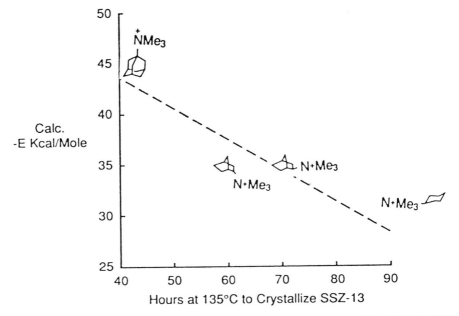

Figure 3.3. Plot of calculated energy of stabilization of organic SDA in SSZ-13 (CHA) vs. experimentally observed crystallization time.

3.5 Molecular modeling

As previously discussed, organic molecules are employed in the synthesis of high-silica zeolites and the organic–inorganic interactions that drive the selection of the zeolite are likely to be the non-bonded van der Waals stabilization energies of the organic molecule by the inorganic silica. Molecular-modeling investigations into the stabilization energies of organic molecules within the channels or cages of zeolites have led to a clearer understanding of the factors that govern zeolite syntheses.

For example, Harris and Zones have used molecular modeling to calculate the van der Waals energy of stabilization of SSZ-13 (CHA) from a variety of organic amine molecules [34]. The calculated van der Waals interaction energies correlate well with the experimental crystallization times (Fig. 3.3) and indicate that the stronger the organic–inorganic van der Waals interaction energy between the organic and the silica, the greater the stabilization of the growing zeolite framework. Small-angle X-ray scattering studies of zeolite crystallization have found that the time between the onset of crystallization (nucleation) and the completion of the crystallization is short compared with the crystal nucleation time [35, 36]. Therefore, the results of the SSZ-13 (CHA) study indicate that the kinetics of zeolite nucleation correlate with the energy of interaction.

Catlow and coworkers have calculated the van der Waals energy of interaction of a series of alkyl amines within a series of zeolite frameworks and compared their modeling results with experimental observations [37]. They correctly calculated the greatest stabilization energy for the experimentally observed quaternized amine/zeolite pairs: zeolite ZSM-11 and tetra-butyl ammonium, zeolite ZSM-5 and tetrapropyl ammonium, and zeolite beta and tetraethyl ammonium. Catlow and coworkers also compared the location of tetrapropyl ammonium in ZSM-5 from the modeling results to the single-crystal X-ray study of as-made ZSM-5 and found that the location of the tetrapropyl ammonium molecule within the ZSM-5 structure was correctly determined from the molecular modeling. These results provide support for the role of molecular modeling in understanding and predicting the organic/inorganic interactions that lead to zeolite formation.

3.5.1 Design of zeolite frameworks

A primary goal of these molecular modeling studies is to gain sufficient insight into the organic–inorganic interactions that lead to zeolite formation in order to rationally design zeolite frameworks. Rational design of zeolites can proceed either by specifying a desired hypothetical framework topology and ascertaining a suitable structure-directing molecule, or by specifying a particular organic molecule and identifying the potential zeolite phases that the organic may direct for.

Lewis and coworkers have developed a procedure in which molecules are computationally 'grown' in desired microporous frameworks in order to determine suitable structure-directing molecules [38]. The computational technique allows the molecules to be 'grown' in the microporous framework subject to space filling restrictions and is referred to as ZEBEDDE (zeolites by *de novo* design). ZEBEDDE proceeds by allowing the user to specify an inorganic host target together with a library of organic molecule fragments from which the SDAs will be constructed. A seed organic fragment is randomly placed into the inorganic framework and the molecule is 'grown' inside the host using a random series of seven actions that employ the user-supplied organic molecule fragment library as a source of new atoms. The allowed actions are 'Build', in which a new frament is bonded to an existing frament; 'Rotate', in which the new fragment is rotated about the new bond; 'Shake' and 'Rock', where the SDA is translated and rotated through the void space; 'Random bond twist', where a randomly selected bond is rotated; and 'Ring Formation', where atoms within a specified cut-off distance are joined to form a ring. Finally, the guest molecule is subjected to an energy minimization either in the gas phase or within the host framework. A cost function is also employed based on the overlap of van der Waals spheres to control the development of new molecules. This cost function is employed to assure that the organic molecule effectively fills the void space of the target microporous framework.

Using this technique, Lewis and coworkers were able to rationalize the efficacy of observed SDAs that have been employed to direct for known zeolite phases and also to predict novel SDAs to direct for the known molecular sieves. In particular, they were able to identify SDA molecules suitable for the synthesis of the previously identified phases: LEV, CHA, MFI, and EUO [38, 39]. Although the technique is effective in these cases it has not been employed to generate a novel zeolite topology.

In contrast to designating a framework and searching for a suitable SDA, zeolite design can occur through specifying a particular SDA that may favor novel molecular sieve phases. We have recently reported the synthesis and structure of three high-silica molecular sieves, SSZ-35 (STF), SSZ-36 (an intergrowth of ITE/RTH), and SSZ-39 (AEI), that are prepared from a library of 37 different cyclic and polycyclic quaternized amine molecules that are used as structure-directing agents (SDAs) [40]. The size and shape of the quaternized amine molecules were purposely designed to obtain novel zeolite structures. To overcome the propensity for SDA molecules to form clathrate structures, Nakagawa originally developed a designed, organic system that is based upon bringing together individual ring structures in a Diels–Alder reaction to build candidate molecules leading to large-pore zeolites [41]. The individual rings themselves, if converted into charged ammonium compounds, would generate clathrate products. Figure 3.4 shows three different-sized norbornane derivatives (based upon Diels–Alder chemistry) and the transition from clathrate to open-pore structure. Many of the rigid, elongated SDA derivatives formed from variations of cyclopentadiene reacting with maleimides produce open-pore, one-dimensional zeolites like MTW or SSZ-31 (shown in Fig. 3.4). These two 12-ring zeolites possess different configurations of the pore opening, with MTW containing a somewhat puckered pore opening [42] while SSZ-31 possesses a very open but highly elliptical pore opening with dimensions ranging from 8.8 Å to 5.5 Å [43].

The initial success of the effort to design the size and shape of the SDA molecules to preclude the formation of the clathrate or 1-D materials resulted from a derivative of the camphor-type molecule (Fig. 3.5) that led to the crystallization of the novel, open-framework zeolites SSZ-35 (STF) and SSZ-36 (ITE/RTH). Subsequently, numerous other zeolite structure-directing molecules were discovered using this design strategy and are presented in reference [41].

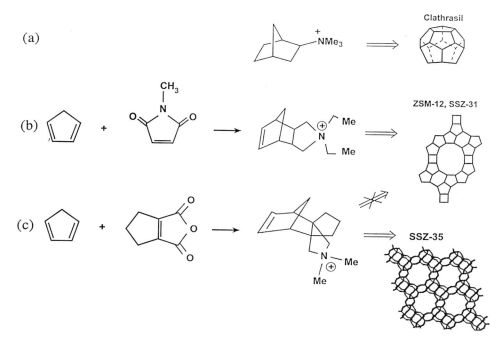

Figure 3.4. Three guest molecules based upon extensions of the Diels–Alder reaction and the types of zeolite generated as the spatial features change. (a) The norbornyl derivative is still small enough to generate cage-centered clathrate structures like nonasil (NON). (b) The tricyclic derivatives with long, central axes produce one-dimensional, large-pore zeolites. (c) The pseudo propellane developed in this guest inhibits the formation of one-dimensional large-pore zeolites and generates zeolites with cavities like SSZ-35. Adapted from [40].

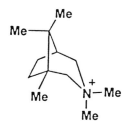

Figure 3.5. A derivative of the camphor-type molecule that led to the crystallization of the novel, open-framework zeolites SSZ-35 (STF) and SSZ-36 (ITE/RTH).

The strategy of designing organic, zeolite directing molecules that are too large or of the wrong geometry to stabilize commonly encountered competing phases has proven successful in developing novel zeolite phases. The primary disadvantage of this process is the lack of control over the framework topology of the zeolite that may form. However, future advances in our understanding of the organic–inorganic interaction and the effects of the inorganic synthesis mixture on molecular-sieve formation may provide greater predictive power over the potential zeolite phases that may form given a particular organic SDA. Next, we provide an example of a strategy for designing a novel zeolite framework.

3.5.2 A rational zeolite design challenge: the design and synthesis of a hypothetical 14 MR zeolite related to the high-silica zeolite SSZ-48

Here, we outline a method for the rational design of a hypothetical extra-large-pore, high-silica zeolite related to the newly discovered high-silica zeolite SSZ-48 [44, 45]. The design strategy employs molecular modeling to sort a set of organic molecule candidates in order to find an appropriate structure-directing agent for the synthesis of the hypothetical, high-silica zeolite.

SSZ-48 is a high-silica zeolite that was synthesized by Zones and coworkers at Chevron using *N,N*-diethyldecahydroquinolinium as the structure-directing agent [44]. We recently solved the structure of SSZ-48 from electron diffraction data and found that the framework structure of this material contains a 1-D, sinusoidal, pore system circumscribed by 12 tetrahedrally coordinated silicon atoms (12 membered-rings, 12 MR) [45]. The pores are elliptical in shape with projected dimensions of 7.91 Å × 10.05 Å (center of oxygen to center of oxygen) and a maximum internal pore diameter of 11.19 Å × 13.65 Å (center of oxygen to center of oxygen). The asymmetric unit contains seven tetrahedrally coordinated silicon atoms (T-atoms) and 14 oxygens resulting in a unit cell content of $[Si_{14}O_{28}]$. The framework density (FD) of the material is 18.7 T-atoms/1000 Å^3. The refined model of SSZ-48 is shown in Fig. 3.6.

The connectivity of the T-atoms in the SSZ-48 structure is similar to the connectivity of the T-atoms in the TON structure that contains 10 MR pores (Fig. 3.7a). Both structures are composed of columns of $[5^4.6^1]$ units as seen in Fig. 3.7. Within these columns the connectivity is identical; however, in the SSZ-48 structure (Fig. 3.7b) the columns are expanded by the insertion of rings of four T-atoms diagonally across the pore. This expansion can occur at the joining of any of the $[5^4.6^1]$ units. The SSZ-48 material is neither disordered nor faulted, and indicates a strong structure-directing effect of the *N,N*-diethyldecahydro-quinolinium SDA for the SSZ-48 topology. If the four T-atom rings are

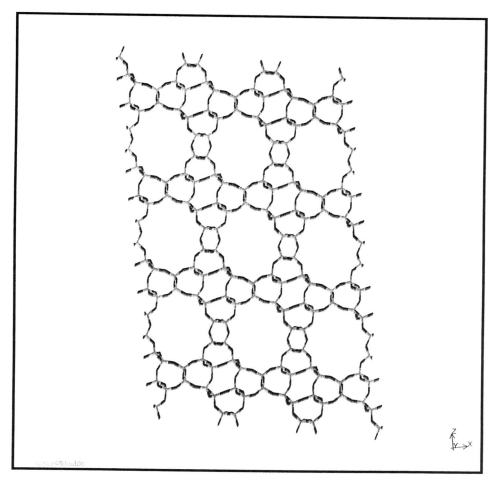

Figure 3.6. Refined structure of SSZ-48.

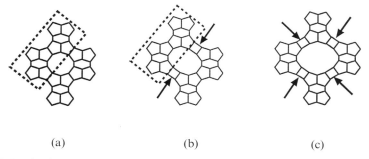

(a)   (b)   (c)

Figure 3.7. Projection of (a) TON structure, (b) SSZ-48 structure, and (c) HYP 14 MR structure. Dotted box indicates [$5^4.6^1$] columns and arrows indicate 4 MR sigma expansion. Adapted from [45].

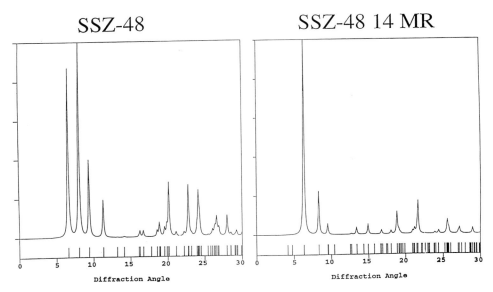

Figure 3.8. Simulated powder X-ray diffraction data for SSZ-48 (left) and HYP-14 MR (right) structures.

inserted at the joining of every $[5^4.6^1]$ unit, a new extra-large pore framework structure is generated that contains 14 T-atom pore openings and has a maximum internal pore diameter of 14.4 Å × 12.8 Å (Fig. 3.7c). The simulated powder X-ray data for SSZ-48 and the hypothetical 14 MR structure are presented in Fig. 3.8.

The 4 MR expansions that relate the zeolites TON (10 MR) – SSZ-48 (12 MR) – HYP (14 MR) have been referred to as sigma expansions, and this type of ring expansion is also observed in the zeolite series of ZSM-48 (10 MR) – SSZ-31 (12 MR) – UTD-1 (14 MR) [46] as illustrated in Fig. 3.9. Within the ZSM-48 sigma expansion series, the zeolites have very similar interconnections of the silica tetrahedra except for the expansion of the columns of $[5^4.6^1]$ units by the 4 MR. Viewed along the pore direction these materials appear to be closely related to the TON sigma expansion series of zeolites; however, the interconnections of the atoms along the channel direction differ significantly. As a result, the TON sigma expansion series possesses sinusoidal channels while the ZSM-48 sigma expansion series possesses straight cylindrical channels.

Although not by design, Balkus and coworkers have successfully synthesized the ZSM-48 – SSZ-31 – UTD-1 sigma expansion series by incrementally increasing the size of a cobaltocenium-derivatized structure-directing agent. They found that Cp_2Co^+ directs for ZSM-48, $(MeCp)_2Co^+$ directs for SSZ-31 and $Cp^*_2Co^+$ directs for the 14 MR zeolite, UTD-1 [47].

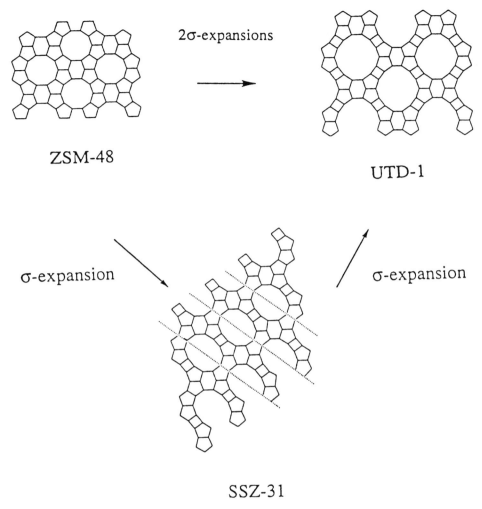

Figure 3.9. ZSM-48 – SSZ-31 – UTD-1 sigma expansion series. Adapted from [43].

In order to find a suitable candidate to direct for the HYP-14 MR material that is related to SSZ-48, a library of bulky decahydroquinolinium derivatized molecules was developed and the van der Waals energy of stabilization of the molecules within the pores of SSZ-48 and the related HYP-14 MR material was calculated using molecular modeling. In addition to stabilizing the HYP-14 MR structure over the SSZ-48 structure, the stabilization of the commonly encountered large-pore zeolite *BEA by the molecules is calculated in order to locate an SDA molecule that also stabilizes the HYP-14 MR structure over *BEA (in high-silica syntheses, often, MFI, MTW and *BEA quickly form [48]; with the organic outlined below, MFI and MTW could not be crystallized because the molecules are too large to reside within their void spaces). The library of derivatized decahydroquinolinium molecules selected for the molecular modeling is shown in Fig. 3.10; the molecules were selected based on their rigidity and hydrophobicity.

The molecular modeling calculations were performed using the Open Force Field module of Cerius^2 [49]. The calculations were carried out by first conducting Monte Carlo sorption simulations at a fixed loading of 1 molecule per unit cell. The sorption simulation probed 50 000 SDA/zeolite configurations and the lowest energy configuration was energy minimized using the Burchart–Universal force field. The framework atoms were fixed during the energy minimization and the coulomb terms of the non-bonded energy calculations were ignored. The calculated stabilization energy represents the non-bonded van der Waals interaction between the framework and the SDA.

As can be seen from Fig. 3.10, the diethyl- (Fig. 3.10a), cyclopentyl- (Fig. 3.10b) and cyclohexyl- (Fig. 3.10c) decahydroquinolinium derivatives all show greater stabilization of the SSZ-48 structure than both the HYP-14 MR structure and *BEA. It is interesting to note that both the cyclopentyl- and the cyclohexyl- derivatives show greater calculated energies of stabilization for SSZ-48 than the N,N-diethyldecahydroquinolinium SDA that was originally used in the synthesis SSZ-48. The synthesis of SSZ-48 from N,N-diethyldecahydroquinolinium requires more than 60 days and produces nanocrystals. If the energy of interaction correlates with the crystallization time for the zeolite, as the work of Zones and Harris suggests [34], then both the cyclopentyl- and cyclohexyl-decahydroquinolinium derivative may prove to be more viable SDAs for the SSZ-48 material than the N,N-diethyldecahydroquinolinium SDA originally used.

From the list of molecules illustrated in Fig. 3.10, it is seen that the 3-methyl-cyclohexyl-decahydroquinolinium derivative (Fig. 3.10d) stabilizes the HYP-14 MR structure over the SSZ-48 structure. This molecule is too large to fit into the SSZ-48 structure as indicated by the positive energy of interaction. However, the calculated energy of interaction of the 3-methyl-cyclohexyl-decahydroquinolinium derivative in the channel system of *BEA indicates that this competing phase is stabilized over the HYP-14 MR structure. These results indicate the importance of modeling SDAs both in the target structure and in potential competing phases in order to locate SDAs that may reasonably direct for the target phase.

From the set of molecules presented in Fig. 3.10, only the dicyclohexyl-decahydroquinolinium derivative (Fig. 3.10e) is calculated to stabilize the HYP-14 MR structure over both *BEA and SSZ-48. The positive energies of interaction between the dicyclohexyl-decahydroquinolinium derivative and

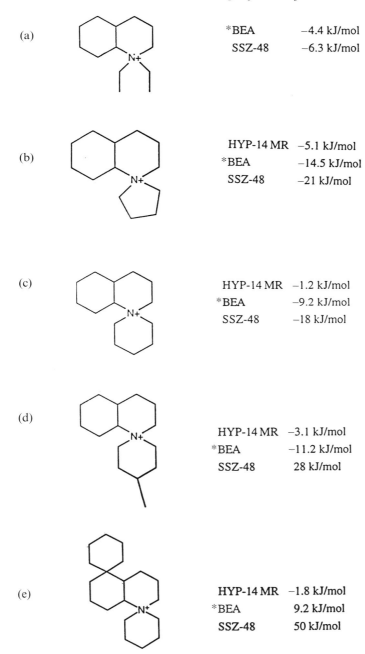

(a)

*BEA   −4.4 kJ/mol
SSZ-48   −6.3 kJ/mol

(b)

HYP-14 MR   −5.1 kJ/mol
*BEA   −14.5 kJ/mol
SSZ-48   −21 kJ/mol

(c)

HYP-14 MR   −1.2 kJ/mol
*BEA   −9.2 kJ/mol
SSZ-48   −18 kJ/mol

(d)

HYP-14 MR   −3.1 kJ/mol
*BEA   −11.2 kJ/mol
SSZ-48   28 kJ/mol

(e)

HYP-14 MR   −1.8 kJ/mol
*BEA   9.2 kJ/mol
SSZ-48   50 kJ/mol

Figure 3.10. Library of candidate organic molecules for HYP-14 MR synthesis and calculated energy of interaction of each molecule in SSZ-48, *BEA, and HYP-14 MR structure.

both BEA and SSZ-48 indicate that this molecule is too large to fit into the channel system of either *BEA or SSZ-48.

Recalling from the early discussion, the experimentally observed properties that tend to make an organic amine a good SDA are the structural rigidity and hydrophobicity of the molecule (C/N^+ ratio optimally between 11 and 14). While the dicyclohexyl-decahydroquinolinium derivative (Fig. 3.10e) is calculated to stabilize the HYP-14 MR structure over SSZ-48 and *BEA and is rigid, i.e. undergoes few conformational fluctuations due to the fused ring structure of the molecule, the high C/N^+ ratio = 19 indicates that the molecule may have solubility problems in the synthesis mixture. To increase the hydrophilicity of the SDA molecule, nitrogen atoms can be substituted into the ring system of the molecule. As seen from the partioning studies of the methylsparteine SDAs (Fig. 3.1), the addition of unquaternized nitrogen atoms into the ring structure of the SDA can increase the hydrophilicity of the molecule, thereby creating an effective zeolite SDA with intermediate hydrophobicity.

3.6 Conclusions

The essential features of the organic molecules that endow them with zeolite structure-directing properties include structural rigidity and intermediate hydrophobicity. Rigid organic molecules are found to favor zeolite formation presumably because molecules that undergo fewer conformational changes are better able to stabilize a smaller number of oxide frameworks through non-bonded organic–inorganic interactions. In addition, as discussed previously, molecules that possess intermediate hydrophobicity (C/N^+ ratio ~11–14) tend to favor high-silica molecular-sieve formation.

Our understanding of the factors that influence the formation of microporous materials has advanced significantly, partly as a result of the dramatic increase in the number of new zeolite phases that have been synthesized and that have had their structures solved. Given the large number of new microporous materials that have been discovered recently, the guest–host relationship of the organic molecules within the inorganic host can be studied in greater detail. In addition, the increased power of computational techniques that probe the organic–inorganic interaction energy has allowed significant insight to be gained into the factors that effect zeolite formation.

Here, we have combined experimental information with computational modeling calculations to outline the design of a new, extra-large-pore, high-silica zeolite. The sigma expansion of the SSZ-48 structure to form the HYP-14 MR structure has been observed in the ZSM-48 – SSZ-31 – UTD-1 zeolite series synthesized from successively bulkier cobaltocenium derived

SDAs. A route to the synthesis of the HYP-14 MR material, which is related to SSZ-48 by a sigma expansion, is outlined. A set of bulky, rigid decahydroqui-nolinium-derivatives was developed and molecular modeling employed to uncover a suitable SDA that stabilizes the desired HYP-14 MR structure over the SSZ-48 structure and over the potential competing large-pore phase, *BEA. An amine analogue of dicyclohexyl-decahydroquinolinium derivative (Fig. 3.10e) is identified as a suitable SDA for the HYP-14 MR structure using computational modeling techniques, and the decahydroquinolinium derivative also possesses the requisite rigidity and hydrophobicity to be an effective SDA.

The extra-large-pore HYP-14 MR possesses a larger pore opening than either of the other 14 MR high-silica zeolites known to exist, UTD-1 [50] and CIT-5 (CFI) [28–30], and could find great utility in the catalysis of molecules of pharmaceutical interest that are too large to enter the pores of the zeolite cata-lysts currently available. This route would represent the first successful example of the directed design of a novel framework zeolite, and highlights the poten-tial of computational techniques in conjunction with experimentally obtained information for designing zeolite structures. The synthesis of the HYP-14 MR material using the zeolite design strategy outlined here could open the possibil-ity for the design of other valuable hypothetical zeolite structures.

3.7 References

1. Feng, P., Bu, X. and Stucky, G. D. (1997). *Nature* **388**, 735.
2. Noble, G. W.,Wright, P. A., Lightfoot, P., Morris, R. E., Hudson, K. J., Kvick, Å. and Graafsma, A. (1997). *Angew. Chem. Int. Ed. Engl.* **36**, 81.
3. Bennett, J. M. and Marcus, B. K. (1998). *Stud. Surf. Sci. Catal.* **37**, 269.
4. Huppertz, H. and Schnick, W. (1997). *Angew. Chem. Int. Ed. Engl.* **36**, 2651.
5. Chen, J., Jones, R. H., Natarajan, S., Hursthouse, M. B. and Thomas, J. M. (1994). *Angew. Chem. Int. Ed. Engl.* **33**, 639.
6. Meier, W. M., Olson, D. H. and Baerlocher, C. (eds.) (1996). *Atlas of Zeolite Structure Types*, 4 edn. Elsevier.
7. McCusker, L. B. (1998). *J. Appl. Cryst.* **21**, 305.
8. Townsend, R. P. (1991). *Stud. Surf. Sci. Catal.* **58**, 259.
9. Kurzendorfer, C.P. *et al.* (1987). *Colloid Polymer Sci.* **265**, 542.
10. Upadek, A. and Krings, P. (1989). *Stud. Surf. Sci. Catal.* **49**, 701.
11. Bellussi, G. and Fattore, V. (1991). *Stud. Surf. Sci. Catal.* **69**, 79.
12. Barrer, R. M. (1982). *Hydrothermal Chemistry of Zeolites*. Academic Press.
13. Barrer, R. M. (1948). *J. Chem. Soc.*, 127.
14. Milton, R.M. (1959). U.S. 2882243.
15. Barrer, R. M. and Denny, P. J. (1961). *J. Am. Chem. Soc.*, 971.
16. Kerr, G. T. and Kokotailo, G. (1961). *J. Am. Chem. Soc.* **83**, 4675.
17. Baerlocher, C. and Meier, W. M. (1970). *Helv. Chim. Acta.* **53**, 1285.
18. Newsam, J. M. *et al.* (1988). *Proc. R. Soc. Lond. A* **420**, 375.
19. Kokotailo, G. T. *et al.* (1978). *Nature* **272**, 437.
20. Kokotailo, G. T. *et al.* (1978). *Nature* **275**, 119.

21. Kubota, Y. *et al.* (1996). *Microporous Mater.* **6**, 213.
22. Goretsky, A. V., Beck, L. W., Zones, S. I. and Davis, M. E. (1999). *Microporous Mater.* **28**, 387.
23. Koller, H., Lobo, R. F., Burkett, S. L. and Davis, M. E. (1999). *J. Phys. Chem. B* **33**, 12588.
24. Franklin, K. R. and Lowe, B. M. (1988). *Stud. Surf. Sci. Catal.* **49**, 174.
25. van Koningsveld, H., van Bekkum, H. and Jansen, J. C. (1987). *Acta. Cryst. B* **43**, 127.
26. Burkett, S. L. and Davis, M. E. (1994). *J. Phys. Chem.* **98**, 4647.
27. Nakagawa, Y. US Patent 5, 271, 922.
28. Lobo, R. F. and Davis, M. E. (1994). *Microporous Mater.* **3**, 61.
29. Wagner, P., Yoshikawa, M., Lovallo, M., Tsuji, K, Tsapatsis, M. and Davis, M. E. (1997). *J. Chem. Soc. Chem. Commun.*, 2179.
30. Yoshikawa, M., Wagner, P., Lovallo, M., Tsuji, K., Takewaki, T., Chen, C.-Y., Beck, L. W., Jones, C. W., Tsapatsis, M., Zones S. I. and Davis, M. E. (1998). *J. Phys. Chem. B* **102**, 7139.
30. Barrett, P. A., Diaz-Cabanas, M. J., Camblor, M. A. and Jones, R. H. (1998). *J. Chem. Soc. Faraday Trans.* **94**, 2475.
31. Tsuji, K., Wagner, P. and Davis, M. E. (1999). *Microporous Mesoporous Mater.* **28**, 461.
32. Lawton, S. L. and Rohrbaugh, W. J. (1990). *Science* **247**, 1319.
33. Lobo, R. F. and Davis, M. E. (1992). *Chem. Mater.* **4**, 756.
34. Harris, T. V. and Zones, S. I. (1994). *Stud. Surf. Sci. Catal.* **84**, 29.
35. de Moor, P.-P. E. A. *et al.* (1997). *Microporous Mater.* **9**, 117.
36. de Moor, P.-P. E. A. *et al.* (1999). *Chem. Mater.* **11**, 36.
37. Bell, R. G. *et al.* (1994). In *Zeolites and Related Microporous Materials: State of the Art 1994* (J. Weitkamp, H. G. Karge, and W. Hoelderich, Eds.), p. 2075. Elsevier: Amsterdam.
38. Lewis, D. W., Willock, D. J., Catlow, R. A., Thomas, J. M. and Hutchings, G. J. (1996). *Nature* **382**, 604.
39. Willock, D. J., Lewis, D. W., Catlow, R. A., Hutchings, G. J. and Thomas, J. M. (1997). *J. Mol. Catal. A.* **119**, 415.
40. Wagner, P., Nakagawa, Y., Lee, G., Davis, M. E., Elomari, S., Medrud, R. C. and Zones, S. I. (2000) *J. Am. Chem. Soc.* **122**, 263.
41. Nakagawa, Y. and Zones, S. I. (1992). In *Molecular Sieves*, Vol. 1 (M. L. Occelli and H. E. Robson, Eds.), p. 222. Van Nostrand Reinhold: New York.
42. La Pierre, R. B., Rohrman, A. C. Jr., Schlenker, J. L., Wood, J. D., Rubin, M. K. and Rohrbaugh, W. J. (1985). *Zeolites* **5**, 346.
43. Lobo, R. F., Tsapatsis, M., Freyhardt, C. C., Chan, I. Y., Chen, C. Y., Zones, S. I. and Davis, M. E. (1997). *J. Am. Chem. Soc.* **119**, 3732.
44. International Patent Appl. WO 98/29336.
45. Wagner, P., Terasaki, O., Ritsch, S., Nery, J. G., Zones, S. I., Davis, M. E. and Hiraga, K. (1999) *J. Phys. Chem. B.* **103**, 8245.
46. Lobo, R. F., Tsapatsis, M., Freyhardt, C. C., Chan, I., Chen, C. Y., Zones, S. I. and Davis, M. E. (1997). *J. Am. Chem. Soc.* **119**, 3727.
47. Balkus, K. J., Rmaeraran, A. R., Szostak, R. and Mitchel, M. (1998). In *Abstracts of the 12th International Zeolite Conference*, 176.
48. Tsuji, K. and Davis, M. E. (1997). *Microporous Mater.* **11**, 53.
49. Cerius2 v3.5, Molecular Simulations Inc., 1997.
50. Freyhardt, C. C., Tsapatsis, M., Lobo, R. F., Balkus, K. J. and Davis, M.E. (1996). *Nature* **381**, 295.

4

Mesoscale self-assembly: the assembly of micron- and millimeter-sized objects using capillary forces

NED BOWDEN, JOE TIEN, WILHELM T. S. HUCK
AND GEORGE M. WHITESIDES

This chapter describes experimental and conceptual issues in mesoscale self-assembly (MESA), using examples from our work in the assembly of millimeter- and micron(micrometer)-sized polyhedral objects using capillary forces. In MESA, objects (from nm to mm in size) self-assemble into ordered arrays through noncovalent forces. Three systems that use capillary forces in MESA are described: these involve the assembly of objects into two-dimensional arrays at the perfluorodecalin/H_2O interface, into three-dimensional arrays at curved liquid/liquid interfaces, and into three-dimensional arrays from a suspension in water. The capillary interactions between objects can be viewed as a type of 'bond' that is analogous to chemical bonds that act between atoms and molecules.

4.1 Introduction

4.1.1 Scope and objectives of the review

Mesoscale self-assembly (MESA) is the self-assembly of objects ranging in size from 3 nm to 10 mm, and interacting through a variety of forces (capillary, magnetic, electrostatic, and light forces). Examples of MESA include the aggregation of gold colloids coated with DNA-terminated self-assembled monolayers (SAMs) using the formation of DNA/DNA duplexes (nanometer-scale self-assembly) [ref. 1]; the crystallization of polystyrene spheres using capillary or electrostatic forces (micron-scale self-assembly) [refs. 2–4]; and the formation of ordered structures from polyhedral objects of polydimethylsiloxane using lateral capillary forces (millimeter-scale self-assembly) [refs. 5, 6]. This chapter describes the area of MESA, with a focus on our work using capillary forces to assemble small objects into ordered arrays or aggregates [refs. 5–15].

This chapter has three parts. The first section outlines the conceptual and experimental issues in MESA: thermodynamics, forces, fabrication of objects, and characterization of arrays. The second section describes studies using MESA that illustrate the level of control over the structure of the arrays that can be achieved in these systems. The third section describes what is known and unknown in MESA and where more work is needed.

4.1.2 Motivation for studying MESA

We study MESA for three reasons. (i) MESA bridges the gap between molecular self-assembly, which has been successful at the nanometer level, and conventional fabrication of machines and parts, which has been successful for scales greater than 100 μm [refs. 16–23]. Few techniques exist to assemble or fabricate objects or arrays in the size region between several nanometers and hundreds of microns, and new techniques in this regime would be welcome. (ii) We wished to develop systems of self-assembly in which we could control the parameters affecting self-assembly more easily than we can with molecules. (iii) We wished to extend the ideas and methods of self-assembly in chemistry and biology to self-assembly on the mesoscale.

4.1.3 Definitions

Self-assembly

Self-assembly is the spontaneous organization of molecules or objects, under steady-state or equilibrium conditions, into stable aggregates, by noncovalent forces; these aggregates are not necessarily at a minimum in energy [refs. 19, 24]. A system that self-assembles can do so either reversibly or irreversibly. If the latter, the components must be able to adjust their positions within the aggregate if a highly ordered structure is to form. Irreversible aggregation without adjustment leads to glasses.

Molecular self-assembly

Molecular self-assembly is defined as the spontaneous condensation of molecules into ordered arrays by noncovalent forces [refs. 20, 24]. For molecular self-assembly to be reversible, the free energy of interaction, $\Delta G°$, between two molecules must be comparable to the average thermal energy, kT, of molecules. Molecular self-assembly ordinarily involves weak interactions (e.g. hydrogen bonds, van der Waals interactions, and hydrophobic interactions). Although one, two, or all three dimensions of the final array may extend into the millimeter scale, the size of the molecules is usually less than 5 nm; notable exceptions are biological macromolecules, especially double-stranded DNA.

Mesoscale self-assembly

Mesoscale self-assembly is the spontaneous organization of *objects* into ordered arrays through noncovalent forces [refs. 7, 12]. We think of an 'object' as a structurally defined entity that is larger than molecules and molecular aggregates: examples include colloids; nanotubes and nanoparticles; viruses and cells; and micron- and millimeter-sized objects made from metals, polymers, or ceramics. Objects involved in mesoscale self-assembly can, in principle, range in size from several nanometers (colloids) to thousands of kilometers (planets and suns), depending on the range of the forces that hold them together.

The word 'mesoscopic' has two separate definitions. The first definition is 'middle'; that is, intermediate in size between small and large [ref. 25]. In our usage of MESA, 'small' is molecular, and 'large' is easily manipulated by hand. We describe objects that are between 3 nm and 10 mm as mesoscopic. A second, complementary, definition of mesoscopic is that the size of the object is comparable to the size of the phenomenon being exploited (or used as a probe) [refs. 26, 27]. In studying light, a mesoscopic object may have dimensions on the order of the wavelength of light ($0.1-10\ \mu m$); in studying electron mobility in a semiconductor, a mesoscopic object may have the dimensions of the mean free path of an electron (10–40 nm); in studying acoustics, a mesoscopic object may have dimensions of the wavelength of sound (0.01–20 m) [refs. 28–30].

Capillary forces

We will describe two mesoscale, self-assembling systems in which the interactions between objects are based on capillary forces. The first is based on polyhedral polydimethylsiloxane (PDMS) objects at a perfluorodecalin (PFD)/H_2O interface. These objects have their faces patterned to be either hydrophobic or hydrophilic, and they assemble via lateral capillary forces that originate from interactions between these faces (Fig. 4.1a). The second system uses polyhedral objects that are suspended in water and have selected faces covered with a water-insoluble liquid – either a hydrophobic organic liquid or a liquid metal solder; these objects assemble via capillary forces into three-dimensional (3D) structures (Fig. 4.1b).

In two-dimensional (2D) MESA, the capillary forces result from interactions between menisci that extend from the faces of the objects [refs. 2–4, 31]. The menisci are the curved parts of the interface near the faces of the objects. In the

N. Bowden et al.

(a)

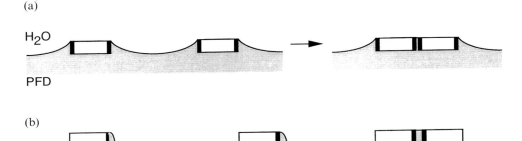

(b)

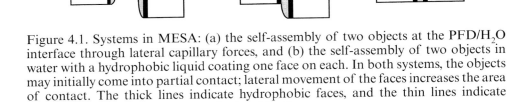

Figure 4.1. Systems in MESA: (a) the self-assembly of two objects at the PFD/H$_2$O interface through lateral capillary forces, and (b) the self-assembly of two objects in water with a hydrophobic liquid coating one face on each. In both systems, the objects may initially come into partial contact; lateral movement of the faces increases the area of contact. The thick lines indicate hydrophobic faces, and the thin lines indicate hydrophilic faces.

work involving polymer plates floating at the PFD/H$_2$O interface, the PFD wets the hydrophobic faces and forms 'positive' menisci that extend above the plane of the interface; water wets the hydrophilic faces and forms 'negative' menisci that extend below the plane (Fig. 4.2a).

There are two components that contribute to the force between objects interacting through overlapping menisci: a change in the surface area of the interface as the objects move relative to one another (capillary force), and a change in the mean level of the PFD/H$_2$O interface as the objects move relative to one another (gravitational force). The capillary force reflects the fact that the molecules at the interface have a higher energy than the molecules in the bulk liquid. When two menisci of the same shape approach one another, both the capillary and gravitational energies decrease, and the objects attract one another (Fig. 4.2b, c). A mismatch in the shape of two menisci decreases the overall change in free energy upon assembly; as a result, the forces between objects with mismatched menisci are less than between objects with matched ones. When a positive meniscus interacts with a negative meniscus, the force between them is repulsive rather than attractive (Fig. 4.2d).

Lateral capillary forces have several useful characteristics. First, the characteristic length for decay of the menisci (from a face several millimeters in width and height) is of the order of a millimeter. Objects of this size are easy to fabricate and to observe. Second, capillary forces are well understood. Their description by the Laplace equation is clear, although often mathematically intractable [ref. 2]. Third, capillary forces are comparable in strength to shear

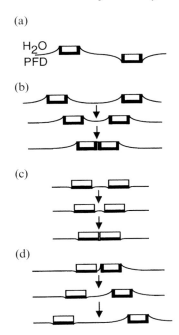

Figure 4.2. There are three types of capillary interaction between objects with positive and negative menisci at the PFD/H$_2$O interface. In (a), we outline nomenclature: one of these objects has approximately the same density as water; the second has approximately that of PFD. The PFD wets the hydrophobic faces (indicated by thick lines) and forms positive menisci; the water wets the hydrophilic faces (indicated by thin lines) and forms negative menisci. In (b), the PFD wets the hydrophobic faces and forms positive menisci. When two faces come together, the net area of the PFD/H$_2$O interface decreases. This decrease is favorable in free energy, brings the pieces together, and holds them in close proximity. In (c), the objects are hydrophilic and sink slightly into the PFD/H$_2$O interface to create small negative menisci. When these objects approach one another, the area of the interface decreases. In (d), a hydrophobic face with a positive meniscus is repelled by a hydrophilic face with a negative meniscus. When these objects move toward one another, the area of the interface *increases*. Positive and negative menisci repel.

forces, the primary form of agitation in our experiments. Fourth, the sense of the menisci at a face (positive or negative) can be changed by varying the hydrophobicity of the face and the densities of the object and the two fluid phases. Fifth, the objects can move relative to one another at the fluid/fluid interface with relatively little resistance; this freedom of movement allows them both to come together and to adjust their positions relative to one another.

In the systems we have used for 3D MESA, the capillary forces act through thin liquid films that coat the faces of the objects [refs. 6, 15]. These films possess a high interfacial free energy in contact with water. When two surfaces coated with liquid come into contact, the films coalesce and decrease the area of the interface, and the free energy of the system decreases. These liquid films act as lubricants that allow lateral motion of the objects relative to one another to optimize orientation of objects. Unlike the lateral capillary forces that act at the PFD/H_2O interface, the capillary forces in 3D MESA act over short separations; the liquid lubricants on the faces of two objects must be in contact.

An advantage of using capillary forces in MESA is that they can be used at a variety of size scales. In previous work lateral capillary forces have been used to cause the assembly of objects ranging from millimeter-sized PDMS objects, to micron-sized polystyrene beads, to nanometer-sized proteins, into 2D arrays [refs. 7, 32, 33]. Although not all of the assemblies have taken place at a PFD/H_2O interface, it should be possible to do so.

We believe that capillary forces used in 3D MESA will be strong enough to assemble nanometer- and micron-sized objects into ordered arrays, and we base this belief on the scaling of forces in MESA. The capillary forces scale with the area of the objects covered with the liquid adhesive. The forces that break apart the arrays scale, in some systems, with the volume of the objects. A decrease of an order of magnitude in the size of the objects corresponds to a decrease of two orders of magnitude in the strength of the capillary forces, and a decrease of three orders of magnitude in the strength of the forces that break apart the arrays. That is, capillary forces become stronger relative to the forces responsible for agitation, at smaller size scales. If capillary forces are strong enough to assemble millimeter-sized objects into ordered arrays, then they should be strong enough to assemble nanometer- and micron-sized objects as well. In a later section, we will show examples of micron-sized objects that have been assembled into ordered arrays using capillary forces.

Another instructive way to view capillary forces in MESA in three dimensions is to consider their analogy with the hydrophobic effect. The capillary forces in some of the systems we describe – especially those at the PFD/H_2O interface – are essentially due to the hydrophobic effect, expressed at the mesoscale. Because the hydrophobic effect is strong on the molecular scale, capillary forces should be able to hold nanometer-sized objects in aggregates against thermal agitation. This view provides a link between molecular self-assembly and MESA using capillary forces.

Capillary bond

By definition, any interaction between objects that tends to hold them together is a bond. In our work, capillary interactions provide the bond between objects. In general, a bond between two molecules or objects has two components: one attractive, and one repulsive (Fig. 4.3). At some distance, these forces are equal

(a)

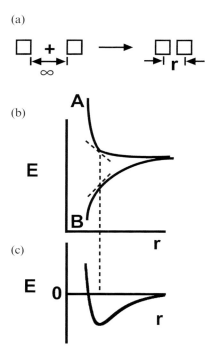

(b)

(c)

Figure 4.3. The potential functions for capillary bonds and for most types of molecular bonds have the same general characteristics. (a) The boxes represent atoms, molecules, or objects that assemble to a finite distance, r (m), apart. (b) There are at least two competing interactions between the objects, one attractive (**B**) and one repulsive (**A**). (c) The net interaction has a minimum in energy at some distance; at this distance, the net force moving the objects relative to one another is zero.

and opposite, and the objects rest at (or oscillate around) a stable separation – that is, the 'bond length'.

We can inquire if there are analogies between capillary bonds – acting on objects – and chemical bonds (whether covalent or noncovalent) – acting on atoms and molecules. In terms of the underlying physics, capillary and atomic/molecular bonds are very different. The capillary bond is largely a result of the high surface tension of the fluid/fluid interface, and as such (and like the hydrophobic effect) is due to the increased order and decreased entropy of water near an interface [ref. 34]. It is largely a collective property of water, and it appears at room temperature as a contribution to ΔG from $T\Delta S$. Atomic and molecular bonds reflect interactions between atoms or molecules and are primarily covalent or electrostatic, and contribute to free energy predominantly through ΔH. Despite these differences, there are also important similarities, at least in form. (i) Both capillary and atomic/molecular bonds are reversible at some level of agitation. (ii) Both capillary and atomic/molecular bonds can be discussed in terms of overlap in space of their components: for capillary bonds, of the menisci; for covalent bonds, of wave functions [ref. 7].

In this model, the capillary bonds holding the objects in contact are analogous to the chemical bonds holding the atoms or molecules in contact. The capillary bonds can be thought of as analogous to the chemical bonds in synthetic organic chemistry (i.e. C–H bonds), or as analogous to the chemical bonds in molecular self-assembly (i.e. hydrogen bonds). Although the physics underlying the bonds may differ, it is still useful to draw these analogies. An important difference between capillary and chemical bonds is that capillary bonds are described by classical mechanics, while chemical bonds are described by quantum mechanics.

4.1.4 Prior work in MESA

Table 4.1 lists several systems that provide examples of MESA [refs. 1–4, 32–58]. Most of the systems involve the formation of close-packed arrays, by using objects (from nanometers to millimeters in size) and forces that are spherical or quasi-spherical in symmetry. The scales of these forces range from ~1 nm (for hydrogen bonds and van der Waals interactions in DNA/DNA duplex formation) to ~10 mm (for capillary bonds and fluid shear).

4.2 Objectives of our work

We developed systems where we could conveniently control the parameters affecting the assemblies and characterize them. These parameters include the shapes, surface properties, densities, and colors of the objects; the directionality of the forces between objects; and the densities and surface properties of the fluids. Some of these systems allow quick examination of tens to thousands of assembling particles. Agitation is normally in the form of fluid shear or gravity. The following sections describe some of the successes and failures in these experiments in self-assembly.

4.2.1 Choice of system

We used objects fabricated from organic polymers with their faces patterned to be hydrophobic or hydrophilic, and attracted by capillarity. The objects were fabricated from polymers – not metals, glass or ceramics – for two reasons. (i) They were easily formed by molding. (ii) The polymers had a density close to that of water; the objects could be suspended in water. We could easily control

Table 4.1. *Examples of self-assembly involving objects that span a range of sizes. The systems are ordered by size*

Description of system	Interaction	Size of the objects	Reference
Au colloids coated with monolayers of thiols terminated with DNA assembled into 3D aggregates either by pairing complementary strands of DNA or by addition of a soluble complementary strand of DNA	Molecular recognition (DNA/DNA)	nm	1, 35, 36, 40
Nanocrystals of Au assembled on a flat substrate by evaporation of a solvent, and linked in a second step with a dithiol	Capillary forces	nm	37
CdSe nanocrystals assembled by evaporation of a suspension (2D array) or by changing the polarity of the solvent (3D array)	Capillary forces	nm	36, 38, 40
Micron-sized polystyrene beads or nanometer-sized Au colloids suspended in water assembled into monolayers or multilayers on indium tin oxide by applying a voltage across two plates	Electrohydrodynamic flow	nm to μm	41, 42
Large proteins or polystyrene beads assembled on a glass plate or Hg surface on evaporation of a solvent	Capillary forces	nm to μm	2–4, 32, 33, 43, 44
Colloidal suspensions of beads crystallized between two plates	Shear and electrostatic forces	nm to μm	45, 46
Particles immersed in a fluid with a different dielectric constant or conductivity (electrorheological fluid) assembled into chains, columns of chains, and glasses in the presence of an electric field	Electric dipole interactions	nm to μm	55
Polystyrene beads assesmbled into close-packed arrays in channels	Capillary forces	nm to μm	47
Charged latex spheres assembled in water	Electrostatic forces	nm to mm	48–50
Latex beads assembled at a step on a surface while suspended in a solvent	Entropy	μm	51

Table 4.1. (*cont.*)

Description of system	Interaction	Size of the objects	Reference
Latex beads and rods suspended in water assembled in periodic optical fields	Light forces	μm	52, 54
Paramagnetic polystyrene beads assembled into 2D arrays at the water/air or water/glass interface by application of a magnetic field. The beads assembled from 6 to 10 particle diameters apart due to dipole–dipole and electrostatic repulsion	Magnetic and electrostatic forces	μm	58
Polystyrene beads assembled into chains, aggregates of chains, and 2D arrays in a ferromagnetic fluid in the presence of an external magnetic field	Magnetic forces	μm	56
Polystyrene beads with multiple ferromagnetic cores suspended in water between two glass walls assembled in the presence of a magnetic field into chains that were loosely correlated to one another	Magnetic and electrostatic	μm	57
Objects in a flowing fluid assembled into grooves on a surface that matched the shape of the objects; fluidic self-assembly	Shear forces and gravity	μm to mm	53

whether a face was hydrophobic or hydrophilic by oxidation in a plasma cleaner. The use of polyhedral objects with tailored surface properties gave us control over the directionality of the capillary forces between objects; this directionality has proved very useful in building complex structures.

We agitated the objects in a dish on an orbital shaker (for 2D MESA) or in a flask rotating around an axis (for 3D MESA). These methods of agitation gave us limited control over the amount of energy that could be supplied to the system and the distribution of energy among the objects. The reliance on fluid shear induced by swirling or rotation to provide agitation energy is one of the current weaknesses of these systems: it is non-isotropic, non-uniform, and difficult to tailor.

4.2.2 *Ease of fabrication and characterization*

Most of our experiments used large (millimeter-sized) polyhedral, polymeric objects that could be easily manipulated by hand. These objects were cast from molds, and the sides were readily differentiated into hydrophobic and hydrophilic sets. By working at a relatively large size scale, we were able to fabricate objects of a complexity difficult to achieve on a smaller, sub-millimeter scale. We could control the colors of the faces of the objects, the densities of the water phase (by adding a salt), and the densities of the objects (by adding a dense metal oxide powder).

We have carried out more limited studies with 10–1000 μm-sized objects. In these studies, we did not have as much control over the shapes that could be fabricated as we did at the millimeter scale. The smaller objects were fabricated using a combination of photolithography, electrochemistry, and evaporation of metals. Differentiating the faces of small objects into hydrophobic and hydrophilic sets was challenging.

The arrays, and the sequence of arrays occurring during self-assembly, have been easy to visualize by direct observation for millimeter-sized objects, or by optical microscopy for sub-millimeter-sized objects. We were able to visualize disordered or partially ordered arrays. Thus, we could study all of the arrays that assembled, not just the ordered ones.

4.3 MESA in two dimensions

4.3.1 *Experimental design*

The PFD/H$_2$O interface

The PDMS objects assembled at the PFD/H$_2$O interface upon agitation using an orbital shaker (Fig. 4.4). We chose this interface for four reasons. (i) PDMS does not swell in contact with either liquid. (ii) The interface has a high surface tension ($\gamma = 0.05$ J m^{-2}); for comparison, the H$_2$O/diethyl ether interface has a surface tension of $\gamma = 0.01$ J m^{-2} [ref. 59]. Capillary forces are, thus, strong. (iii) PDMS ($\rho = 1.05$ g cm^{-3}) has a density between those of PFD ($\rho = 1.91$ g cm^{-3}) and water ($\rho = 1.00$ g cm^{-3}); the objects floated at the interface. (iv) A thin film of PFD remained between the faces of objects when they assembled. This film acted as a lubricant and allowed the objects to move laterally relative to each other, and thus to minimize the free energy of the system.

A number of other liquid/liquid and liquid/vapor systems might be used for MESA. In principle, one could use any immiscible pair of liquids, provided that the objects that were assembling were not swelled by either, and that the

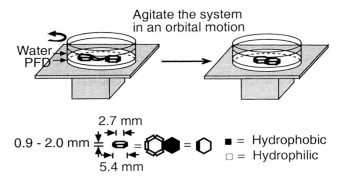

Figure 4.4. The size of the objects and how the arrays were assembled.

system showed a sufficiently high interfacial free energy to generate the required capillary forces.

We briefly examined the self-assembly of PDMS objects at the air/water interface, but have not pursued this system because the results were not entirely reproducible: the water edge on the faces of the objects tended to pin irreproducibly (i.e. we did not have control over whether the whole object was recessed into or floating above the interface). With sufficient attention to detail, this system might be made to work. Our instinct and experience, however, is that the liquid/liquid interface is a more forgiving and reproducible environment for MESA than is the liquid/air interface.

Fabrication

Long (~ 10 cm) hexagonal rods of PDMS were cast from molds (Fig. 4.5) [ref. 5]. PDMS was used because it has a low surface free energy, low cost ($60 per kg), a high surface free energy for the plasma-oxidized surface, and a density that is bracketed by those of PFD and water [refs. 39, 60]. To form a mold, a brass rod was cast in PDMS, and then removed. The mold was oxidized in an oxygen plasma and placed under vacuum with a fluorosilane $Cl_3SiCH_2CH_2(CF_2)_6CF_3$. PDMS prepolymer was added to the PDMS mold, cured, and removed to generate a replica of the brass rod. Rods of various sizes and shapes could be fabricated using this method; soaking in CH_2Cl_2 with Sudan red 7B or crystal violet dyed the rods to aid in visualization. The faces of the final objects that were to remain hydrophobic were covered with tape or ink (from a Sharpie© Marker), and the faces that were to be hydrophilic were left exposed. The objects were oxidized in an oxygen plasma for 5 min. After oxidation, the tape or ink was removed, and we cut the rods into thin plates by hand using a razor blade. The plates were placed at the PFD/H_2O interface with tweezers. The key advantages in this process were the ease of fabricating a

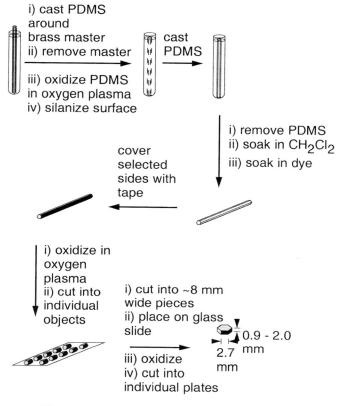

Figure 4.5. Fabrication of the PDMS objects.

range of hexagonal plates with different dimensions, and the ease of controlling the wettability of faces.

Agitation; shear forces

The objects were placed in a circular dish (14.5 cm in diameter) and agitated on an orbital shaker. The orbital shaker moved in a circular path (with a diameter of 2.5 cm) parallel to the plane of the stationary PFD/H_2O interface; this circular movement caused the liquid to swirl, and generated shear forces that separated the objects from one another. We have not studied these shear forces in detail, but in general, higher rotation speeds corresponded to stronger agitation.

Characterization of individual hexagons and arrays

The objects were easily viewed by eye. We used an optical technique to measure the tilt of the objects relative to the interface; the tilt angles provided

N. Bowden et al.

Number of
Hydrophobic Sides:

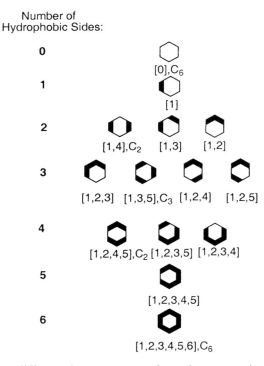

Figure 4.6. Fourteen different hexagons can form by permuting the location and number of hydrophobic faces. The hydrophobic faces are labeled by placing the location of the face in square parentheses. Centrosymmetric hexagons with C_n symmetry perpendicular to the hexagonal face are noted; these hexagons float parallel to the PFD/H_2O interface.

information about the contours of the menisci. Photographs and micrographs of the menisci from the side provided views of their cross sections.

The arrays were characterized by photography: after the assembly was complete (usually 30–60 min), the agitation was stopped, and a picture was taken of the array.

4.3.2 Nomenclature

In the figures that follow, we indicate the hydrophobic faces with thick lines and the hydrophilic faces with thin lines (Fig. 4.6). The positions of the hydrophobic faces are numbered by viewing the objects from above and counting in a clockwise direction; the hydrophobic faces are placed in square parentheses such as [a,b,c, . . .]. For instance, a hexagon with four adjacent hydrophobic faces and two hydrophilic faces is called a [1,2,3,4] hexagon. When we discuss a hydrophobic face, we will place parentheses around the number (i.e. the [3] face

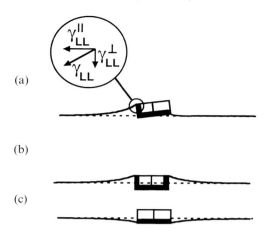

Figure 4.7. Vertical capillary forces acting on the hydrophobic faces pull the hexagons into the PFD/H_2O interface. The dashed line indicates the level of the interface far from the objects. (a) Hexagons with an unbalanced distribution of vertical capillary forces float with a tilt relative to the plane of the PFD/H_2O interface. The surface tension, γ_{LL}, can be separated into vertical, $\gamma^{\perp}_{LL}$, and horizontal, $\gamma^{\parallel}_{LL}$, components. (b, c) Hexagons with a balanced distribution of vertical capillary forces float parallel to the plane of the interface. Thick and thin lines indicate hydrophobic and hydrophilic faces, respectively.

on a [1,2,3,4] hexagon). When we discuss a hydrophilic face we will not use parentheses (i.e. the 6 face on a [1,2,3,4] hexagon).

4.3.3 Objects with a noncentrosymmetric pattern of hydrophobic faces are tilted at the interface

Objects with an asymmetric distribution of hydrophobic faces have a corresponding asymmetric distribution of vertical capillary forces that causes these objects to tilt at the interface (Fig. 4.7). This tilt shapes the contours of the menisci, because the meniscus on a tilted face is not symmetric about the center of the face. Two faces on different plates are held together more strongly when the contours of their menisci match than when they do not match. Thus, tilting can regulate the strength of attractive interactions between objects.

4.3.4 Results

Hexagonal plates

We have examined the arrays that assembled from the 14 different types of hexagonal plates formed by permuting the number and location of the hydrophobic faces (Fig. 4.6) [ref. 7]. The objective of this initial study was to establish

Table 4.2. *The 14 different hexagons and examples of the arrays they assembled into. We note whether the arrays were ordered, partly ordered, or disordered*

Centrosymmetric hexagons	Noncentrosymmetric hexagons

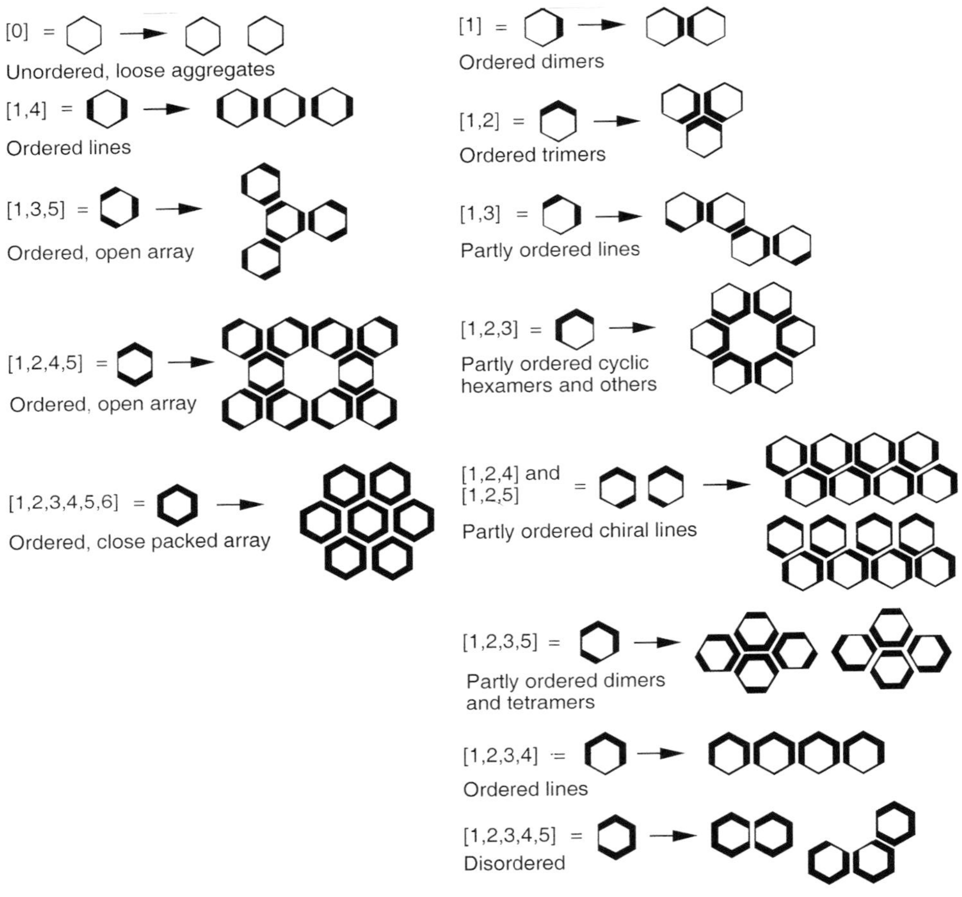

the rules that governed self-assembly in one particular system. The hexagons assembled into a variety of ordered, partially ordered, and disordered arrays (Table 4.2).

Centrosymmetric hexagons float parallel to the interface The hexagons with a centrosymmetric pattern of hydrophobic faces floated parallel to the plane of the PFD/H_2O interface. These hexagons, with the exception of the [0] hexagons, assembled into ordered arrays (Table 4.2). The [0] hexagons had only

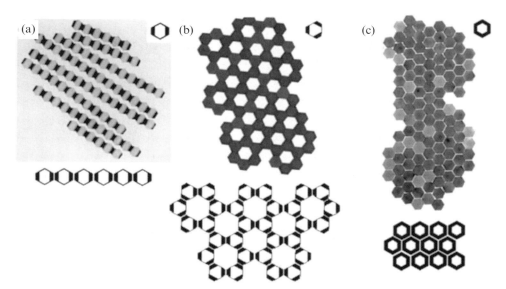

Figure 4.8. The arrays that assembled from the (a) [1,4], (b) [1,3,5], and (c) [1,2,3,4,5,6] hexagons. All the hexagons had a centrosymmetric pattern of hydrophobic faces and they floated parallel to the PFD/H$_2$O interface. In (a) the dark faces were hydrophobic and the light faces were hydrophilic.

small negative menisci, were weakly attracted to one another, and assembled into a loose aggregate. The [1,4], [1,3,5], and [1,2,3,4,5,6] hexagons assembled into arrays that would be predicted by opposing hydrophobic faces (Fig. 4.8).

The [1,2,4,5] hexagons could assemble into either an open or close-packed array, both of which juxtapose the hydrophobic faces; we only observed formation of the open array, however (Fig. 4.9). To study why only the open array formed, we assembled a close-packed array of hexagons by hand and agitated it (Fig. 4.9b). This close-packed array was stable to the agitation used to assemble the open array of hexagons. On the other hand, two hexagons, assembled by hand into the orientation that could lead to the close-packed array, spontaneously dissociated upon agitation (Fig. 4.9c). From these observations, we believe that the preference for the open array reflected the greater stability of the dimer of [1,2,4,5] hexagons that leads to an open array over that of the dimer that leads to a close-packed array. It is unclear whether the open array is actually energetically favored over the close-packed array.

Noncentrosymmetric hexagons are tilted at the interface Hexagons with a non-centrosymmetric pattern of hydrophobic faces floated with a tilt relative to the plane of the PFD/H$_2$O interface, in some cases, burying the faces and vertices in the interface. Table 4.3 lists the tilt angles.

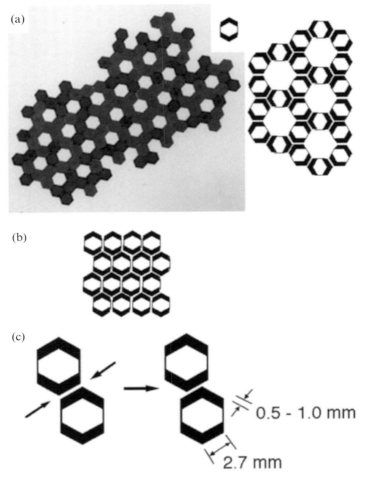

Figure 4.9. (a) The [1,2,4,5] hexagons assembled into an open array. (b) A closed array does juxtapose the hydrophobic faces, but it was not observed as a product of the agitation. (c) Two hexagons, when placed in proximity to one another, spontaneously shifted laterally. This motion was due to the proximity of the positive and negative menisci originating on the hydrophobic and hydrophilic faces. The two arrows indicate where the negative and positive menisci are close enough to interact. The exponential decay length of the menisci – the distance from the face where the height of the menisci has decreased by a factor of 1/e of its maximum height – is ~1.2 mm (or, in the plates shown here, approximately one-half the width of a face). Thus, the menisci on the faces can interact even when in close proximity.

Table 4.3. *The measured tilt angles, α, for 1.2 mm and 2.0 mm thick hexagons*

Hexagon	α (°)[a]	
	1.2 mm	2.0 mm
⬡	7 ± 2	
⬡	3 ± 1	
 ⬡	14 ± 1	26 ± 2
⬡	3 ± 2	
⬡	6 ± 2	
 ⬡	15 ± 1	30 ± 1
⬡	5 ± 2	
 ⬡	14 ± 1	
 ⬡	9 ± 2	

Note:
[a] The uncertainties were standard deviations of at least 25 measurements on different hexagons.

Tilted hexagons assembled to match the contours of their menisci. For example, although the [1,2] hexagons (1.2 mm thick) could assemble in two different ways to juxtapose their hydrophobic faces – trimers and parallel lines (Fig. 4.10a, b, c) – only trimers formed. This selection in structure occurred because hexagons that assembled into trimers matched the contours of their menisci; hexagons that assembled into the parallel lines did not match the contours of the menisci (Fig. 4.11a). Parallel lines assembled by hand were stable to agitation. We conclude that the preference for trimers lies in the kinetics of formation and stabilities of the dimers of the [1,2] hexagons.

In the course of our experiments with the [1,2] hexagons, we noticed a tendency for the trimers to aggregate weakly with one another. The trimers aggregated with one another based on interactions between the small positive menisci on the edges of the exposed tilted hydrophilic faces. To increase the strength of the interaction between the trimers, we assembled [1,2] hexagons that were 2.0 mm thick (Fig. 4.10d, e, f). These hexagons rapidly formed trimers at high ($\omega = 1.5\ s^{-1}$) rates of agitation; the trimers then weakly aggregated into

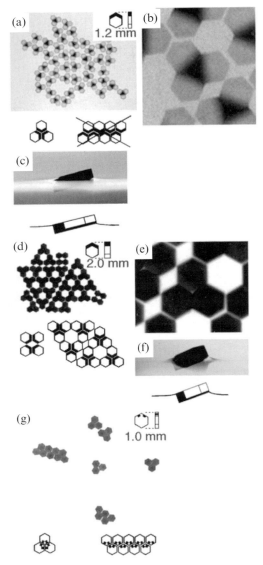

Figure 4.10. Assemblies of [1,2] hexagons. (a) The 1.2 mm thick [1,2] hexagons assembled into trimers that weakly aggregated when the agitation was stopped. (b) The hexagons in trimers assembled into close contact; the trimers were separated by a more noticeable distance apart. (c) The optical micrograph of a 1.2 mm thick [1,2] hexagon from the side shows the tilt of the hexagon at the interface. (d) The 2.0 mm thick hexagons assembled into trimers at high rates of agitation, and the trimers further assembled into the arrays shown at lower rates of agitation. The tetramer was also a stable product of the assembly. (e) The trimers assembled a noticeable distance apart. (f) The optical micrograph of a 2.0 mm thick [1,2] hexagon from the side shows the tilt and large positive menisci at the intersection of the 4 and 5 faces. (g) The 1.0 mm thick $[1_{1:2:1}, 2_{1:2:1}]$ hexagons assembled into trimers and lines.

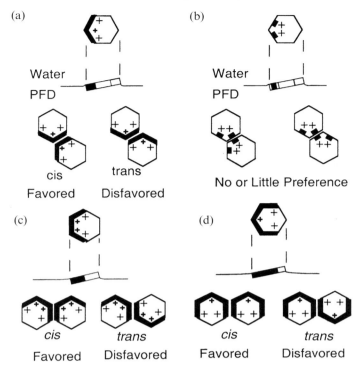

Figure 4.11. The tilt of the hexagons determined the contour of the menisci. The shape of the meniscus is indicated by the + symbol; the larger this symbol, the higher the meniscus at that point on the face. There are at least two ways to form dimers of the [1,2], [$1_{1:2:1}, 2_{1:2:1}$], [1,2,3], and [1,2,3,4] hexagons. In (a), (c), and (d), the favored *cis* configuration matches the contours of the menisci well; the disfavored *trans* configuration matches the contours of the menisci less well. In (b), the [$1_{1:2:1}, 2_{1:2:1}$] hexagons float almost parallel to the interface, and there is little or no preference between the *cis* and *trans* configurations.

the arrays shown in Fig. 4.10d at low ($\omega = 0.75$ s^{-1}) rates of agitation. This assembly thus showed a hierarchy of strengths in its capillary bonds.

To test the influence of the tilt on the assembly of the [1,2] hexagons, we fabricated hexagons with two adjacent sides hydrophobic that floated with a *small* tilt at the interface. The [1,2] faces on these hexagons were only half-hydrophobic: the central halves of the faces were hydrophobic and the edges hydrophilic. We call these objects [$1_{1:2:1}, 2_{1:2:1}$] hexagons. The numbers in subscripts indicate the pattern of the area of the face that is hydrophilic and hydrophobic (indicated by underlining the number). The [$1_{1:2:1}, 2_{1:2:1}$] hexagons tended to float with a smaller tilt than the [1,2] hexagons for two reasons. (i) The total vertical capillary forces on the half-hydrophobic faces were weaker than on the fully hydrophobic faces. (ii) The tilt of the hexagons pulled the hydrophilic areas of the faces into the PFD/H$_2$O interface and created substantial negative menisci that resisted the tilt. The smaller tilt of the [1$_{1:2:1}$,2$_{1:2:1}$] hexagons allowed them to assemble into both the *cis* and the *trans* dimers readily (Fig. 4.11b). The [1$_{1:2:1}$,2$_{1:2:1}$] hexagons assembled into a mixture of trimers and lines (Fig. 4.10g).

The same experiments with [1,2,3] hexagons yielded a variety of structures. Dimers, trimers, tetramers, cyclic arrays, and lines formed from 1.2 mm thick [1,2,3] hexagons (tilt angle: 15°) (Fig. 4.12). These hexagons assembled with the [1] face in contact with the [3] face. This configuration matched the contours of the menisci (Fig. 4.11c). The line motif was remarkable because the hexagons assembled with contacts at vertices rather than along faces. Thicker (2.0 mm thick) [1,2,3] hexagons (tilt angle: 30°) assembled into similar arrays (Fig. 4.12d, e). The 2.0 mm thick hexagons tended to assemble into longer, more stable lines than the 1.2 mm thick hexagons. Assemblies of the [1$_{1:2:1}$,2$_{1:2:1}$,3$_{1:2:1}$] hexagons (tilt angle: 5°) resulted from interactions between all three of the half-hydrophobic faces (Fig. 4.12f). The difference in the arrays that assembled from the [1,2,3] hexagons and the [1$_{1:2:1}$,2$_{1:2:1}$,3$_{1:2:1}$] hexagons show the importance of tilting.

The [1,2,3,4] hexagons assembled into lines through interactions between the [1] and [4] faces (Fig. 4.13). These lines showed four interesting characteristics. (i) The lines assembled so that the [1] faces were in contact with the [4] faces. The contours of the menisci on the faces were matched when two hexagons assembled with the [1] and [4] faces in contact; contacts between two [1] faces, or between two [4] faces mismatched the contours of the menisci on the faces (Fig. 4.11d). (ii) The [2] and [3] faces were buried into the interface and had small, positive menisci. Thus, the lines did not assemble into contact at high rates of agitation, but did assemble under weak agitation and low shear. (iii) The lines assembled into a loose, parallel array, with individual lines a small distance apart. Since the menisci on the juxtaposed [2] and [3] faces were mismatched, the lines assembled with a small distance between them. (iv) The vertex between the 5 and 6 faces had a small positive meniscus because it was pulled out of the interface (Fig. 4.13). This meniscus caused the hydrophilic faces to attract the exposed hydrophobic faces weakly.

Modeling receptor/ligand interactions

We wished to develop a macroscopic model of the interactions between molecular ligands and receptors. Molecular recognition is a broad subject that describes selective assembly in chemistry and biology, with examples from DNA–protein complex formation to asymmetric catalysis. The principle behind molecular recognition dictates that the molecules that mate have complementary shapes and interfacial characteristics. Our extension of this principle to the mesoscale involved the self-assembly of objects that matched both

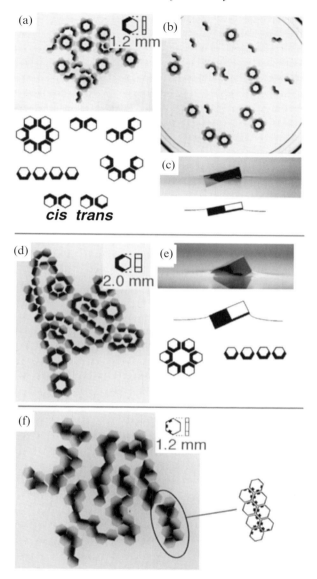

Figure 4.12. Assemblies of the [1,2,3] hexagons. (a) The 1.2 mm thick [1,2,3] hexagons assembled into the arrays that aggregated further when the intensity of agitation was decreased. (b) The arrays were separated from one another by hand to show which was stable under high agitation. (c) The optical micrograph of a 1.2 mm thick [1,2,3] hexagon shows the tilt relative to the PFD/H_2O interface. (d) The 2.0 mm thick [1,2,3] hexagons assembled into the arrays shown. (e) The optical micrograph of the hexagon from the side shows the tilt and the positive menisci on the 5 face. (f) The $[1_{1:2:1}, 2_{1:2:1}, 3_{1:2:1}]$ hexagons assembled into a mixture of arrays. Some hexagons assembled with a half-hydrophobic face in contact with a hydrophilic face.

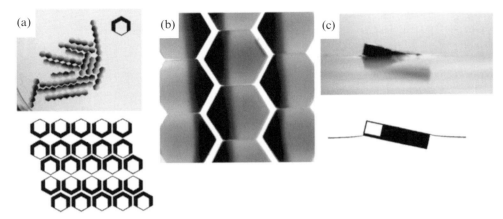

Figure 4.13. Assemblies of [1,2,3,4] hexagons. (a) The [1,2,3,4] hexagons assembled into lines that aggregated weakly when the agitation was stopped. (b) A close-up of the lines. The dark faces on the hexagons in (a) and (b) are hydrophobic; the light faces are hydrophilic. (c) The optical micrograph of a [1,2,3,4] hexagon from the side shows the tilt of the hexagon at the interface.

the shapes of the juxtaposed faces and of the menisci at those faces. These systems might be used as physical models for processes that occur on the molecular level, and in methods of selective recognition and assembly of components.

Figure 4.14 describes three systems where recognition is size- and shape-selective [ref. 9]. In one system, two [1,3] hexagons assembled with a receptor in the presence of [1,4] or [1,2,3] hexagons (Fig. 4.14). In a second system, two ligand-receptor pairs (2a and 1a; 2b and 1b) formed in the presence of each other. The pattern of the hydrophobic faces on the ligands and receptors was chiral, and the receptors and ligands assembled in a way that juxtaposed enantiomeric chiral faces. In a third system, one [1,4] hexagon selectively assembled with a receptor. Receptors that selectively assembled two or three [1,4] hexagons were also fabricated.

These systems are imperfect models for molecular recognition, since both the potential functions and energy distributions that describe the interactions in MESA are different from those at the molecular level. Moreover, the encounter frequencies between objects in MESA (10^{-3}–10^{-2} s^{-1}) are much smaller than those between molecules (10^2–10^3 s^{-1}, for micromolar concentrations). Despite these differences, our model manages to exhibit the salient characteristics of molecular recognition: assembly depends on the shapes and interfacial properties of the faces that recognize one another.

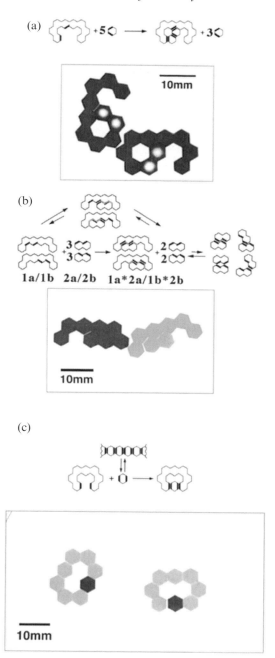

Figure 4.14. MESA can be used to model receptor/ligand interactions. In (a), two [1,3] hexagons – with dark edges and clear centers – assemble into the receptor. In (b), the ligands, 2a and 2b, assemble into the receptor, 1a and 1b, based on the chirality of the pattern of the hydrophobic faces. In (c), the dark [1,4] hexagon assembles into the light receptor.

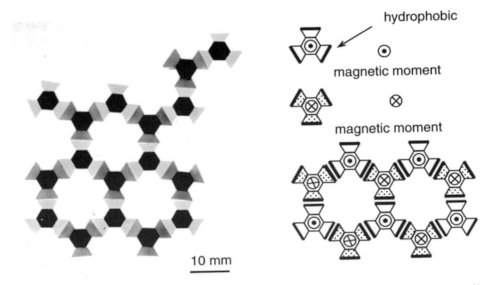

Figure 4.15. Assemblies of 'propellers' at the PFD/H$_2$O interface that combine capillary and magnetic interactions. The direction of the magnetic dipole of the central (black) hexagonal plate is perpendicular to the plane of the interface. The light, grey-colored parts are made of dyed, non-magnetic PDMS.

MESA using magnetic forces

The introduction of forces other than capillarity extends the complexity of the assemblies that can be assembled. We chose to explore magnetic interactions to complement capillary forces in MESA, because (i) they can be both repulsive and attractive; (ii) their strength can be easily tailored during fabrication of the objects (see below); and (iii) capillary and magnetic forces can act simultaneously and independently (that is, they have similar ranges).

We made magnetic hexagonal plates by adding powdered Fe$_3$O$_4$ to liquid PDMS and curing the polymer in a strong magnetic field. The assembling units consisted of central magnetic hexagonal plates that were attached to non-magnetic, colored PDMS pieces. All of the magnetic dipoles were perpendicular to the hexagonal face (and perpendicular to the PFD/H$_2$O interface). Self-assembly proceeded in the absence of an external magnetic field, and resulted solely from interactions between the magnetic dipoles and capillary forces between the hexagonal plates. Because the long, exposed edges of the colored PDMS pieces were hydrophobic, the capillary interactions are always attractive, regardless of the polarity of the magnetic core. The magnetic interactions were either repulsive or attractive, depending on the alignment of the magnetic dipoles (parallel or anti-parallel). The combination of capillary and magnetic interactions resulted in arrays that alternated pieces with the direction of magnetization down and up (light and dark pieces in Fig. 4.15).

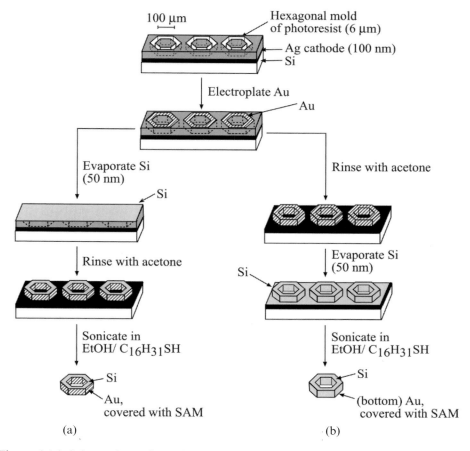

Figure 4.16. Schematic outline of the process used to fabricate hexagonal metal rings. (Reprinted with permission from *J. Am. Chem. Soc.*, 1998, **120**, 8267–8268. Copyright 1998 American Chemical Society.)

4.4 MESA on curved interfaces

MESA at a liquid/liquid interface is not limited to two dimensions. Objects placed on a curved interface will assemble into a pseudo-3D system. The capillary forces between the objects are the same as those described for 2D MESA [ref. 5]. The curved interface is made by putting a drop of an immiscible liquid in another, isodense liquid; the result is a sphere. The method can easily be extended to different shapes (catenoid, cylinder, cone) by distorting the shape of the drop [ref. 39]. The spherical assemblies in these experiments are made of metallic hexagons (100 μm sides; 6 μm thick) with hydrophobic and hydrophilic sides [ref. 11]. Electrodeposition through lithographically defined molds fabricated metallic hexagonal rings (Fig. 4.16). Hydrophilic faces were introduced

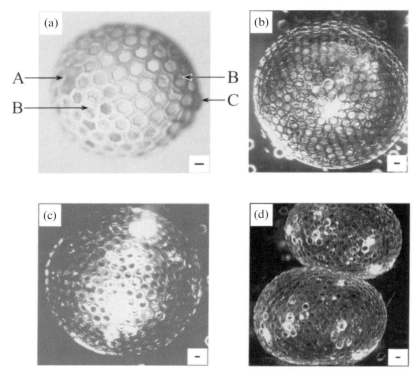

Figure 4.17. (a) Porous sphere composed of self-assembled hexagonal rings welded by electrodeposition of Ag. The rings had approximately 100 μm sides and were assembled on a drop of chlorobenzene in an aqueous silver-plating solution. Defects are indicated by arrows: **A**, void; **B**, double layer; **C**, point at which the electrode was attached. (b) Hexagons assembled on a drop of water in heptane. (c) Hexagons assembled on a drop of PFD in water. (d) Two spheres pressed against each other do not fuse; the assemblies have substantial stability even without welding to connect the particles. Scale bars are 100 μm. (Reprinted with permission from *J. Am. Chem. Soc.*, 1998, **120**, 8267–8268. Copyright 1998 American Chemical Society.)

by evaporating a thin layer (50 nm) of Si onto the gold rings; this silicon oxidizes spontaneously, and forms a hydrophilic, native SiO_2 film. The remaining gold faces were covered with a hydrophobic SAM of hexadecanethiol. Two types of hexagonal ring were prepared: hexagonal rings with all faces, except for the bottom, hydrophilic; and rings with all faces hydrophobic, except for the hydrophilic top.

   Drops of water in heptane, or drops of PFD or chlorobenzene in water, were covered with the appropriate hexagonal rings. The rings assembled at the interfaces and gentle swirling induced packing of the objects into a hexagonal array (Fig. 4.17). These rings acted as mesoscale surfactants that stabilized the drops against fusion when they were pressed together. In all experiments, an excess of hexagons was used; the hexagons not adsorbed at the interface remained suspended in the continuous phase.

The assemblies were welded into a permanent form by electrodeposition of silver. The rings were allowed to assemble around a drop of chlorobenzene in an aqueous silver-plating solution. Chlorobenzene is approximately isodense with the silver-plating solution; this similarity in density minimizes the distortion of the drop from a sphere. The drop, covered with the assembled metallic rings, was touched with an electrode (a very thin gold wire), a second electrode was placed in the solution, and a power supply connected to these wires. Silver deposition started near the electrode, but quickly spread over the whole sphere, because the conductive hexagonal rings were in contact. When all the rings were welded together (after ~1 min), the whole sphere was pulled from the bath; liquids drained from the structure spontaneously. These free-standing, porous metallic structures were mechanically strong enough to be handled in air.

The hexagonal rings did not assemble into perfect crystalline arrays: because the rings have low aspect ratios and a large central hole, they overlap and form double layers. Beyond this issue of overlap, hexagonal rings can, in any event, never completely cover the surface of a sphere. We did not attempt to solve this problem by adding pentagonal rings.

4.5 MESA in three dimensions

Our efforts in the self-assembly of three-dimensional (3D) arrays (Table 4.4) evolved to answer three questions: first, and perhaps most importantly, *how* does one assemble in three dimensions? Here, 'assembly in 3D' refers to self-assembly that does not take place at an interface, but rather in the bulk of a liquid. Second, what sorts of 3D structure can be self-assembled at the mesoscale? Rods, crystals, and sheets certainly, but what about spirals, sponges, or ropes? Third, how small are the dimensions to which this type of self-assembly can be scaled? For some of the applications of 3D structures in optics that we envision, it is desirable to have a method to make 3D arrays of ~1-μm-sized objects.

4.5.1 Experimental design

One answer to the first question – how to assemble in 3D – is simple: use capillary forces. A strategy for self-assembly based on capillarity requires a method of fabricating objects, coating them with thin liquid films (either completely or only on selected areas), and agitating the liquid-coated objects so that they collide to allow coalescence of liquid films (Fig. 4.18). Among the lubricants

Table 4.4. *Structures obtained with 3D MESA, and the experimental parameters involved in the self-assembly of each*

Assembling piece	Number of pieces per aggregate	Size of piece	Lubricant
	2	5 mm	
	3	5 mm	
	4	5 mm	Alkanes, photocurable methacrylate
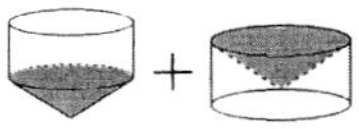	2	5 mm	
	8–1000	5 mm	
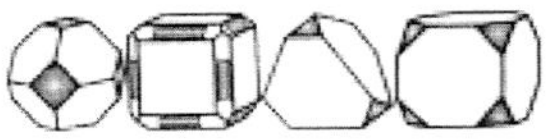	50–200	1 cm	
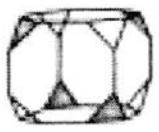	12	1 cm	Liquid metal
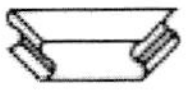	~20	1 cm	
10 μm-sized plates	10^4–10^5	10 μm	Alkanes, photocurable methacrylate

we tried were alkanes, photocurable methacrylates, and liquid metals. The liquid metals were particularly useful in the self-assembly of objects in which only a small fraction of the total surface area was coated with liquid, since the surface tension of a liquid metal is high (~ 200 mN/m for Bi-containing alloys, compared to ~ 20 mN/m for alkanes).

Fabrication of pieces

In contrast to the pieces used in 2D MESA, the pieces used in 3D MESA were made of polyurethane rather than PDMS. Objects made of PDMS aggregate in water, regardless of whether they are coated with liquid or not: because it is compliant, contacting surfaces come into intimate contact; it is also hydrophobic. The choice of polymer arose from purely practical considerations: fabrication

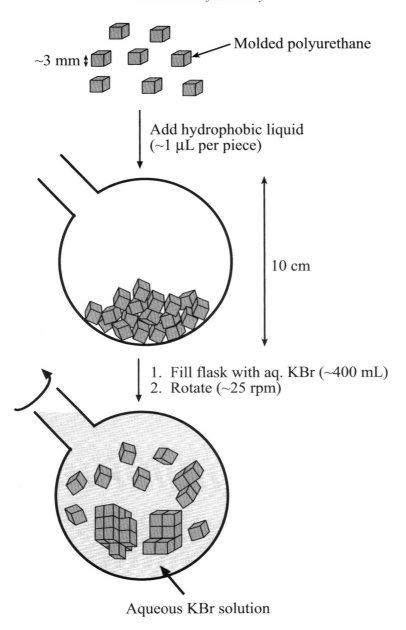

~3 mm ↕ Molded polyurethane

Add hydrophobic liquid
(~1 μL per piece)

10 cm

1. Fill flask with aq. KBr (~400 mL)
2. Rotate (~25 rpm)

Aqueous KBr solution

Figure 4.18. Schematic sketching the experimental procedure used in 3D mesoscale self-assembly. Molding of a polyurethane prepolymer in a PDMS master generated polyhedra. Lubricant was added to the polyhedra in a water-filled Morton flask; the use of liquid solder as a lubricant required the covering of selected faces with solder-coated copper tape. Axial rotation of the flask provided the agitation needed to cause collisions between liquid-coated pieces. The schematic depicts the formation and self-assembly of hydrophobic cubes.

of 3D objects involved the molding of a prepolymer into a PDMS master, and the prepolymer of polyurethane that we used (Norland Optical Adhesive 73, Norland) was easy to mold. The PDMS masters themselves were cast against machined metal pieces (made of aluminum or brass). It was not necessary to polish the metal to a mirror-like finish.

A typical assembly experiment consisted of placing molded pieces of polyurethane in a flask, coating the pieces with liquid, and rotating the flask to tumble the pieces. Deposition of liquid films required adding a mixture of ethanol and the lubricant to cover the pieces. Gradual addition of water to an ethanol solution containing the lubricant then precipitated the lubricant onto the pieces. In both cases, we carefully minimized the introduction of bubbles that could interfere with assembly.

Agitation

The use of rotation in an aqueous medium to provide agitation made it possible to adjust the strength of agitation. High rotation speeds, as in 2D MESA, caused stronger collisions between objects. Increasing the density of the aqueous medium by addition of salts (KBr or KI) decreased the agitation strength by decreasing the speed with which the objects fell in the flask. Adjustment of the tilt of the flask, from nearly horizontal or nearly vertical to an angle in between, provided yet another route to tailor the intensity of agitation.

Characterization

In general, all of the arrays formed within a few hours (or at most a few days) after beginning agitation. After formation of the array, the lubricant was solidified by irradiation (for the photocurable methacrylate) or cooling (for the liquid metal). The array could then be removed from the flask for examination.

4.5.2 *Results and discussion*

Table 4.4 describes the structures obtained with 3D MESA. The following sections address each case in turn.

Finite structures of mm-sized objects

Our first attempt in 3D MESA focused on the self-assembly of structures that consisted of a finite number of pieces (Fig. 4.19) [ref. 6]. The surfaces of these pieces were either hydrophilic or hydrophobic. The shapes of these pieces, and the locations of the hydrophobic surfaces, were chosen so that aggregates, when formed, did not expose any hydrophobic surfaces – after assembly, only hydrophilic surfaces were exposed to the aqueous phase. In most cases, the assembling

Figure 4.19. Pictures of self-assembled, finite structures. (a) A torus. (b) An ellipsoid. (c) A spheroid. (d) A cylinder, from two different pieces with complementary shapes. (Reprinted with permission from *Nature*, 1997, **386**, 162–164. Copyright 1997 Macmillan Publishers Ltd.)

units were identical: for example, two half-donuts assembled into a whole donut. In one case, we allowed objects with complementary shapes to self-assemble: some objects had protruding hydrophobic surfaces, while others had indented ones (Fig. 4.19d). These objects assembled as dimers, and demonstrated rudimentary shape recognition in 3D MESA. Assembly was fast in all cases.

Extended arrays of mm-sized objects

Extension of these results to arrays that formed extended lattices in 3D required modification of the pieces. In particular, the liquid-coated films on each piece were placed so that the arrays always possessed at least one exposed liquid-coated face as they were forming. Thus, growth of the aggregate did not stop after the array consisted of a certain number of objects.

Self-assembly of arrays that could tile 3D space relied on pieces that were completely coated with lubricant (Fig. 4.20) [ref. 15]. The sequence of steps for self-assembly of these arrays is similar to that seen in molecular-scale crystallization: a few blocks initially bond and form a nucleus; these initial small

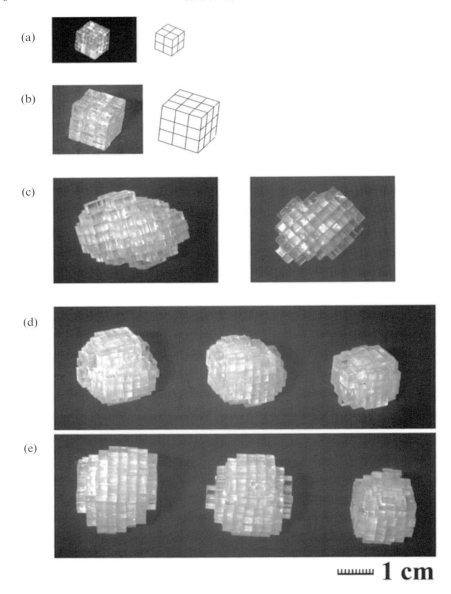

Figure 4.20. Pictures of self-assembled crystals of mm-sized objects. (a) An array of 8 cubes. (b) An array of 27 cubes. (c) Arrays obtained from assembly of ~1000 cubes. (d) An array of 36 hexagonal plates. (e) An array of ~100 cubes. Each cube in (e) has three intersecting, perpendicular, cylinder channels machined through it. (Reprinted with permission from *J. Am. Chem. Soc.*, 1998, **120**, 12670–12671. Copyright 1998 American Chemical Society.)

aggregates grow by addition of individual pieces. As in molecular crystallization, 3D MESA favors the formation of a few large arrays over the formation of many small ones. It also favors the formation of arrays with crystalline external morphologies through collisions between arrays. These collisions 'polish' the sides of the arrays by dislodging pieces that comprise incomplete layers. If we induced the formation of one monolithic array by reducing the agitation strength, then the external morphology of this array was oblate (Fig. 4.20c). These results hold for other shapes, such as hexagonal prisms and cubes with holes drilled through them.

The formation of *open* 3D structures required the selective wetting of the faces of the assembling pieces [ref. 8]. Here, the lubricant was a liquid, low-melting-point solder that had a high surface tension [ref. 61]. The high surface energy compensated for the fact that lubricant covered only a fraction of the total surface area of a piece. Lattices that possessed simple cubic, body-centered cubic, or face-centered cubic symmetries self-assembled from truncated octahedra or truncated cubes (Fig. 4.21). Other geometries formed more intricate structures: a trapezoidal piece with complementary S-shaped ends self-assembled into a helix (Fig. 4.22) [ref. 8]. Rods with notches formed a structure reminiscent of a log cabin (Fig. 4.23) [ref. 62]. Hexagonal rods could form open lattices that resembled elongated honeycombs [ref. 62]. The sizes of nearly all of these structures appeared to be limited only by the number of pieces used in each experiment.

Only in the case of the helix did fundamental limitations arise. The helix grew to ~2–3 turns (~20 pieces) and did not grow further. The helices were apparently unstable to mechanical agitation. Because there was no adhesion between turns, the helices expanded like a spring during agitation, and therefore broke easily.

Arrays of 10-μm-sized plates

The third question in 3D MESA – the dimension of the smallest pieces that can self-assemble – motivated experiments with metallic, polyhedral, 10-μm-sized plates [ref. 63]. The fabrication of 10-μm-sized polyhedra is difficult, so these experiments examined the self-assembly of flat plates, which were easily obtained at this scale. Photolithography, electrodeposition, and electron beam evaporation provided a convenient way to form plates having surfaces whose properties could be modified selectively. The surfaces of these plates consisted of either gold or chromium. Treatment with a long-chained alkanethiol formed a hydrophobic SAM on the gold; with a phosphonic acid-terminated thiol, a hydrophilic one. Similarly, a phosphonic acid-terminated thiol formed a hydrophobic SAM on chromium (through coordination of the acid with surface Cr

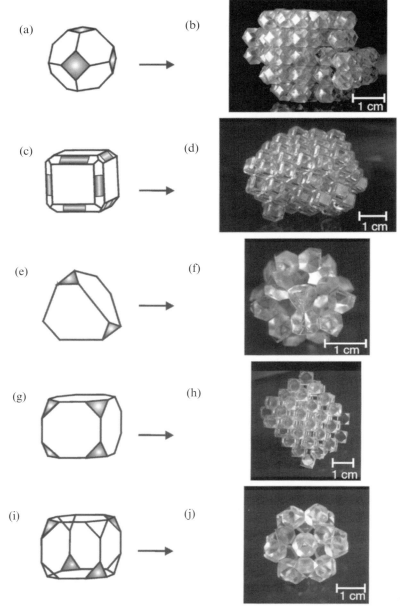

Figure 4.21. Pictures of self-assembled crystals with open structure. (a, b) An array of truncated octahedra, with simple cubic symmetry. (c, d) An array of truncated cubes, with fcc symmetry. (e, f) An array of truncated tetrahedra. (g, h) An array of truncated cubes, with bcc symmetry. (i, j) An array of truncated cubes. Each cube in (i, j) has solder on only four of its eight triangular faces. (Reprinted with permission from *Science*, 1999, **284**, 948–951. Copyright 1999 American Association for the Advancement of Science.)

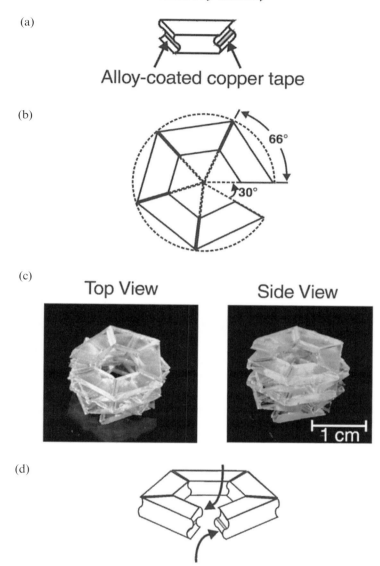

(a)

Alloy-coated copper tape

(b)

66°

30°

(c)

Top View   Side View

1 cm

(d)

Figure 4.22. Self-assembly of a helix. (a, b) The design of the assembling pieces. (c) Pictures of a self-assembled helix. (d) Proposed explanation of how chiral helices form: first, five pieces assemble into an open ring; then pieces attach to the ring by approaching along the directions indicated by arrows. Pieces that do not approach along these directions cannot attach to the ring because they are blocked by overhanging 'lips' that protrude out from the S-shaped ends of the ring. (Reprinted with permission from *Science*, 1999, **284**, 948–951. Copyright 1999 American Association for the Advancement of Science.)

(a)

(b)

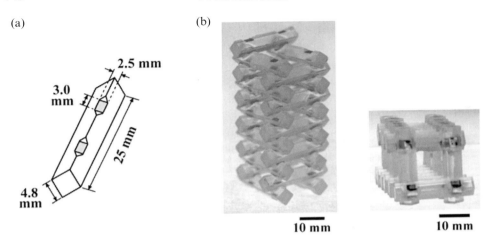

2.5 mm

3.0
mm

25 mm

4.8
mm

10 mm   10 mm

Figure 4.23. Self-assembly of an open, rectangular structure. (a) Design of the pieces. (b) Pictures of self-assembled structures.

oxides). Precipitation of hydrophobic lubricant onto the hydrophobic faces of each plate took place from addition of water to a suspension of plates in a mixture of ethanol and lubricant.

The important conclusion from these experiments is that 3D MESA *does* scale down to the 10-μm-scale. It is possible to generate well-ordered arrays with crystalline morphologies (Fig. 4.24). These arrays have enough order to function as diffraction gratings. The location of the hydrophobic faces determines the external morphology of the resulting array, and a rough correlation exists between the growth rate along a certain axis and the amount of hydrophobic surface area perpendicular to that axis.

These results have not identified any limitation on the lower limit to sizes that can be assembled using capillary forces. In fact, since capillary forces should become stronger relative to other forces (primarily gravitational and electrostatic) at smaller size scales [ref. 64], we believe 3D MESA based on capillarity should apply to even smaller pieces (<0.5 μm). The inference that capillary forces may be appropriate for self-assembly in systems of very small components is compatible with other evidence (Table 4.1).

Defects

As with all self-assembled structures, these 3D arrays possess defects. They may take the form of misalignment ('slip') between planes in an array, omitted members of the arrays (vacancies), or misaligned individual pieces. The simplest way to remove defects due to slip is to make the hydrophobic faces nonplanar. For instance, assemblies of deformed cubes, where three faces of each

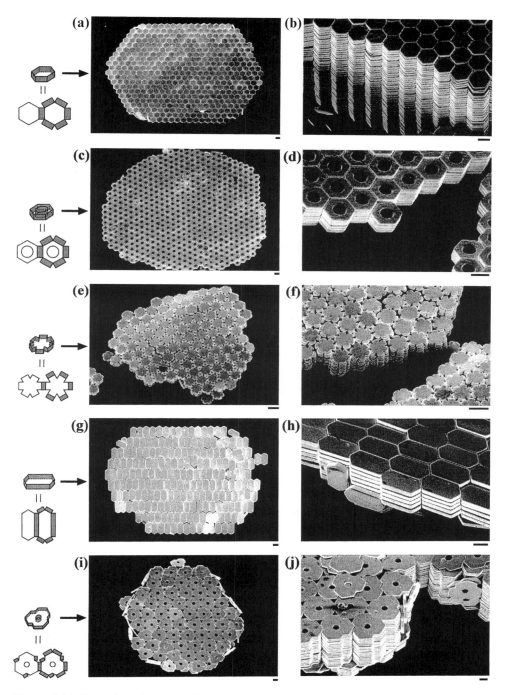

Figure 4.24. Scanning electron micrographs of self-assembled arrays from 10-μm-sized objects.

cube are convex and three are concave, should not possess slip planes. The other types of defect – vacancies and misplaced pieces – are inherent to self-assembly that relies on only one force to effect assembly. It is certainly possible to minimize these defects by slow crystallization, but how to eliminate them is unclear (as it is unclear in crystallization of atoms or molecules). We estimate that the number of defects in these arrays is ~ 0.1–1% of the total number of pieces.

4.6 Discussion

4.6.1 MESA using capillary forces can generate a variety of arrays

MESA can be used to generate a wide variety of structures: two-dimensional, three-dimensional, open, closed, linear, micron or millimeter in size, or chiral. MESA also offers a simple model for molecular recognition. Most of our work has been with millimeter-sized polyhedral objects, although MESA also applies to nanometer- and micron-sized objects. We are limited in the size, shape, and patterns of arrays we can explore by limitations in our methods of fabricating the individual objects. The fabrication of structurally complex micron-sized objects in a way that makes it possible to control properties of the faces individually is difficult or, for some designs, impossible with existing methods. New methods that generate regular geometric shapes with patterned faces would be welcome. Because the capillary forces for both two-dimensional and three-dimensional MESA scale into the micron and nanometer size range, we believe that even more complex arrays can be readily assembled using these methods and ideas.

4.6.2 Thermodynamics and kinetics are different in mesoscale and molecular self-assembling systems

The concepts of 'thermodynamics' and 'kinetics' are poorly defined in MESA. We do not understand the distribution of energies, the rates of encounter, or of order/disorder in the assemblies. The distribution of energy among the objects is unknown, and may vary from system to system. We have not measured the rates of encounter or the rates of collision within each encounter complex, and do not know how to measure the entropy of the system.

There are two experimental advantages to these systems that may bring a better level of understanding to their kinetics and thermodynamics. (i) Because the objects are relatively large, we can assemble them by hand into any configuration of interest and test their stability to the agitation. We can, as a result, determine if arrays are formed because of the kinetics of their assembly or because of their stability. (ii) These systems may be amenable to computer simulation because the interactions involved in systems of objects interacting strongly through capillarity are classical.

4.6.3 *MESA extends the ideas and methods of molecular self-assembly and biology from the molecular scale to the mesoscale*

Biology and organic chemistry are replete with examples of self-assembly. Examples include lipid bilayers, the DNA duplex, proteins in their correctly folded forms, self-assembled monolayers, and crystals. In MESA, we extend ideas abstracted from molecular recognition – shape recognition, chirality, directional interactions, hierarchy of bonds, and hydrophobicity – to the mesoscale.

4.6.4 *Future directions*

MESA is the beginning of a new area of research. Now that its feasibility has been established, it can go into several directions. New areas to explore include catalysis, self-replicating systems, introduction of new forces, assembly of small electrically active components, and energy dissipating systems. A common direction in MESA will be to develop strategies that allow large arrays to assemble with high levels of precision and perfection.

Acknowledgments

We thank Tricia Breen for helpful comments. We also thank Francisco Arias, Tricia Breen, Jeff Carbeck, Insung Choi, Tom Clark, Noo Li Jeon, Tanja Hadzic, Rustem Ismagilov, Scott Oliver, Abe Stroock, Andreas Terfort, Marcus Weck, and Hongkai Wu who have carried out many of the experiments in this area, and who have provided many hours of helpful discussions and thoughtful ideas. Financial support was provided in part by the NSF (grant number CHE9901358), NIH (grant number GM30367), and DARPA.

References

1. Mirkin, C. A., Letsinger, R. L., Mucic, R. C. and Storhoff, J. J. (1996). *Nature* **382**, 607.
2. Kralchevsky, P. A., Paunov, V. N., Denkov, N. D., Ivanov, I. B. and Nagayama, K. (1993). *J. Colloid Interface Sci.* **155**, 420.
3. Yamaki, M., Higo, J. and Nagayama, K. (1995). *Langmuir* **11**, 2975.

4. Kralchevsky, P. A. and Nagayama, K. (1994). *Langmuir* **10**, 23.
5. Bowden, N., Terfort, A., Carbeck, J. and Whitesides, G. M. (1997). *Science* **276**, 233.
6. Terfort, A., Bowden, N. and Whitesides, G. M. (1997). *Nature* **386**, 162.
7. Bowden, N., Choi, I. S., Grzybowski, B. and Whitesides, G. M. (1999). *J. Am. Chem. Soc.* **121**, 5373.
8. Breen, T. L., Tien, J., Oliver, S. R. J., Hadzic, T. and Whitesides, G. M. (1999). *Science* **284**, 948.
9. Choi, I. S., Bowden, N. and Whitesides, G. M. (1999). *J. Am. Chem Soc.* **121**, 1754.
10. Choi, I. S., Bowden, N. and Whitesides, G. M. (1999). *Angew. Chem. Int. Ed. Engl.* **38**, 3078.
11. Huck, W. T. S., Tien, J. and Whitesides, G. M. (1998). *J. Am. Chem. Soc.* **120**, 8267.
12. Isaacs, L., Chin, D. N., Bowden, N., Xia, Y. and Whitesides, G. M. (1999). *Self-Assembling Systems on Scales from Nanometers to Millimeters: Design and Discovery* (Reinhoudt, D. N., Ed.), vol. 4, p. 309. John Wiley & Sons: New York.
13. Terfort, A. and Whitesides, G. M. (1998). *Adv. Mater.* **10**, 470.
14. Tien, J., Terfort, A. and Whitesides, G. M. (1997). *Langmuir* **13**, 5349.
15. Tien, J., Breen, T. L. and Whitesides, G. M. (1998). *J. Am. Chem. Soc.* **120**, 12670.
16. Madou, M. (1997). *Fundamentals of Microfabrication*. CRC Press: New York.
17. MacDonald, J. C. and Whitesides, G. M. (1994). *Chem. Rev.* **94**, 2383.
18. Conn, M. M. and Rebek Jr., J. (1997). *Chem. Rev.* **94**, 2383.
19. Lehn, J.-M. (1995). *Supramolecular Chemistry: Concepts and Perspectives; a Personal Account*. VCH: Weiheim.
20. Lehn, J.-M. (1998). *Angew. Chem. Int. Ed. Engl.* **27**, 89.
21. Bain, C. D. and Whitesides, G. M. (1989). *Angew. Chem., Int. Ed. Engl.* **28**, 506.
22. Merritt, M. V., Mrksich, M. and Whitesides, G. M. (1997). *Using Self-Assembled Monolayers to Study the Interactions of Man-Made Materials with Proteins*, 2nd edn. (Lanza, R. P., Chick, W. L. and Langer, R., Eds.), p. 211. R. G. Landes Co.: Austin, TX.
23. Mrksich, M. (1997). *Curr. Opin. Colloid Interface Sci.* **2**, 83.
24. Whitesides, G. M., Mathias, J. P. and Seto, C. T. (1991). *Science* **254**, 1312.
25. Namba, S. (1994). *J. Phys. Soc. Jpn.* **63**, 224.
26. Groma, I. (1997). *Phys. Rev. B: Condens. Matter.* **56**, 5807.
27. Rosen, A. (1998). *Adv. Quantum Chem.* **30**, 235.
28. Freeman, M. H. (1990). *Optics* (10th edn.), Butterworths: London.
29. Kelly, M. J. (1995). *Low-Dimensional Semiconductors, Materials, Physics, Technology, Devices*. Oxford Science Publications, Clarendon Press: Oxford.
30. Meyer-Arendt, J. R. (1995). *Introduction to Classical and Modern Optics*. Prentice-Hall, Inc.: Englewood Cliffs, New Jersey.
31. Makse, H. A., Havlin, S., King, P. R. and Stanley, H. E. (1997). *Nature* **386**, 379.
32. Nagayama, K., Takeda, S., Endo, S. and Yoshimura, H. (1995). *J. Appl. Phys.* **34**, 3947.
33. Lazarov, G. S., Denkov, N. D., Velev, O. D., Kralchevsky, P. A. and Nagayama, K. (1994). *J. Chem. Soc. Faraday. Trans.* **90**, 2077.
34. Israelachvili, J. (1992). *Intermolecular and Surface Forces*. Academic Press Inc.: San Diego.

35. Elghanian, R., Storhoff, J. J., Mucic, R. C., Letsinger, R. L. and Mirkin, C. A. (1997). *Science* **277**, 1078.
36. Alivisatos, A. P. (1997). *Endeavour* **21**, 56.
37. Kolagunta, V. R., Janes, D. B., Bielefeld, J. D., Andres, R. P., Osifchin, R. G., Henderson, J. I. and Kubiak, C. P. (1996). *Proc. Electrochem. Soc.* **95**, 56.
38. Murray, C. B., Kagan, C. R. and Bawendi, M. G. (1995). *Science* **270**, 1335.
39. Kim, E. and Whitesides, G. M. (1995). *Chemistry of Materials* **7**, 1257.
40. Alivisatos, A. P., Johnsson, K. P., Peng, X., Wilson, T. E., Loweth, C. J., Bruchez Jr., M. P. and Schultz, P. G. (1996). *Nature* **382**, 609.
41. Trau, M., Sankaran, S., Saville, D. A. and Aksay, I. A. (1995). *Langmuir* **11**, 4665.
42. Trau, M., Saville, D. A. and Aksay, I. A. (1996). *Science* **272**, 706.
43. Paunov, V. N., Kralchevsky, P. A., Denkov, N. D. and Nagayama, K. (1993). *J. Colloid Interface Sci.* **157**, 100.
44. Velev, O. D., Furusawa, K. and Nagayama, K. (1996). *Langmuir* **12**, 2385.
45. Weiss, J. A., Oxtoby, D. W., Grier, D. G. and Murray, C. A. (1995). *J. Chem. Phys.* **103**, 1180.
46. Murray, C. A. and Grier, D. G. (1996). *A. Rev. Phys. Chem.* **47**, 421.
47. Park, S. H., Qin, D. and Xia, Y. (1998). *Adv. Mater.* **10**, 1028.
48. Crocker, J. C. and Grier, D. G. (1998). *MRS Bull.* **23**, 24.
49. Grier, D. G. (1998). *Nature* **393**, 621.
50. Weiss, J. A., Larsen, A. E. and Grier, D. G. (1998). *J. Chem. Phys.* **109**, 8659.
51. Adams, M., Dogic, Z., Keller, S. L. and Fraden, S. (1998). *Nature* **393**, 349.
52. Burns, M. M., Fournier, J. M. and Golovchenko, J. A. (1989). *Phys. Rev. Lett.* **63**, 1233.
53. Tu, J. K., Talghader, J. J., Hadley, M. A. and Smith, J. S. (1996). *Electronic Letters* **31**, 1448.
54. Dufresne, E. R. and Grier, D. G. (1998). *Rev. Sci. Instrum.* **69**, 1974.
55. Martin, J. E., Odinek, J. and Halsey, T. C. (1992). *Phys. Rev. Lett.* **69**, 1524.
56. Skjeltorp, A. T. (1983). *Phys. Rev. Lett.* **51**, 2306.
57. Hwang, Y. H. and Wu, X.-L. (1994). *Phys. Rev. E* **49**, 3102.
58. Dimitrov, A. S., Takahashi, T., Furusawa, K. and Nagayama, K. (1996). *J. Phys. Chem.* **100**, 3163.
59. Markina, Z. N., Bovkun, O. P., Zadymova, N. M., Roskete, E., Shchukin, E. D., Makarov, K. N. and Gervits, L. L. (1988). *Zh. Vses. Khim. O-va. im. D. I. Mendeleeva* **33**, 346.
60. Morra, M., Occhiello, E., Marola, R., Garbassi, F., Humphrey, P. and Johnson, D. (1990). *J. Colloid Interface Sci.* **137**, 11.
61. Howie, F. H. and Hondros, E. D. (1982). *J. Mater. Sci.* **17**, 1434.
62. Oliver, S., Bowden, N. and Whitesides, G. M., unpublished results.
63. Tien, J., Clark, T. D., Duffy, D. and Whitesides, G. M., unpublished results.
64. Maboudian, R. and Howe, R. T. (1997). *J. Vac. Sci. Technol. B* **15**, 1.

5

Design of amphiphiles for the modulation of catalytic, membranous and gelation properties

SANTANU BHATTACHARYA

5.1 Introduction

Amphiphiles are molecules in which there exist a distinct polar segment and a lipophilic segment. Existence of segments of such opposing polarities makes these molecules self-organize upon solubilization in water in such a manner that the polar portion of these molecules remain exposed to water while the hydrophobic segments remain virtually buried [1]. Thermodynamically driven supramolecular aggregates of amphiphilic molecules, including those of natural origin (known as lipids), constitute some of the most widespread organizations in biological systems. These make up nearly 30–70% (by weight) of the components of the cell membrane, and are among the most intriguing type of self-assembly known [2]. The exceptional properties of amphiphilic molecules have, since the early development of chemistry, attracted the curiosity of many scientists, including those whose interests are at the interface of biology and materials science.

This chapter will present a few selected instances of design of synthetic amphiphiles including some from the author's own laboratory that help in the understanding of various fundamental issues that pertain to the aggregation properties of these amphiphiles. Therefore rather than a review of the topic, this is an attempt to highlight certain recent developments involving designed molecular assemblies. References to review articles rather than to original papers are given wherever possible to limit the otherwise long list of references.

5.2 Catalyses and reactions in amphiphilic aggregates

Catalytic processes are part of modern strategies for the fast, inexpensive production of selective chemicals and are central to most modern chemical manufacturing industries. This is often used in finding applications, which span the range from large-scale petrochemical to small-scale biochemical processes. Catalysis may be intelligently employed in the decontamination of pollutants and undesired by-products, as well as providing a means for designing novel reaction systems that do not require environmentally hazardous compounds [3].

The ability of micelles or related aggregates to alter reaction rates and selectivity has been an area of active research for the past several decades. Reactants are partitioned into the aggregates by coulombic and hydrophobic interactions; the observed rate accelerations are largely a result of the increased localization of the reactants and also of the typical physicochemical properties of the micellar environment, which are significantly different from those of the bulk solvents. This unique ability of the aggregate systems has therefore prompted several scientists to employ micellar media for catalytically carrying out specific reactions.

Micellar catalysis has been demonstrated to be effective for a large number of reactions, resulting in an extensive (although often empirical) knowledge base for predicting the behavior of general classes of reaction with different types of surfactant. For instance, the importance of micelles in reactive systems has been demonstrated during the initial stage in emulsion polymerization, as potential replacements for volatile and toxic organic solvents, as templates in the syntheses of new molecules and materials, and for developing rudimentary mimics for enzymes.

Earlier studies of numerous research groups have considerably helped to establish a basic understanding of the mechanism of various types of micellar catalysis. Several excellent accounts of prior work have been published [4–8]. Pseudophase treatments of reactivity in association colloids provide a chemically satisfying approach with predictive power [9]. The present section of this article focuses on a few examples from recent literature that explore the catalytic effects observed from aqueous micellar or mixed-micellar aggregates.

Attempts to develop new aggregate-based recipes for catalytic decontamination of areas tainted with phosphorus-based persistent chemicals continue to receive attention. Most of these studies employ a simulant phosphotriester, *p*-nitrophenyldiphenyl phosphate (PNPDPP), as a test substrate and examine the efficacy of each formulation developed. Moss *et al.* described a potent esterolytic reagent based on 9-iodoso-10-phenanthroate in micellar media [10].

Bhattacharya and Snehalatha reported the utility of a cationic surfactant aggregate-based formulation of monoperoxyphthalate (MPP) as an inexpensive, 'off-the-shelf' reagent for catalytic decontamination [11]. They also presented evidence of the formation of the acylated or phosphorylated monoperoxyphthalate intermediates during the phosphorolysis and esterolysis

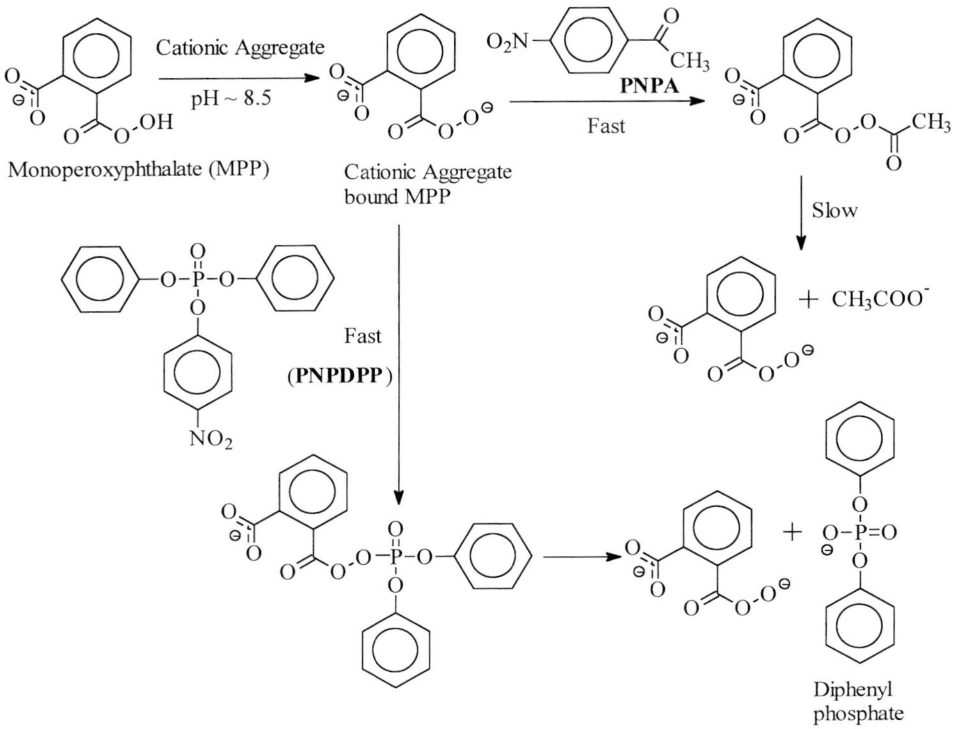

Scheme 5.1

reactions of appropriate activated substrates in mildly basic pH induced by MPP under micellar conditions. On the basis of this work, the authors also propose a mechanism consistent with their experimental findings (Scheme 5.1).

Khan and Arifin reported a study involving the kinetics and mechanism of an intramolecular general base-catalyzed methanolysis of phenyl salicylate in cationic micellar media [12]. The authors showed the participation of methanol in the reaction via its partitioning into the micellar pseudophase and thereby modifying the overall aggregate microenvironment.

Owing to the uncertainties associated with substrate or reagent partitioning into various sub-sites of micelles, the quantitative predictions about the course of reactions and observed changes in the rates of such reactions often become difficult. Consequently attention has been directed toward the design and synthesis of amphiphiles in which either substrate or reagent functions are covalently attached.

5.2.1 *Functionalized surfactant micelles*

Covalent introduction of a residue into an amphiphile, often at the headgroup level, results in the generation of functional surfactants (Fig. 5.1). These molecules also aggregate on their own upon solubilization in water or could be comicellized with a supporting, yet functionally inert, surfactant. Functional surfactants containing reactive nucleophilic or substrate groups participate chemically in the reaction. The principal advantage that functional surfactants offer over ordinary non-functional surfactants is in defining the location of the reactive substrates and nucleophiles at the interfaces of the micellar aggregate. This also brings about significant and often favorable changes in the pK_a values of certain functional groups at this location. Because micelles are dynamic entities, microheterogeneous in character, and possess morphological features that are superficially similar to globular proteins, functional micelles offer a good starting point for the design of primitive analogues of enzymes.

Consequently, new investigations dealing with reactions in micellar solutions composed of functional surfactants, or mixtures of inert and functional surfactants, continue to appear in the literature. An interesting study of acid-catalyzed hydrolysis of 2-(*p*-tetradecyloxyphenyl)-1,3-dioxolane (*p*-TPD) in aqueous sodium dodecyl sulfate (SDS) solutions has been reported [13]. In this case, *p*-TPD behaves as a non-ionic functional surfactant and apparently forms non-ideal mixed micelles with the anionic surfactant (SDS). Based on the observed kinetic data, the authors propose that, at elevated temperatures, the thermodynamic non-ideality results in the manifestation of two populations of micelles, one rich in SDS and the other rich in *p*-TPD.

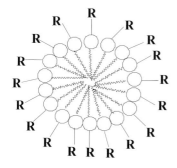

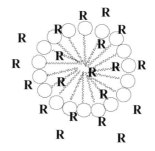

Covalently functionalized micelle with reactant (R) units

Micelle to which reactants (R) are added

Figure 5.1. Idealized presentation of functionalized micellar aggregate.

Boyer *et al.* reported [14] on the comparison of performances of anionic and cationic surfactants on inducing reduction of ketones by $NaBH_4$ in two-phase water–heptane systems. Although cationic surfactants are more efficient, the surfactants of either charge type exhibit discernible catalytic effect. This prompted the authors to propose the role of surfactants as promoters of the transfer of lipophilic reactants from heptane to the aqueous phase.

4,4-Dialkylaminopyridine (DAAP) and its derivatives have been widely employed as highly efficient super-nucleophilic acyl-transfer catalysts in the 1980s [15–17]. A large number of new polymeric and amphiphilic systems bearing the DAAP residue have been developed to potentiate the catalytic effects associated with the chemistry of DAAP [18–21]. Most of these studies addressed the abilities of various DAAP systems to confer strong hydrolytic reactivity against a set of test substrates in water, or in mixed aqueous–organic solvents. These studies provide a basic understanding of the origins of catalytic efficiency and selectivity in man-made systems when compared to biological catalysts.

In this area, recent unrelated efforts of the groups of Bhattacharya and Fife toward the development of new aggregate and polymer-based DAAP catalysts deserve mention. Bhattacharya and Snehalatha [22] report the micellar cataly-sis in mixtures of cetyl trimethyl ammonium bromide (CTAB) with synthetic anionic, cationic, nonionic, and zwitterionic 4,4′-(dialkylamino)pyridine functional surfactant systems, **1b–c** and **2a–b**. Mixed micelles of these func-tional surfactants in CTAB effectively catalyze cleavage of various alkanoate and phosphotriester substrates. Interestingly these catalysts also conform to the Michaelis–Menten model often used to characterize the efficiency of natural enzymes. These systems also demonstrate superior catalytic activity as compared to the ones previously developed by Katritzky and co-workers (**3** and **4**).

1

a: R = CH_3

b: R = $(CH_2)_2CONHC_{18}H_{37}$

c: R = n-$C_{18}H_{37}\overset{\oplus}{N}Me_2(CH_2)_3NHCO(CH_2)_2$, $\overset{\ominus}{Br}$

2

a: R = n-$C_{18}H_{37}$

b: R = n-$C_{18}H_{37}\overset{\oplus}{N}Me_2(CH_2)_3NHCO(CH_2)_2$, $\overset{\ominus}{Br}$

Bhattacharya *et al.* have also reasoned the lack of substantial rate enhancements with the surfactants, **3** and **4**. In amphiphile **3**, the DAAP residues are probably confined in orientations that are 'inaccessible' to the substrates as the DAAP unit is attached to the surfactant head group via a rigid, inflexible piperazinium moiety. On the other hand in **4**, the DAAP moieties are far away from the headgroups. Long hydrophobic chain spacing between DAAP and the headgroup charge promote looping of the DAAP units into the aggregate interior. Under these circumstances the substrates probably do not have optimal access to the 'buried' DAAP residues. In contrast, in the DAAP catalysts **1** and **2**, the DAAP residues are attached closer to the headgroup region via shorter, flexible spacer chains. The presence of amide-type linkages in some of these amphiphiles, (such as in **1b**) between the hydrophobic chain and the headgroup might help form inter-molecular hydrogen bonded networks to further assist holding the DAAP units around the Gouy–Chapman layer. Among these, the anionic carboxylate ion bearing DAAP catalyst **2a** is the most effective. In the mixed micelles of **2a** with excess CTAB, the ion-pairing $(NMe_3^+ \ldots {}^-OOC-)$ at the headgroup region should impose the location of DAAP for the optimal reaction with the substrate.

3

a: R = CH$_3$
b: R = CH$_2$COOH

4
R = *n*-Bu
a: X = OH
b: X = OSO$_3$H
c: X = OPO$_3$H$_2$
d: X = COOH

Bhattacharya and Snehalatha also examined the reactivities of both functionalized DAAP surfactants and simple non-amphiphilic DAAP analogues in microemulsion droplets and found that their reactivities were quite altered in such aggregates when compared with their reactivities in micellar media [23–24].

Fife and co-workers described a macromolecule, **5** containing 4–DAAP and a bis-(trimethylene) disiloxane backbone that exhibited enzyme-like substrate selectivity for the esterolysis of *p*-nitrophenyl alkanoates, **6** (Scheme 5.2). This synthetic polymer showed highest levels of activity toward substrate **6** ($n = 14$), when it was used as a nucleophilic catalyst for the solvolysis of a series of

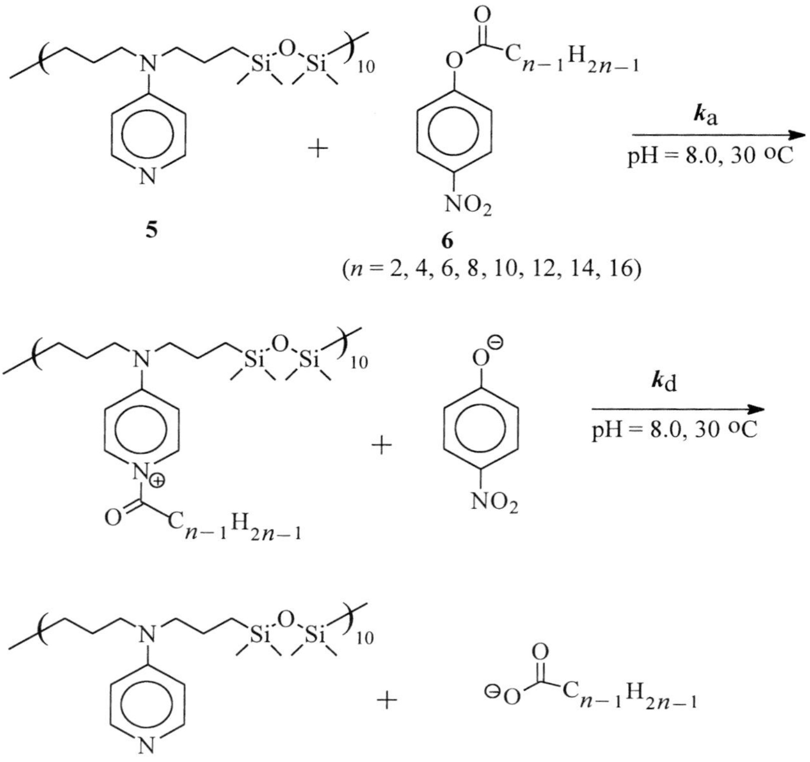

Scheme 5.2

homologues of **6** in 1:1 (v/v) MeOH–water solution [20]. Interestingly the rates of such catalytic reactions were shown to be enhanced significantly by the presence of an anionic surfactant SDS. Both the substrate and the catalyst become confined within the micellar aggregates of SDS in this situation. Fife and coworkers reported an instance of a direct salt effect type of interaction between a buffering agent, tris(hydroxymethyl)ammonium ion, and catalyst **5**, which selectively showed highest levels of activity toward the substrate **6** ($n = 6$), for the esterolysis of a series of **6** that varies in terms of chain length [25].

5.3 Metallomicelles

Many proteolytic enzymes utilize *metal ions* in their active site for catalyzing specific hydrolytic reactions [26]. Consequently, quite a number of electron-rich ligands for the complexation of specific metal ions [27] have been designed. Hydrolytic reactions mediated by such complexes have also been examined in order to obtain insight into the mechanisms responsible for this kind of catalysis [28]. These studies provided a basis for understanding certain actions of the metalloproteases, which often use water molecules or internal alcoholic side-chain residues (e.g. serine or threonine) as nucleophiles. Coordination of the nucleophilic residues or water molecules to the ligated metal ion enhances their acidity, facilitating deprotonation at physiological pH and speeding the reaction with electrophilic substrates, e.g. amides, esters, or phosphates. In addition, the electrophilicity of a substrate molecule is often enhanced through its coordination with the metal ions present at 'active' sites.

In this context the functional amphiphiles based on metal ion-complexing ligands offer advantages in the pursuit of designing metalloprotease mimics because they aggregate in aqueous media. Such supramolecular ensembles provide hydrophobic binding sites for the substrates and induce catalytic effects through appropriate functional groups [29]. These aggregates are known as 'metallomicelles' [30].

Scrimin and co-workers have shown that the hydrolysis of α-amino acid esters can be substantially accelerated by transition metal ions, such as Cu^{2+}, in the presence of appropriately designed ligand amphiphiles [31]. Aggregates made of the amphiphilic Cu^{2+}-complex (**7**) were found to accelerate the hydrolytic activity. The esterolytic reaction was found to be faster by nearly three orders of magnitude in micellar aggregates than in their monomeric form with the *p*-nitrophenyl picolinate (**PNPP**) under the conditions $[(7)]$ or $[(8)] = [Cu^{2+}] = 1.3 \times 10^{-4}$ M. This also resulted in an approximately 100-fold acceleration of the reaction in metallomicelles with respect to the effects of Cu^{2+} alone and nearly 10-fold retardation by the water-soluble complex. This difference in behavior was attributed to the shifting of the equilibrium via coordination of the alcoholic side arm of the ligand in the micellized state. In the water-soluble, monomeric complex (**8**), however, the OH arm was not involved in coordination to the Cu^{2+}-ion. Because the coordination was effective only in micellar media, this system was in a position to 'exploit' the nucleophilicity of the Cu^{2+}-bound alcohol group (Fig. 5.2).

Scrimin and co-workers have extensively studied the ester-cleavage abilities of various Cu(II)-chelating *bidentate* ligand [(2-hydroxymethyl)pyridine] amphiphile, **9**, and related bolaphile, **10**, in micellar media [32]. The corresponding metallomicelles are powerful catalysts for the cleavage of substrates, e.g. *p*-nitrophenyl alkanoates, that do not coordinate with the metal-complex core. Subsequent studies demonstrated that *tridentate* ligand amphiphiles such as **9**,

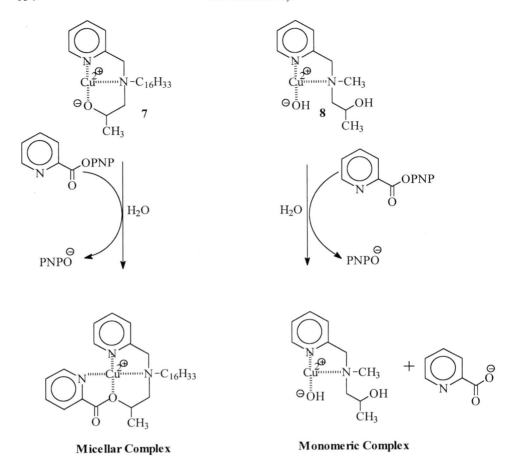

Micellar Complex **Monomeric Complex**

Figure 5.2. Monomeric complex **8** hydrolyzes substrate PNPP directly, as the $-CH_2OH$ sidearm in the ligand is not involved in the Cu^{2+}-ion coordination. However, the micellar complex **7** is acylated during the hydrolysis of PNPP through its $-CH_2OH$ sidearm, which is also involved in the Cu^{2+}-ion binding.

form Cu(II)-chelating micelles that are capable of potentiating ester hydrolysis even when the esters cannot participate in the complex formation [33].

Bhattacharya *et al.* have recently developed some Cu^{2+}-chelating dialkyl-aminopyridine-bearing amphiphiles (**11–13**) and investigated their ester hydrolyzing abilities in aqueous mixed micellar aggregates with CTAB [34]. These ligand amphiphiles conferred somewhat better catalytic activities at pH 7.6, 25 °C, than the catalytic system based on micellar **9**/Cu^{2+}-complex developed by Scrimin [33]. These amphiphiles formed 2:1 ligand/Cu^{2+}-complexes which were also kinetically competent. All of these systems comprising the co-aggregates of **11**, **12** and **13** and CTAB demonstrated turnover in the presence of excess substrate. Nevertheless these findings suggested that an increase in the π-electron density on the N-atom of the pyridine ring in the DAAP ligand did not significantly contribute to any major rate enhancement in the hydrolytic reactions relative to Cu^{2+}-complexes of **9** under micellar conditions.

Kimura *et al.* employed novel Zn^{2+}-based metallomicellar systems that were effective agents for deacylation and dephosphorylation reactions [35]. The authors provided a qualitative analysis of the species present in mixtures of Zn^{2+}-ion and macrocycles.

Mechanistic knowledge of enzyme-catalyzed reactions is being applied increasingly to incorporate minimalistic structural units to synthetic analogues. Some of the efforts described above have successfully recreated at least a fraction of the activities of natural enzymes. This biomimetic approach will be sustained as long as there is a need for developing stable, inexpensive catalysts that can survive the conditions in which the natural enzymes get denatured and do not sustain their activity.

5.4 Designer lipids for the modulation of membranous properties

Lipids and related amphiphiles are important building blocks of biological membranes [2]. Several naturally occurring lipids have been isolated from sources as diverse as plants, mammalian tissues, bacteria and marine organisms [36]. However, owing to the practical difficulties associated with their isolation, effective purification and characterization, most of the physical approaches to the understanding of membrane behavior at a molecular level have been largely devoted to the analogues of lipids of defined molecular structure. Early pioneering studies in this direction demonstrated the feasibility of exploring various membrane-forming properties from molecules of totally synthetic, non-biological origin [37].

Reports describing the design of new lipid architectures continue to appear in the literature. A few instances of case studies involving newer approaches toward lipid design for achieving control of membrane level properties will be discussed in the present section.

5.4.1 Ion-paired lipids

Spontaneous formation of specific supramolecular patterns or folding motifs could often be rationalized on the basis of the stabilizing interactions that are non-covalent in character. Ion-pairing or salt-bridge formation brings about significant stabilization. Thus, ion-pairs are frequently encountered inside different protein molecules [38a]. Even in membrane proteins such as rhodopsin [38b] the existence of such salt-bridges inside the membrane-embedded region has been demonstrated. Ion-pairing is crucial in other biological events as well. Implications of ion-pairing through anionic lipid mediated release of nucleic acid during cationic lipid induced gene transfection have already been described [39].

When two amphiphiles of identical charge are mixed, micellar aggregates are formed [5] but spontaneous lamellae, vesicle or nanodisc formation is observed when amphiphiles with *oppositely* charged head groups are mixed [40]. The resulting systems are called 'ion-paired amphiphiles'. Presumably the electrostatic interactions at or near the headgroup region and hydrophobic association of the non-polar chains drive the formation of 'tight' complexes that eventually organize into lamellae or bilayers.

These amphiphiles are important as they provide a 'bridge' to the supramolecular gap that exists between single- and double-chain amphiphiles. This fact, together with the ready availability and low cost of single-chain surfactants as compared with their double-chain counterparts, provides considerable stimulus for the design and exploitation of these types of lipids as lamellar material. The following examples illustrate how specific ion-pairing influences membranous aggregate formation.

Regen *et al.* [41] investigated the aggregation and polymerization behavior of the ion-paired lipid, **14**. In another study, Regen *et al.* also compared the barrier-forming properties of the bilayers derived from ion-paired lipid, cetyltrimethylammonium palmitate (CTAP, **15**) with that of 1,2-dipalmitoyl-*sn*-glycero-3-phosphocholine, DPPC, a naturally occurring lipid of identical chain length [42].

14

15

Kobuke and co-workers [43] synthesized the ion-paired lipids composed of glycolate ethers, **16a–16d**. These lipids exhibited novel properties characteristic to the formation of single ion channels, upon insertion into a bilayer lipid membrane.

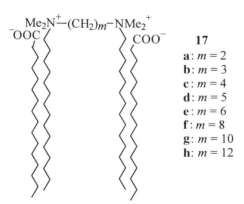

16a ($n = 2, m = 18$)
16b ($n = 2, m = 1$)
16c ($n = 3, m = 18$)
16d ($n = 3, m = 1$)

Bhattacharya and De introduced [44] a novel class of amphiphiles (**17a–h**) in which the gemini surfactant is ion-paired with palmitate ions at the level of the head groups. The resulting amphiphiles differ in terms of the length of a spacer polymethylene chain that has been used to covalently connect the two segments of the gemini. These amphiphiles not only formed vesicular aggregates, as evidenced from dye entrapment experiments, electron microscopy (TEM) and dynamic light scattering (DLS), but their vesicular properties could also be adjusted conveniently by the choice of the spacer chain length. Thus the bilayers of **17a** undergo a solid gel-to-fluid thermal phase transition at ~ 74 °C, while the bilayers of **17h** melted at ~ 39 °C. Similarly the vesicles of **17a** were impermeable and those of **17h** were too leaky. The rate of transmembrane permeation increased progressively with the increase in *m*-value (**17b–17g**). The origin of such control of the membrane level properties has been rationalized on the basis of differences in their packing modes in the resulting bilayer membranes.

$$\text{Me}_2\overset{+}{\text{N}}{-}(\text{CH}_2)_m{-}\overset{+}{\text{N}}\text{Me}_2$$
$$\overline{\text{OOC}} \qquad\qquad \text{COO}^-$$

17

a: $m = 2$
b: $m = 3$
c: $m = 4$
d: $m = 5$
e: $m = 6$
f: $m = 8$
g: $m = 10$
h: $m = 12$

Bhattacharya *et al.* have also shown [45] that it is possible to generate vesicles via ion-pairing of two cationic amphiphiles with one bis-anionic *bolaform* amphiphile (see below). To explore whether subtle structural differences in the starting ion-pairs are reflected in the properties of the supramolecular assemblies, Bhattacharya *et al.* synthesized four hybrid bolaphile/amphiphile ion-pairs **18a–d**. Three of these ion-paired compounds, **18a–c**, contain the following molecular modules. The bolaphilic parts in these ion-pairs retain a central aromatic core from which two lipophilic chains are connected, and the termini of these chains contain carboxylate residues. The disubstituted (1,2–, 1,3–, 1,4–) central aromatic units in these molecules are isomeric with respect to each other and act as structural anchors in these bolaphiles. Thus depending on the position of the covalent linkage of the undecanoate chains to the central core, these lipophilic chains propagate in directions dictated by the manner in which they are connected to the respective aromatic ring. The $-CO_2^-$ residues at the termini are finally ion-paired with the cetyl trimethyl ammonium (CTA^+) cations. To put these results into proper perspective, the vesicular properties of the corresponding amphiphile (monopolar) anion/cation pair, cetyltrimethylammonium 11-phenoxy undecanoate, **18d**, and CTA palmitate, **15**, were also examined.

TEM, DLS and dye entrapment studies were carried out with these aggregates that confirm the vesicle formation from **18a–d**. Remarkably the permeability of these vesicular aggregates was found to be highly influenced by the geometry of the bola-amphiphilic counterion. The T_m values were very different although these ion-pairs were isomeric and were found to depend strongly on the architecture of the starting ion-pairs. X-ray diffraction studies of the cast multilayered films produced from each of these ion-pairs were also performed to investigate the possible nature of the packing and organization in these membranes. Most of the lipid suspensions of this class showed polymorphism in their organizations. Although cast films of the vesicles **18a** and **18b** showed tilted bilayer and interdigitated modes of packing, that of **18c** showed unit layer thickness close to the length of the ion-pair in a typical extended bolaform conformation. It is evident that by judicious incorporation of central, isomeric, disubstituted aromatic units as structural anchors into different bolaphiles one can modulate the properties of the resulting vesicles.

18a

18b

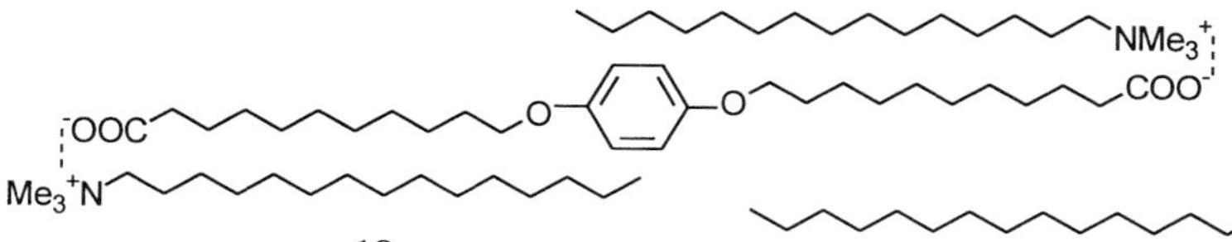

18c

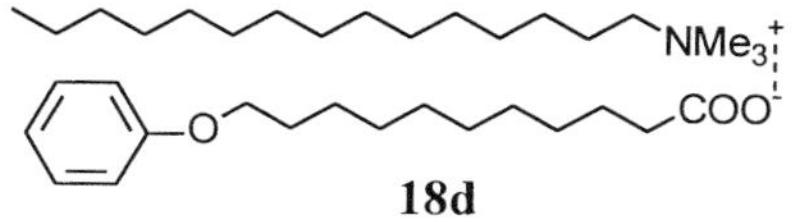

18d

Jaeger *et al.* described [46] some novel double-chain vesicle-forming cleavable amphipiles which were synthesized via ion-pairing of two oppositely charged head groups. These amphiphiles formed giant vesicles which cleave at the labile carbon–nitrogen double bond around pH ~ 3.

5.4.2 Bolaform lipids

Membranes in thermophilic and acidophilic bacteria can sustain their normal biological function even in extreme physiological conditions ($>85\,^{\circ}C$ or very low pH). Owing to these exceptional properties, there has been a lot of interest in the structure–function studies of the types of lipids that occur in these organisms. The widths of the membranes that occur in archaebacteria are considerably thinner than those typically found in the eukaryotic cells. This can be explained on the basis of their differences in the lipid molecular structures of the respective biological species. Because of the presence of two polar head groups at each terminus of a lipophilic segment (bolaform amphiphiles) in the archaebacterial lipids, these aggregates tend to form 'monolayer' membranes (MLMs) in water [47]. In contrast, the lipids that are found in mammalian cells are typically double-chain, monopolar amphiphiles such that their aggregation in aqueous media results in the formation of bilayer membranes (BLMs). Synthetic bolaform amphiphiles also represent an important class of biomimetic materials and provide insight into membrane stabilization strategies without requiring polymerization. Several reports describing the design and synthesis of symmetric bolaamphiphiles have appeared in literature of which a few instances will now be described.

Thompson *et al.* have developed a few symmetric, tetraether bolaamphiphiles **19a**, **19b**, with anionic headgroups [48]. The aqueous dispersions of these lipids behave as good mimics of archaebacterial membranes.

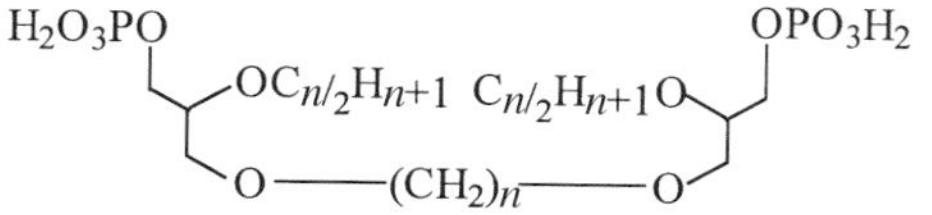

$n = 20/16$

19a

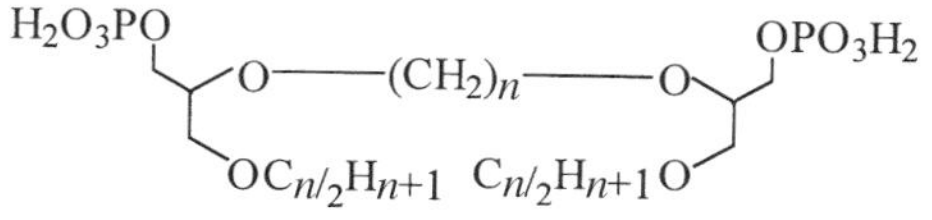

$n = 20$

19b

Moss introduced the idea of using synthetic lipid analogues bearing cleavable ester functionalized head groups [49a−d]. These lipids were employed extensively for kinetically differentiating the rates of ester hydrolyses at the outer and the inner surfaces of vesicles and also for examining the transbilayer dynamics (flip-flop) in membranes. In this connection Moss *et al.* reported the synthesis of a novel group of biscationic bolaphiles, **20**, and examined the transbilayer dynamics in bilayer liposomes [49e].

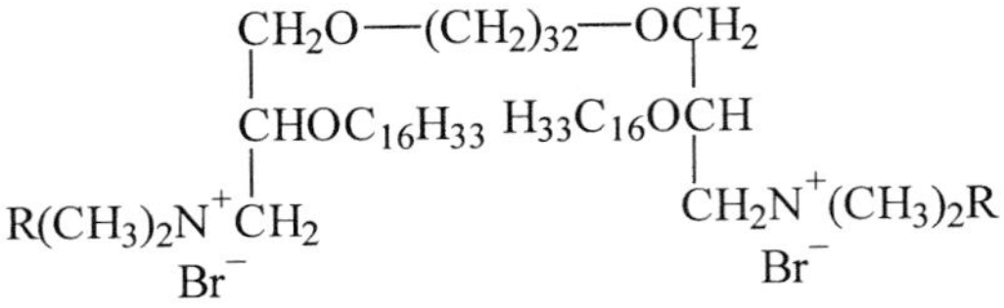

$$CH_2O\!-\!(CH_2)_{32}\!-\!OCH_2$$
$$CHOC_{16}H_{33} \quad H_{33}C_{16}OCH$$
$$R(CH_3)_2N^+CH_2 \qquad\qquad CH_2N^+(CH_3)_2R$$
$$Br^- \qquad\qquad\qquad\qquad Br^-$$

20

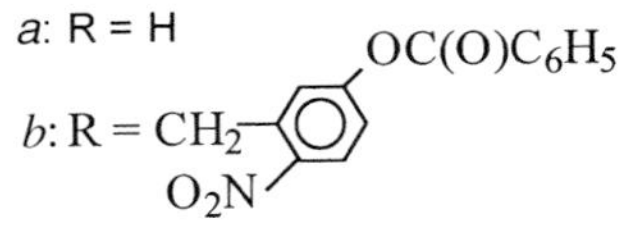

a: R = H

b: R = CH$_2$— with OC(O)C$_6$H$_5$ and O$_2$N

A new family of crown-ether-based bolaamphiphiles, **21**, that aggregate into a previously unknown type of bolaamphisome was prepared by Gokel and co-workers [50]. Evidence was presented on vesicle formation from the aqueous suspension of such bolaform amphiphiles. Several reports describing the synthesis of various macrocyclic models related to archael compounds have also appeared in the literature [51]. Fyles *et al.* prepared a novel series of asymmetric bis-macrocyclic bolaphiles, **22**, and evaluated their transport activities in vesicles and planar bilayer membranes [52].

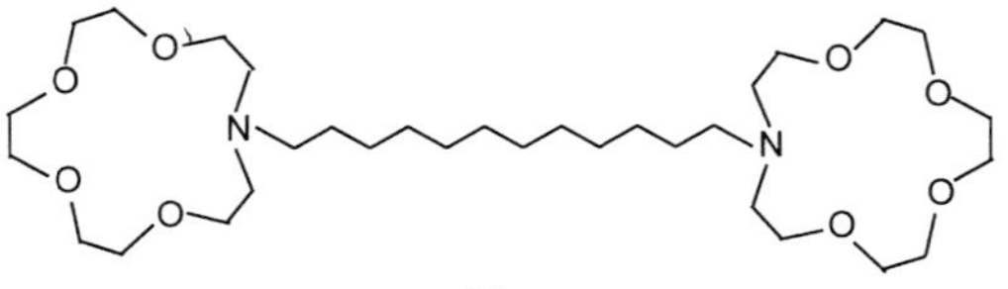

21

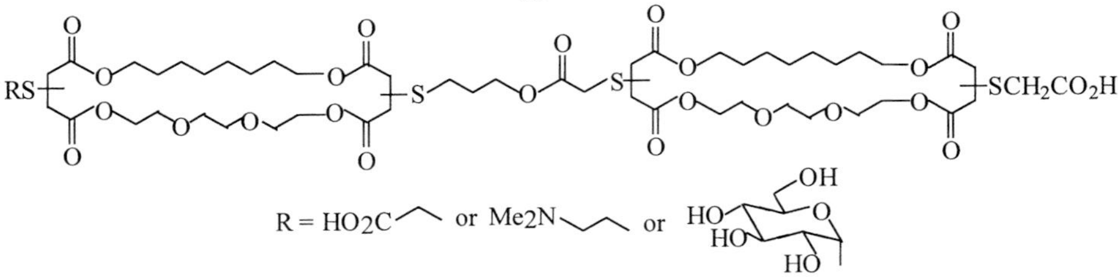

R = HO$_2$C— or Me2N— or HO-sugar

22

Certain bolaform systems provide opportunities to differentiate between the outer and inner surfaces of the MLMs. Bolaamphiphiles bearing two head groups of different sizes, that vary in terms of charges or charge types, may allow the formation of totally asymmetric membranes, especially if one of the head groups can be 'forced' to aggregate selectively on one surface. Their stability range is appropriate for their applications in cellular transfection, surface derivatization, immobilization of integral membrane protein receptors, etc.

Generally, however, bolaform amphiphiles with two different head groups usually give vesicles with a statistical distribution of both head groups on the inner and outer surfaces [47]. Fixation of the charged head group on the inner surface can be achieved by attachment to polymeric, colloidal or solid materials with opposite charges. Herein lies the significant potential for the application of bolaamphiphiles and MLMs. They can be 'engineered' to protect surfaces, stabilize colloids, and can be used to bring in contact with inorganic and organic materials with very high surface areas per unit mass of material [53]. One such example is the long-chain monoester monoamide bolaphile, **23**, with a carbohydrate head group and a tetra-alkylammonium head group. This has been used to entrap polynucleotides as well as inorganic colloids [54].

$$ RS\text{-}\diagdown\diagdown\diagdown\diagdown\diagdown\diagdown\diagdown\overset{\displaystyle O}{\underset{\displaystyle O}{C}}\diagdown\diagdown\diagdown\diagdown\diagdown\diagdown\diagdown\overset{\displaystyle H}{\underset{\displaystyle O}{N}}\diagdown\diagdown N^{+}(CH_3)_3 \quad I^{-} $$

R = glucose, mannose, galactose

23

5.4.3 Lipid dimerizations

Dimerization provides a convenient way by which one can control the packing and the organization of amphiphilic supramolecular assemblies. A large number of micelle-forming dimeric amphiphiles (gemini) have been developed [55]. Impressive alterations in their physical properties have been observed, particularly with the dimeric amphiphiles with short spacer chains [56]. Because of their very low critical micellar concentration, high viscoelasticity, better wettability, pronounced antibacterial properties, and their enhanced tendency towards lowering of surface tension at the oil–water interfaces, these materials are receiving a lot of current attention [57].

The corresponding dimeric analogues of lipids are also important, owing to their occurrence in biological membranes. For example cardiolipins, which constitute a class of complex dimeric phospholipids that occur mainly in the heart and skeletal muscles, show high metabolic activity [58]. The association of these type of acidic phospholipid in cell membranes with Ca^{2+} often triggers membrane fusion as Ca^{2+} binding induces formation of dimeric or higher order lipid complexes [59]. Similarly, glycolipid A_2, another complex multimeric lipid, is composed of two phosphate head groups and as many as seven hydrophobic chains. It serves as an immunomodulator in mediating selective recognition of toxins, bacteria or viruses at the cell surface [60].

Therefore a detailed study of the influence of dimerization of well-defined synthetic lipids seems to be the most direct approach to achieve an understanding of the role of these complex amphiphilic molecules in living cells. In addition, new lipidic materials based on dimerization might be of practical value. Based on these expectations, several reports describing the design, synthesis and characterization of a number of new dimeric lipids have appeared in recent literature. Most of these lipids are made up of two identical monomeric lipid moieties connected at the level of the head groups or very close to these by a spacer unit, which can be either hydrophilic or hydrophobic, or even be either rigid or flexible.

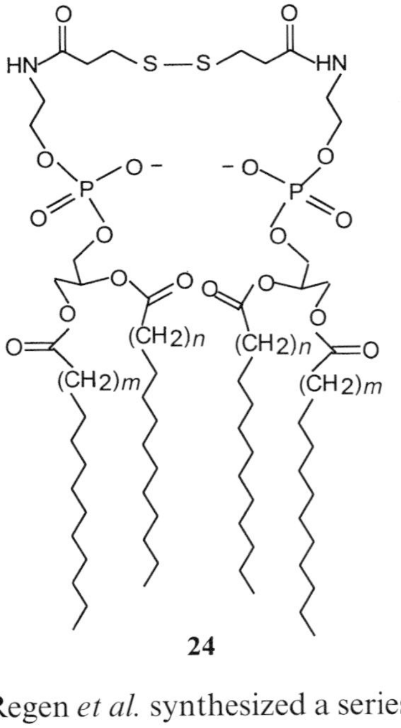

24

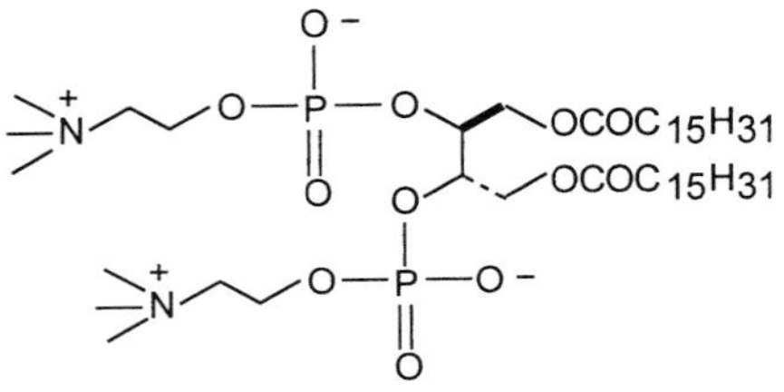

25a

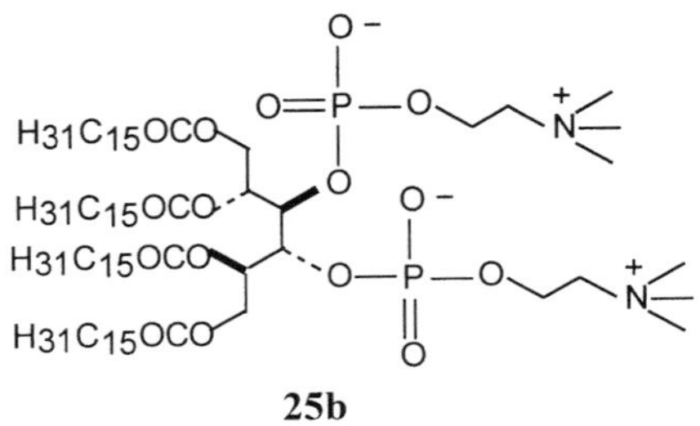

25b

Regen *et al.* synthesized a series of disulfide-linked phospholipid dimers, **24**, to study the lipid mixing and domain formation in natural phospholipid membranes [61]. Lennox *et al.* developed two novel bis-phosphatidylcholine lipids, **25a** and **25b**. These were employed to understand the interfacial activation mechanism of the membrane bound phospholipase A_2 enzymes [62].

Nolte and co-workers synthesized three different stereoisomers of a novel phosphatidic acid analogue bearing two phosphate groups, **26a**, **26b** and **26c**. They examined the aggregation behavior of these lipids in their aqueous suspensions and also studied their fusion and fission processes upon vesicle formation [63].

$$Na_2O_3PO\text{---}\overset{*}{\underset{*}{\,}}\text{---}O\text{---}C(=O)\text{---}n\text{-}C_{17}H_{35}$$
$$Na_2O_3PO\text{---}O\text{---}C(=O)\text{---}n\text{-}C_{17}H_{35}$$

26a (*R, R*)
26b (*S, S*)
26c (*R, S*)

Bhattacharya *et al.* developed a family of dimeric lipid systems that are based on ion-pairing [44] or pseudoglyceryl skeleton [64]. These dimeric pseudoglyceryl lipids are structurally similar to the naturally occurring glycerol bridged, dimeric phosphatidic acid known as cardiolipin (**27**), except that the former is cationic and cardiolipin is negatively charged. To put these results into appropriate perspective, Bhattacharya *et al.* also examined the properties of the membranes formed from the corresponding monomeric lipid, **28** (*m* = 0) and based on that rationalized the membrane level properties of the series of dimeric lipids, **29a–29h**.

$$CH_2\text{---}O\overset{O}{\overset{\|}{P}}OCH_2CHCH_2O\overset{O}{\overset{\|}{P}}O\text{---}CH_2$$

with O^-, OH, O^- substituents, connecting to:

$$CH_2\text{---}CH \qquad CH\text{---}CH_2$$
$$O \quad O \qquad\qquad O \quad O$$
$$C_{16}H_{33} \quad C_{16}H_{33} \qquad C_{16}H_{33} \quad C_{16}H_{33}$$

CARDIOLIPIN (27)

The presence of an intervening $-(CH_2)_m-$ spacer at the level of headgroup in gemini pseudoglyceryl lipids, **29**, strongly affects the properties of these amphiphiles at the membrane level. Thus at low and high *m*-values (*m* = 3–4 or 20–22), the vesicles were found to be nearly impermeable and showed high thermal gel → liquid crystalline phase transition temperatures. On the other hand the vesicles formed from the lipids with *m*-value 5–12 showed little differences in their melting temperatures, although the transmembrane permeation rates increased with the increase in *m*-values. Possibly because of the increasing headgroup bulk and spacer chain looping with the increase in *m*-value up to

$m \leq 16$, the membranes from the corresponding lipids showed poorer hydro-carbon chain packing. This was also manifested in the increase in the trans-membrane permeability with the rise in m-value. This probably also enabled these vesicles to undergo facile fusion relative to their low ($m = 3$–4) or high ($m = 20$–22) m-value counterparts. The membranes derived from the lipids with m-value 20–22 were thermally more resistant to melting. X-ray diffraction studies on the cast multilayered films of these vesicles confirmed distinct differences in their organizational and packing arrangements as a function of m-value. The evidence of interdigitation with lipids of high m-value was shown unambiguously from ESR measurements [64b].

CH$_2$—NMe$_3^+$, Br$^-$
/
CH$_2$—CH
/   \
O   O
| |
C$_{16}$H$_{33}$ C$_{16}$H$_{33}$

$m = 0$
28

CH$_2$—NMe$_2^+$(CH$_2$)$_m$NMe$_2^+$—CH$_2$
/   Br$^-$ Br$^-$ \
CH$_2$—CH   CH—CH$_2$
/   \   /   \
O   O   O   O
| | | |
C$_{16}$H$_{33}$ C$_{16}$H$_{33}$ C$_{16}$H$_{33}$ C$_{16}$H$_{33}$

$m = 3, 4, 5, 6, 12, 16, 20, 22$
29a - 29h

In summary, the dimeric lipids (**29a–29h**) with low m-value (3–4) and high m-value (20–22) showed exceptional thermal, lipid-packing and cholesterol-association properties. Obviously the introduction of a polymethylene spacer chain at the level of headgroup brought about dramatic effect on the aggregation behavior, membrane organization and lipid packing of **29**.

The findings described above with unrelated systems reveal the widespread effects of lipid dimerization. Clearly the covalent connection of two '*monomeric*' lipids brings about an effect at the membrane level that is much more pronounced than one would expect from seemingly trivial structural modifications at the lipid level. These findings emphasize the continuing needs for newer design of synthetic lipid structures to expand our understanding of their behavior upon membrane formation.

5.5 Gelation by amphiphile association

Impressive materials properties can also be achieved by the solvent-specific aggregation properties of appropriately designed amphiphilic materials. A brief account on the organogel or hydrogel formation by amphiphilic molecules is now presented, citing a few examples from the current literature. Both types of gel offer potential application in such diverse fields as cosmetics, implants, pharmacology, medicine, hardeners of toxic solvent spills, and even in environmental clean-up [65].

A gel is a form of matter that is intermediate between a solid and a liquid. Hence these are wet and soft materials and, therefore, capable of enduring large deformation. Living organisms are largely made of gels. Most of the animal tissues are gelatinous and are generally composed of protein and polysaccharide based polymeric networks. A number of synthetic polymeric systems also show excellent gelation properties [66].

Recently small molecule (MW < 1000) induced gelation of organic solvents is receiving increasing attention [67]. One reason for this interest lies in the fact that the gels made from small organic molecules may offer some advantages over traditional polymeric gels. This is because of the reversible nature of the non-covalent interactions that hold these small-molecule-based assemblies together. Because these materials are made up of low-molecular-mass systems, they can be made to flow like small molecules in dilute solutions or at elevated temperatures [68]. Solvent-selective gelation is thought to proceed via the formation of highly fibrous aggregates in which the gelator molecules are held together by non-covalent interactions that produce a three-dimensional, highly entangled 'network' in which the specific solvent molecules can be entrapped. These three-dimensional networks are often built through several hydrogen bonding and/or van der Waals contacts.

Molecules of widely different structures may produce such aggregation behavior. Hence compounds as diverse as anthracene alkyl ethers [69a], cyclic depsipeptides [69b], gluconamides [69c], steroid analogues [69d], and polar compounds containing long apolar moieties [69e], all show gelation abilities in different solvents.

Fuhrhop *et al.* described the formation of aqueous gels from N-alkylgluconamide amphiphiles [70]. Although these gels were unstable, the corresponding admixtures with sodium dodecyl sulfate (SDS) afforded gels with shelf-lives up to several months. Electron microscopy revealed the presence of fibers of bimolecular thickness of ~4 nm in these gels.

Yamada *et al.* described the formation of reverse micellar fibers in organic solvents using aggregates of tripeptide containing amphiphiles [71]. The tripeptide units in these aggregates were shown to possess a parallel chain β-sheet structure that was present not only in water but also in carbon tetrachloride.

Oda *et al.* described the use of gemini surfactant/tartaric acid admixtures which could trap up to 1200 solvent molecules per molecule of the gelator [72]. In these gelators, the biscationic geminis contain chiral counterions and they associate to form gels with both water and organic solvents such as chloroform

upon their assembly into helical aggregates, which have similar organizations in
either media.

30

Bhattacharya *et al.* [73] reported gelation behavior of several L-phenylalanine-
based amphiphiles. To explore the impact of molecular structures on gelation
of L-phenylalanine derivatives, Bhattacharya and co-workers synthesized as
many as twelve L-phenylalanine-based mono- and bi-polar derivatives and
solubilized each of these in selected solvents. The formation of gel was found to
depend on the concentration of the gelling agent, solvents and the temperature.
The SEM and TEM studies suggested the formation of intertwined threads
and fibers juxtaposed by slender filaments, which also produced a network with
pores, which probably held the solvent molecules due to surface tension in the
gel.

The specific role played by each structural subunit in inducing gelation from
the above derivatives was examined in detail. Clearly, the hydrogen-bonding
interactions between the N—H and C=O of the urethane as well as the amide
connector, the inter-aromatic π-stacking interactions and the van der Waals
interactions of the long hydrocarbon segments are the essential driving forces
for the formation and the mechanical stability of gels. One intriguing requirement was that the core *phenylalanine* backbone should be in its enantiomeri-
cally pure form. Presumably all of these structural parameters in concert drive
the entanglement of amphiphiles into a matrix that is capable of gelation.

Conversion of monopolar L-phenylalanine-based amphiphiles to their bolaform counterparts did not result in the enhancement of gelation capacities.
However, the bipolar analogues manifested novel adhesion properties.

In another study, Ragunathan and Bhattacharya described the formation of
gels and fibrous assemblies from L-serine-based amphiphiles in chloroform [74]. In this instance also, chiral purity was essential for the observed gelation. The microscopic examinations of these fibers reveal the existence of
braided fibers and helices in these aggregates.

It is apparent that the solubilization of amphiphilic molecules with chiral head
groups produces aggregates such as braided fibers, helices, ribbons, rods, or
tubules. When these organizations are sufficiently extended, they start to get
entangled and develop the ability to entrap solvent molecules in the resulting
reticulum. Many such instances eventually lead to the formation of gels. This phe-

nomenon is fascinating as it opens up new opportunities in aggregate chemistry. However, predictable molecular design of gelator molecules still remains a challenge. Substantial effort will have to be directed toward the design of new gelators based on supramolecular association for specific applications in the near future.

5.6 Conclusions

Results obtained from various research initiatives in the area of self-organizing amphiphilic aggregates in the past two decades or so have contributed to our better understanding of the issues that govern the reactivity, structure, dynamics and morphology of the aggregates. The focus has now shifted toward the design of properties based on information coded at the level of the molecular structure of the amphiphiles. New types of material such as mesoporous solids and nano-structured materials are already showing ways to interesting applications.

Significant improvements in the properties can also be achieved by using the various types of functionalized surfactant aggregates. Scientific endeavors for the design of enzyme mimics will continue until natural enzymes are 'engineered' such that they can participate in the processes such as decontamination of toxic substances.

New intelligent efforts will also help us realize novel biomimetic systems for specific tasks and man-made molecular systems with attractive material properties. Recent efforts toward synthesizing mesoporous materials with uniform and controllable pore morphology employ surfactants. This area represents an exciting new territory for exploiting defined surfactant aggregates. Another notable emerging area is in the design of gene delivery agents involving cationic surfactants.

The present account confines itself to a few examples, which demonstrate the ingenious exploitation of amphiphilic aggregate-based systems for specific purposes. It is clear that a collection of the 'right' molecules, properly assembled, can accomplish much more than an equal number of the same molecules functioning in isolation. Remarkable advances have been made in recent years in the physical techniques by which one can characterize the above type of aggregates and utilize them for various materials applications. State-of-the-art characterization coupled with clever molecular design should hold the key for the development of 'smart' systems.

Acknowledgment

Much of this work was supported by grants from DST and JNCASR. It is a pleasure to acknowledge the contributions of many present and former coworkers whose names appear in the literature citations. I also thank Ms Y. Krishnan and Mr Dileep P.V. for their help in the preparation of this manuscript.

5.7 References

1. (a) Berezin, I. V., Martinek, K. and Yatsimirskii, A. K. (1973). *Russ. Chem. Rev.* **42**, 787. (b) Lindman, B. and Wennerstrom, H. (1980). *Top. Curr. Chem.* **87**, 1.
2. (a) Merz, K. M. Jr. and Roux, B. (1996). *Biological Membranes. A Molecular Perspective from Computation and Experiment*; Birkhäuser, Boston. (b) Slater, J. L. and Huang, C.-H. (1992). In *The Structure of Biological Membranes* (P. Yeagle, Ed.), CRC Press: Boca Raton, FL, pp. 175–210. (c) Marsh, D. (1990). *CRC Handbook of Lipid Bilayers*. CRC Press; Boca Raton, FL.
3. Haber, J. (1994). *Pure Appl. Chem.* **66**, 1597.
4. Cordes, E. (1973). *Reaction Kinetics in Micelles*. New York: Plenum Press.
5. Fendler, J. and Fendler, E. (1975). *Catalysis in Micellar and Macromolecular Systems*. New York: Academic Press.
6. (a) Bunton, C. A. (1991). In *Kinetics and Catalysis in Microheterogeneous Systems* (M. Gratzel and K. Kalyanasundaram, Eds.), p. 13. New York: Marcel Dekker Inc. (b) Tascioglu, S. (1996). *Tetrahedron* **52**, 11113.
7. Scrimin, P. (1996). 'Supramolecular Control of Structure and Reactivity', In *Perspectives in Supramolecular Chemistry* (A.D. Hamilton, Ed.), vol. **3**, p. 101. New York: John Wiley & Sons.
8. Rathman, J. F. (1996). *Curr. Opin. Colloid Interface Sci.* **1**, 514.
9. (a) Romsted, L. S., Bunton, C. A. and Yao, J. (1997). *Curr. Opin. Colloid Interface Sci.* **2**, 622. (b) Bunton, C. A., Nome, F., Quina, F.-H. and Romsted, L. S. (1991). *Acc. Chem. Res.* **24**, 357.
10. Moss, R. A., Bracken, K. E. and Emge, T. J. (1995). *J. Org. Chem.* **60**, 7739.
11. Bhattacharya, S. and Snehalatha, K. (1997). *J. Org. Chem.* **62**, 2198.
12. Khan, M. N. and Arifin, Z. (1995). *Langmuir* **11**, 2393.
13. Ruzza, A. A., Nome, F., Zanette, D. and Romsted, L. S. (1995). *Langmuir* **11**, 2393.
14. Boyer, B., Betzer, J. F., Lamaty, G., Leydet, A. and Roque, J. P. (1995). *New J. Chem.* **19**, 807.
15. Delaney, E. J., Wood, L. E. and Klotz, I. M. (1982). *J. Am. Chem. Soc.* **104**, 799.
16. Vaidya, R. A. and Mathias, J. (1986). *J. Am. Chem. Soc.* **108**, 5514.
17. Deratani, A., Darling, G. D., Horac, D. and Frechet, J. M. J. (1987). *Macromolecules* **20**, 767.
18. Cei, G. and Mathias, J. (1990). *Macromolecules* **23**, 4127.
19. Rubinstajn, S., Zeldin, M. and Fife, W. K. (1991). *Macromolecules* **24**, 2682.
20. Fife, W. K., Rubinstajn, S. and Zeldin, M. (1991). *J. Am. Chem. Soc.* **113**, 8535.
21. Katritzky, A. R., Duell, B. L., Seiders, R. P. and Durst, H. D. (1987). *Langmuir* **3**, 976.
22. Bhattacharya, S. and Snehalatha, K. (1995). *Langmuir* **11**, 4653.
23. Bhattacharya, S. and Snehalatha, K. (1997). *Langmuir* **13**, 378.
24. Bhattacharya, S. and Snehalatha, K. (1996). *J. Chem. Soc. Perkin Trans 2*, 2021.
25. Wang, G.-J., Ye, D. and Fife, W. K. (1996). *J. Am. Chem. Soc.* **118**, 12536.

26. See, for example, Kabunde, T., Strater, N., Tucker, P., Witzel, H. and Krebs, B. (1995). *Science* **268**, 1489.
27. (a) Young, M. J. and Chin, J. (1995). *J. Am. Chem Soc.* **117**, 10577. (b) Kimura, E., Kodama, Y., Koike, T. and Shiro, N. (1995). *J. Am. Chem. Soc.* **117**, 8304. (c) Kady, I. O., Tan, B., Ho, Z. and Scarborough, T. (1995). *J. Chem. Soc. Chem. Commun.,* 1137. (d) Hendry, P. and Sargeson, A. M. (1990). *Prog. Inorg. Chem.* **38**, 201.
28. For example, see (a) Chin, J. (1991). *Acc. Chem. Res.* **24**, 145. (b) Fife, T. H. (1991). In *Perspectives in Bio-inorganic Chemistry* (R.W. Hay, Ed.). JAI Press: London.
29. For a comprehensive review, see Murakami, Y., Kikuchi, J. I., Hisaeda, Y. and Hayashida, O. (1996). *Chem. Rev.* **96**, 721.
30. For further discussion on metallomicelles, see Tagaki, W. and Ogino, K. (1985). *Top. Curr. Chem.* **128**, 144.
31. Scrimin, P., Tecilla, P. and Tonellato, U. (1992). *J. Phys. Org. Chem.* **5**, 619.
32. Scrimin, P., Tecilla, P., Tonellato, U. and Vendrame, T. (1989). *J. Org. Chem.* **54**, 5988.
33. Scrimin, P., Tecilla, P. and Tonellato, U. (1991). *J. Org. Chem.* **56**, 161.
34. Bhattacharya, S., Snehalatha, K. and George, S. K. (1998). *J. Org. Chem.* **63**, 27.
35. Kimura, E., Hashimoto, H. and Koik, T. (1996). *J. Am. Chem. Soc.* **118**, 10963.
36. (a) Grunner, S. M. and Jain, M. K. (1985). *Biochim. Biophys. Acta* **818**, 352. (b) Bittman, R., Clejan, S., Jain, M. K., Deroo, P. W. and Rosenthal, A. F. (1981). *Biochemistry* **20**, 2790. (c) Friedberg, S. J. and Holpert, M. (1978). *J. Lipid Res.* **19**, 57. (d) Langworthy, T. A. (1985). In *Bacteria: The Treatise on Structure and Function* (C. R. Woese and R. S. Wolfe, Eds.). Academic: Orlando, FL, pp. 459–497. (e) De Rosa, M., Gambacorta, A. and Gliozzi, A. (1986). *Microbiol. Rev.* **50**, 70. (f) Setter, K. O. (1982). *Nature* **300**, 258. (g) Brock, T. D., Brock, K. M., Belley, R. T. and Weiss, R. L (1972). *Arch. Mikrobiol.* **84**, 54.
37. (a) Kunitake, T. (1992). *Angew. Chem. Int. Ed. Engl.* **31**, 2137. (b) Fendler, J. H. (1984). *Science* **223**, 888.
38. (a) Schultz, G. E. and Schirmer, R. H. (1988). *Principles of Protein Structure* (C. R. Cantor, Ed.). Springer-Verlag: New York, pp. 28, 41. (b) Sakmar, T. P., Franke, R. R. and Khorana, H. G. (1989). *Proc. Natn. Acad. Sci. USA* **86**, 8309.
39. Bhattacharya, S. and Mandal, S. S. (1998). *Biochemistry* **37**, 7764.
40. (a) Kaler, E. W., Murthy, A.K., Rodriegg, B. E. and Zasadzinski, J. A. N. (1989). *Science* **245**, 1371. (b) Fukuda, H., Kawata, K., Okuda, H. and Regen, S. L. (1990). *J. Am. Chem. Soc.* **112**, 1635. (c) Kondo, Y., Uchiyama, H., Yoshino, N., Nishiyama, K. and Abe, M. (1995). *Langmuir* **11**, 2380. (d) Zemb, T., Dubois, M., Demé, B. and Guilik-Krzywicki, T. (1999). *Science* **283**, 816.
41. Hirano, K., Fukuda, H. and Regen, S. L. (1991). *Langmuir* **7**, 1045.
42. Chung, Y.-C. and Regen, S. L. (1993). *Langmuir* **9**, 1937.
43. Kobuke, Y., Ueda, K. and Sokabe, M. (1993). *J. Am. Chem. Soc.* **114**, 7618.
44. (a) Bhattacharya, S. and De, S. (1995). *J. Chem. Soc. Chem. Commun.,* 651. (b) Bhattacharya, S. and De, S. (1999). *Langmuir* **15**, 3400.
45. (a) Bhattacharya, S. and De, S. (1996). *J. Chem. Soc. Chem. Commun.,* 1288. (b) Bhattacharya, S., De, S. and Subramanian, M. (1998). *J. Org. Chem.* **63**, 7640.
46. Jaeger, D.A., Li, B. and Clark, J. T. (1996). *Langmuir* **12**, 4314.
47. Escamilla, G. H. and Newkome, G. R. (1994). *Angew. Chem. Int. Ed. Engl.* **33**, 1937.

48. Thompson, D. H., Wong, K. F., Humphry-Baker, R., Wheeler, J. J., Kim, J.-M. and Rananavare, S. B. (1992). *J. Am. Chem. Soc.* **114**, 9035.

49. (a) Moss, R. A., Bhattacharya, S. and Chatterjee, S. (1989). *J. Am. Chem. Soc.* **111**, 3680. (b) Moss, R. A., Ganguli, S., Okumura, Y. and Fujita, T. (1990). *J. Am. Chem. Soc.* **112**, 6391. (c) Moss, R. A. and Li, J.-M. (1992). *J. Am. Chem. Soc.* **114**, 9227. (d) Moss, R. A. (1994). *Pure Appl. Chem.* **66**, 851. (e) Moss, R. A., Fujita, T. and Okumura, Y. (1991). *Langmuir* **7**, 2415.

50. Munoz, S., Mallen, J., Nakano, A., Chen, Z., Gay, I., Echegoyen, L. and Gokel, G. W. (1993). *J. Am. Chem. Soc.* **115**, 1705.

51. (a) Menger, F. M., Chen, X. Y., Brocchini, S., Hopkins, H. P. and Hamilton, D. (1993). *J. Am. Chem. Soc.* **115**, 6600. (b) Yamauchi, K., Sakamoto, Y., Moriya, A., Yamada, K., Hosokawa, T., Higuchi, T. and Kinoshita, M. (1990). *J. Am. Chem. Soc.* **112**, 3188.

52. Fyles, T. M., Loock, D., van Staraten-Nijenhuis, W. F. and Zhou, X. (1996). *J. Org. Chem.* **61**, 8866.

53. Fuhrhop, J.-H. and Bach, R. (1992). In *Advances in Supramolecular Chemistry* (G. W. Gokel, Ed.), Vol. 2, pp. 25–63. JAI Press: Greenwich, CT.

54. Fuhrhop, J. H. and Tank, H. (1987). *Chem. Phys. Lipids* **43**, 193.

55. (a) Menger, F. M. and Littau, C. A. (1991). *J. Am. Chem. Soc.* **113**, 1451. (b) Menger, F. M. and Littau, C. A. (1993). *J. Am. Chem. Soc.* **115**, 10083. (c) Zana, R. and Talmon, Y. (1993). *Nature* **362**, 228. (d) Karaborni, S., Esselink, K., Hilbers, P. A. J., Smit, B., Karthauser, J., van Os, N. M. and Zana, R. (1994). *Science* **266**, 254.

56. (a) De, S., Aswal, V. K., Goyal, P. S. and Bhattacharya, S. (1996). *J. Phys. Chem.* **100**, 11664. (b) Aswal, V. K., De, S., Goyal, P. S., Bhattacharya, S. and Heenan, R. K. (1998). *Phys. Rev.* **57E**, 776. (c) De, S., Aswal, V. K., Goyal, P. S. and Bhattacharya, S. (1998). *J. Phys. Chem.* **102**, 6152. (d) Oda, R., Huc, I. and Candau, S. J. (1997). *J. Chem. Soc. Chem. Commun.*, 2105. (e) Duivenvoorde, F. L., Feiters, M. C., van der Gaast, S. J. and Engberts, J. B. F. N. (1997). *Langmuir* **13**, 3737.

57. (a) Zana, R. (1996). *Curr. Opin. Colloid Interface Sci.* **1**, 566. (b) Rosen, M. J. (1993). *Chemtech*, 30.

58. (a) Hübner, W., Mantsch, H. H. and Kates, M. (1991). *Biochim. Biophys. Acta* **1066**, 166. (b) Grunner, S. M. and Jain, M. K. (1985). *Biochim. Biophys. Acta* **818**, 352. Other examples of dimeric lipids are also known, see for example: DeRosa, M. and Gambacorta, A. (1988). *Prog. Lipid Res.* **27**, 153.

59. (a) *Molecular Mechanisms of Membrane Fusion* (1988). (S. Okhi, D. Doyle, T. D. Flanagan, S. W. Hui and E. Mayhew, Eds.) Plenum: New York. (b) Ohnishi, T. and Ito, T. (1974). *Biochemistry* **13**, 881.

60. Lockhoff, O. (1991). *Angew. Chem. Int. Ed. Engl.* **30**, 1611, and references cited therein.

61. Vigmond, S. J., Dewa, T. and Regen, S. L. (1995). *J. Am. Chem. Soc.* **117**, 7838.

62. Klotz, P., Cuccia, L. A., Mohamed, N., Just, G. and Lennox, R. B. (1994). *J. Chem. Soc. Chem. Commun.*, 2043.

63. Sommerdijk, N. A. J. M., Hoeks, T. H. L., Synak, M., Feiters, M. C., Nolte, R. J. M. and Zwanenburg, B. (1997). *J. Am. Chem. Soc.* **119**, 4338.

64. (a) Bhattacharya, S., De, S. and George, S. K. (1997). *J. Chem. Soc. Chem. Commun.*, 2287. (b) Bhattacharya, S. and De, S. (1999). *Chem. Eur. J.* **5**, 2335.

65. (a) Dagani, R. (1995). *Chem. Eng. News*, pp. 53. (b) Walker, G. (1995). *New Scientist*, 6 May, 24.

66. (a) Osada, Y. and Gong, J.-P. (1998). *Adv. Mater.*, 827. (b) Eyne, D. R. (1980). *Science* **207**, 1315. (c) Smith, K. A. and Balazs, E. A. (1968). *Biopolymers* **6**, 677. (d) Yin, Y.-L, Prud'homme, R. K. and Stanky, F. (1992). In *Polyelectrolyte Gels* (R. S. Harland and R. K. Prud'homme, Eds.), ch. 6. American Chemical Society: Washington.
67. (a) Terech, P. and Weiss, R. G. (1997). *Chem. Rev.* **97**, 3133. (b) van Esch, J. H., and Feringa, B. L. (2000) *Angew. Chem. Int. Ed. Engl.* **39**, 2263.
68. (a) Pourcain, C. B. and Griffin, A. C. (1995). *Macromolecules* **28**, 4116. (b) Zimmerman, N., Moore, J. S. and Zimmerman, S. C. (1998). *Chem. Ind.*, 604.
69. (a) Brotin, T., Utermohlen, R., Fages, F., Bouas-Laurent, H. and Desvergne, J.-P. (1991). *J. Chem. Soc. Chem. Commun.*, 416. (b) de Vries, E. J. and Kellogg, R. M. (1993). *J. Chem. Soc. Chem. Commun.*, 238. (c) Fuhrhop, J.-H., Schneider, P., Rosenberg, J. and Boekma, E. (1987). *J. Am. Chem. Soc.* **109**, 3387. (d) Hafkemp, R. J. H., Feiters, M. C. and Nolte, R. J. M. (1994). *Angew. Chem. Int. Ed. Engl.* **33**, 986. (e) Shinkai, S. and Murata, K. (1998). *J. Mater. Chem.* **8**, 485.
70. Fuhrhop, J.-H., Svenson, S., Boettcher, C., Rossler, E. and Vieth, H.-M. (1990). *J. Am. Chem. Soc.* **112**, 4307.
71. Yamada, N., Koyama, E., Imai, T., Matsubara, K. and Ishida, S. (1996). *J. Chem. Soc. Chem. Commun.*, 2297.
72. Oda, R., Huc, I. and Candau, S. J. (1998). *Angew. Chem. Int. Ed. Engl.* **37**, 2689.
73. (a) Bhattacharya, S., Acharya, S. N. G. and Raju, A. R. (1996). *J. Chem. Soc. Chem. Commun.*, 2101. (b) Bhattacharya, S. and Acharya, S. N. G. (1999) *Chem. Mater.* **11**, 3121.
74. Ragunathan, K. G. and Bhattacharya, S. (1995). *Chem. Phys. Lipids* **77**, 13.

6

Nanofabrication by the surface sol-gel process and molecular imprinting

IZUMI ICHINOSE, SEUNG-WOO LEE AND
TOYOKI KUNITAKE

6. 1 Introduction

Fabrication and nanostructural control of metal oxide thin films have been
playing important roles in various areas of materials science. A thin surface
oxide layer can give rise to excellent properties for materials. For example,
improvement of abrasion resistance or lubricativity, chemical and mechanical
stabilization, enhancement of hydrophilicity or water repellency can be
achieved by forming thin oxide surfaces. Thin oxide layers with nano-
controlled structures are known to display advantages for designing unique
chemical functions like specific adsorption of organic molecules, improved catalytic properties and capture of light energy.

High-vacuum dry-processes, such as chemical vapor deposition (CVD) and
molecular beam epitaxy (MBE), have made it feasible to control precisely the
thickness of metal oxide thin films. In these techniques, the preparative conditions like pressure and substrate temperature can be widely varied, and the elemental composition in individual atomic layers is controllable by sequential
supply of precursor gases [1]. The dense, defect-less oxide films thus prepared
are frequently used as underlayers of microelectronics devices.

On the other hand, the progress of wet-processes as preparative techniques
of metal oxide films has been remarkable. The so-called soft solution process
that provides oxide layers by means of electrochemical oxidation of a metal
surface is expanding as a synthetic method of various mixed metal oxides with
controlled thickness [2]. The two-dimensional (2D) sol-gel process based on the
hydrolysis of metal alkoxides at the air/water interface has been reported as a
preparative technique of ultrathin oxide films (Fig. 6.1a) [3]. It is also known
that LB films of metal complexes of long-chain alkyl carboxylic acid can be
converted to metal oxide films after removal of organic component by oxygen
plasma [4] and UV-ozone treatments (Fig. 6.1b) [5]. Preparation of metal oxide

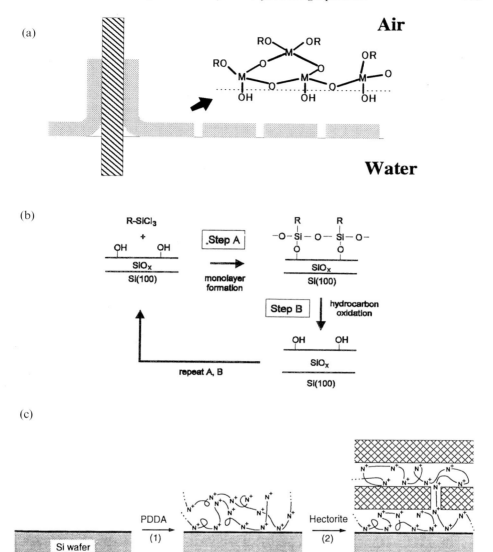

Figure 6.1. Wet processes for the preparation of ultrathin metal oxide thin films. (a) 2D sol-gel process at the air/water interface [3], (b) repeated adsorption and UV-ozone oxidation of alkylsiloxane monolayer ($R = C_{18}H_{37}$) [5], and (c) two-step adsorption cycle of alternate electrostatic adsorption of linear polymers and inorganic sheets [7] ((b) is with permission of American Chemical Society and (c) is with permission of *Science*, copyright AAAS (1994)).

films based on the stepwise adsorption from solution may be divided into two kinds. One employs formation of metal complexes at solid surfaces [6], and another is alternate adsorption of metal oxides and organic polymers via electrostatic attraction (Fig. 6.1c) [7]. In the latter case, the metal oxide layers are derived from a large variety of inorganic particles [8], clays [9] and poly-nuclear complexes [10]. These techniques are conceived as an important meth-odology of organizing molecules, and the wet fabrication techniques attract much interest. For example, molecular composites of organic dyes on oxide electrodes are useful for efficient light energy conversion [11]. The wet process makes possible ready combination of the inorganic moieties with small mole-cules, polymers and biomaterials. This opens wide chemical applications of oxide thin films.

We have developed the surface sol-gel process based on stepwise adsorption of metal alkoxides [12]. It can be widely employed for the fabrication of novel ultrathin films. In this technique, the metal alkoxides are chemisorbed on a solid substrate modified with surface hydroxyl group, and the substrate is washed in adequate organic solvents. The oxide gel films with molecular thick-ness are formed after hydrolysis of chemisorbed alkoxides. At this point, active hydroxyl groups are reproduced on the surface, and are used for further adsorption of alkoxides to grow metal oxide gel films. The films obtained by the surface sol-gel process generally contain some unhydrolyzed alkoxide groups. This gel is not a glass-like solid and is quite different from oxide crystals. However, as described later, the structural flexibility of the gel gives useful properties for forming precise nano-composites with organic compounds and for its use in molecular imprinting. In this chapter, we describe structural and physicochemical characteristics of ultrathin oxide gel films prepared by the surface sol-gel process, their extension to organic/inorganic nano-composite films, and their various applications. Furthermore, use of the gel films as a molecular recognition matrix is described.

6.2 Surface sol-gel process

6.2.1 *Stepwise adsorption of metal alkoxides*

Preparation of metal oxide thin film by means of stepwise absorption of metal alkoxide has been carried out in the past for the activation of heterogeneous catalysts [13]. For example, Asakura *et al.* prepared one-atomic layer of niobium oxide by repeating chemisorption of $Nb(OEt)_5$ on silica beads. The catalyst obtained by immobilizing platinum particles on a niobum oxide layer showed improved reactivity for hydrogenation of ethylene in comparison with

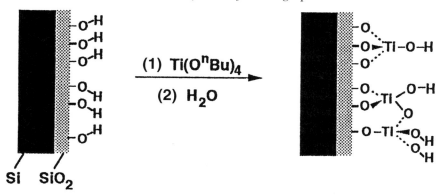

Figure 6.2. Schematic illustration of the chemisorption and hydrolysis of $Ti(O^nBu)_4$ to form an ultrathin TiO_2 film [15].

platinum particles on bare silica beads [14]. Kleinfeld and Ferguson prepared an ultrathin TiO_2 film by repeated adsorption of $Ti(O^nBu)_4$ onto silicon wafers with oxidized surface (Fig. 6.2) [15]. The substrate was dipped into the alkoxide solution (5 mM, 30s) in a glove box filled with dry N_2 gas, sufficiently rinsed with dry solvent and with pure water. Thickness growth of a TiO_2 thin film estimated from ellipsometry measurement was 3–4 Å for the first adsorption cycle and about 1 Å for the following adsorption cycles. In this case, the amount of the chemisorbed Ti $(O^nBu)_4$ was independent of adsorption time (30 s to 4 days), and was limited by the number of surface hydroxyl groups. Though the adsorbed thickness per cycle was smaller than thickness of TiO_2 monolayer (3.4 Å), the film grew regularly in proportion to the number of adsorption cycles.

We examined stepwise adsorption of various metal alkoxides on the solid surface for the purpose of obtaining uniform oxide films. Fig 6.3 shows an example where a gold-coated quartz crystal microbalance (QCM) was used as a substrate. The electrode modified by mercaptoethanol was immersed in a metal alkoxide solution, rinsed in an adequate organic solvent, and the alkoxide molecule chemisorbed on the surface was hydrolyzed by water or by moisture in the air. QCM frequency was measured after drying the electrode by flushing with nitrogen gas, and the adsorption, rinsing, and hydrolysis procedures were repeated. It is possible to carry out all these operations in the ambient atmosphere. Figure 6.4a shows how the QCM frequency change depends on adsorption time for the adsorption of $Ti(O^nBu)_4$. The adsorption was saturated within 1 min in toluene/ethanol = 1/1, and frequency change was about 50 Hz. In our system with a 9 MHz resonator, frequency decrease of 1 Hz corresponds to mass increase of about 0.9 ng. And a frequency change of 50 Hz corresponds

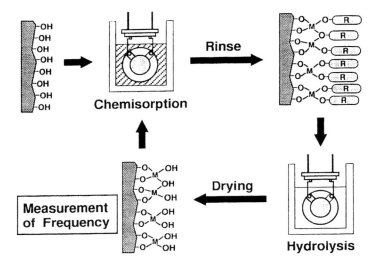

Figure 6.3. Schematic illustrations of sequence of the surface sol-gel process. A QCM elecrode is used as a substrate.

to thickness increase of 8 Å from equation (6.1) [16], using the density of bulk TiO_2-gel (1.7 g/cm^3).

$$2d\,(\text{Å}) = -\Delta F\,(\text{Hz})/1.83\,\rho\,(\text{g/cm}^3) \tag{6.1}$$

This frequency shift varies somewhat with the concentration of alkoxide and the structure of the electrode surface. In the case of a gold-coated QCM resonator (USI System, Fukuoka) that was subjected to smoothing treatment by an electrode, a frequency change of about 30 Hz (thickness increase; 5 Å) was observed under the experimental conditions. The QCM frequency changes during sequential adsorption of $Ti(O^nBu)_4$ and $Al(O^nBu)_3$ are shown in Fig. 6.4b. The linear frequency shifts observed in both cases clearly indicate regular growth of oxide gel films on the electrode. In the case of $Al(O^nBu)_3$, the frequency change per 1 cycle (272 ± 90 Hz) is larger than that of $Ti(O^nBu)_4$. This is probably due to the fact that $Al(O^nBu)_3$ formed oligomers in stock solution, and that they were adsorbed on the hydrolyzed surface without de-aggregation. In this case, the frequency shifts readily vary if the alkoxide and rinsing solvents are changed.

The averaged QCM frequency shifts and adsorption conditions for various metal alkoxides are listed in Table 6.1. In the cases of $Al(O^nBu)_3$, $Sn(O^iPR)_4$, $InSn_3(OR)_x$, and $VO(O^iPr)_3$, the alkoxides do not adsorb at room temperature, owing to their lower reactivities, and it is necessary for all cases to raise the adsorption temperature to 45–50°C. In particular, $Si(OMe)_4$ needs to be adsorbed at high concentrations for longer adsorption times. In contrast, in the

(a)

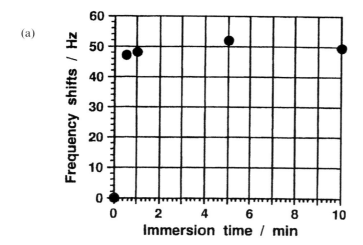

(b)

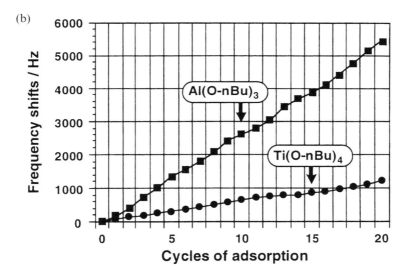

Figure 6.4. (a) Immersion time dependence of QCM frequency shift for Ti(O^{n}Bu)$_4$ adsorption (100 mM toluene/ethanol = 1/1, 16 °C). (b) QCM frequency decrements at each adsorption cycle. (●) Ti(O^{n}Bu)$_4$ (100 mM in toluene/ethanol = 1/1, 18 °C), (■) Al(O^{n}Bu)$_3$ (88 mM in ethanol/carbon tetrachloride = 2/3. 45 °C).

case of Zr(O^{n}Pr)$_4$, regular frequency changes (109 ± 57Hz) were observed only at low temperature. Adsorption conditions for In(OCH$_2$CH$_2$OCH$_3$)$_3$ were close to that of Ti(O^{n}Bu)$_4$, and the average frequency change was 35 ± 20Hz. The frequency change in each cycle tends to increase gradually for Nb(O^{n}Bu)$_5$. This increase seems to arise from hydrolysis of the alkoxide solution by the humidity in the air, as the ΔF value again decreases when a fresh solution is

Table 6.1. *Condition and stepwise adsorption and QCM frequency changes*

compd	adsorption		hydrolysis		frequency change
	concn/solvent[a]	immersion time/temp	solvent	immersion time/temp	
Zr(O^{n}Pr)$_4$	100mM/T-P	3 min/10°C	water	1 min/10°C	109 ± 57 Hz
Ti(O^{n}Bu)$_4$	100mM/T-E	3 min/18°C	water	1 min/18°C	61 ± 37 Hz
Al(O^{n}Bu)$_3$	88mM/E-C	10 min/45°C	water	1 min/45°C	272 ± 90 Hz
B(O^{n}Et)$_4$	300mM/benzene	5 min/20°C			517 Hz[b]
Ti(acac)$_2$	20mM/methanol[b]	5 min/60°C	aq. NaOH (pH 12)	1 min/20°C	48 ± 35 Hz
Si(OMe)$_4$	500mM/T	20 min/30°C	water	10 min/30°C	30 ± 18 Hz
In(OC$_2$H$_4$OMe)$_3$	100mM/methoxyethanol	3 min/30°C	water	1 min/30°C	35 ± 20 Hz
Sn(O^{n}Pr)$_4$	10mM/iP	5 min/50°C	water	5 min/50°C	145 ± 63 Hz
InSn$_3$(OR)$_x$	30mM/iP-M	5 min/50°C	water	5 min/50°C	32 ± 19 Hz
BaTi(OR)$_x$	10mM/T-E	5 min/30°C	water	5 min/30°C	79 ± 39 Hz
VO(O^{i}Pr)$_4$	200mM/iP	10 min/50°C	water	5 min/50°C	43 ± 17 Hz

Notes:

[a]T-P, toluene/propanol 1/1; T-E, toluene/ethanol 1/1; E-C, ethanol/carbon tetrachloride 2/3; iP-M, isopropanol/methanol 1/1.
[b]Methanol containing 300mM of acetic acid. [c]Average frequency change for 20 cycles. [d]Water-saturated toluene.

used. The condensed niobium butoxide may adsorb on the hydroxyl surface more readily than the monomer species. $B(OEt)_3$ chemisorbed on a surface is easily desorbed upon hydrolysis in water. Instead, the chemisorbed surface can be hydrolyzed with moisture in the air, and the stepwise adsorption is achieved by repeating chemisorption and rinsing. Elution of barium ion from the chemisorbed layer of $BaTi(OR)_x$ double alkoxide can be also suppressed by using water-saturated toluene for hydrolysis.

Precursors useful in the surface sol-gel process are not restricted to alkoxides. The requirements as precursors are chemisorption on surface hydroxyl groups and regeneration of the hydroxyl groups after hydrolysis. For example, $TiO(acac)_2$ repeatedly adsorbs, when acid and alkali are added to the adsorption and hydrolysis media, respectively. Except for the case of $Nb(O^nBu)_5$, all the compounds listed in Table 6.1 show linear frequency shifts. Adsorption conditions such as concentration, temperature, and immersion time are dependent on the solubility, reactivity, and the ease of hydrolysis of alkoxides. These conditions are varied as the structure of alkoxide units is changed. For example, $Ti(O^iPr)_4$, which exists as a monomer in solution [17], requires conditions different from $Ti(O^nBu)_4$, which tends to form oligomer species.

It is possible to control the adsorption thickness of alkoxides by intentionally forming the oligomer species in stock solution. Figure 6.5a shows frequency shifts for $Ti(O^nBu)_4$ in the presence of water. The adsorption tends to be excessive after several cycles, but, by using a fresh alkoxide solution at the cycle marked with the asterisk, the original adsorption is recovered. The average frequency shift increased with the mole ratio (n) of water against $Ti(O^nBu)_4$, and the thickness increase per cycle became 46 Å ($n = 2/3$), 72 Å ($n = 1$) and 300 Å ($n = 2$). Figure 6.5b shows a scanning electron micrograph of the cross section of TiO-gel film ($n = 2/3$) prepared on a gold-coated QCM resonator. This thin film has a thickness of constant 90 ± 10 nm, and the surface is smooth over a large area. The density is estimated to be 1.6 g/cm^3, by substituting the total frequency shift (5221 Hz) during film growth and the thickness (90 nm) in equation (6.1). Bulk TiO_2-gel, which was obtained by adding excessive water into titanium butoxide solution and sufficient drying of the precipitates formed, gives a density of 1.7 g/cm^3. The density of an ultrathin film formed by the surface sol-gel process (1.6 g/cm^3) is very close to that of bulk TiO_2-gel.

6.2.2 Structure and property

Ultrathin oxide gel films prepared by the surface sol-gel process include many nano-sized voids, and the film density is generally low. Their compositions and

(a)

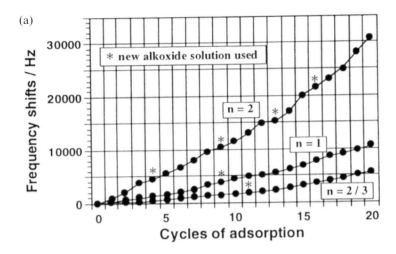

(b)

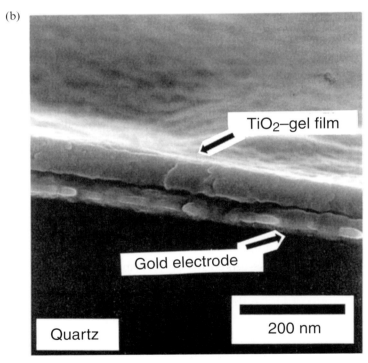

Figure 6.5. (a) Stepwise adsorption of $Ti(O^nBu)_4$ in the presence of water; $n = H_2O/Ti(O^nBu)_4$. (b) Scanning electron micrograph of TiO_2-gel film on Au-coated resonator. Total frequency shift: 5221 Hz.

estimated structures estimated from XPS measurements are shown in Fig. 6.6. Atomic ratio of TiO_2-gel formed from 100 mM $Ti(O^nBu)_4$ in toluene/ethanol = 1:1 is $Ti:C:O = 1.0:5.1:4.2$. Judging from the large carbon ratio and the FT-IR data, intact butoxide units (or ethoxide units) must be partially contained in this film. When 20 mM $Ti(O^nBu)_4$ in toluene/methanol = 5:1 is used, the element ratio is $Ti:C:O = 1.0:2.1:2.6$. Assuming that the carbon content is derived from solvent methanol, the observed ratio (C/Ti = 2.1) indicates that at least two methanol moieties coordinate on one Ti atom. These TiO_2-gel films probably have a chain structure of $-Ti(OR)_2-O-Ti(OR)_2-O-$ units which must be partially cross-linked. For 100 mM $Zr(O^nBu)_4$ in toluene/propanol = 5:1, the element ratio obtained from XPS measurements is $Zr:C:O = 1.0:3.7:4.0$. The ZrO_2-gel film probably has a three-dimensionally bridged structure $(-O-ZrO(O^nPr)-O-)$ unlike the TiO_2-gel film, as judged from the higher oxygen content and the use of propanol solvent. In the case of $Nb(O^nBu)_5$, which is readily hydrolyzed by water, three-dimensional cross-linking appears abundant in the gel film, and the niobium atom seems to be cross-linked with three or four surrounding niobium atoms via an oxygen bridge. The reactivity of $Al(O^nBu)_3$ is low, however, and the film prepared at 45 °C shows a considerably higher degree of condensation. $In(OCH_2CH_2OCH_3)_3$ also seems to have a highly cross-linked structure, because the composition is $In:C:O = 1.0:2.9:2.6$.

The elemental composition reflects the degree of hydrolysis of precursor alkoxides, that is, the degree of condensation via oxygen bridges. Other important film characteristics are thickness, smoothness, and porosity. The film thickness is confirmed by scanning electron microscopy, and the adsorbed film mass is estimated by QCM measurement. The film density can be calculated from thickness and mass. For example, a TiO_2-gel processes alkoxide units and nano-sized voids in the interior, and the density is 1.7 g/cm³. This density is half to one third of that of TiO_2 crystal (3.8–4.3 g/cm³). The density of an Al_2O_3-gel film, which has a highly cross-linked structure, is greater: 2.2–2.3 g/cm³. In contrast, In_2O_3-gel forms a highly cross-linked, non-uniform film, which is made of randomly filled particles, and the density becomes 1.0 g/cm³: only 1/7 of In_2O_3 crystal (7.18 g/cm³).

The electrical conductivity of the ultrathin oxide gel films prepared on an ITO electrode has been measured upon deposition of the aluminium electrode [18]. TiO_2-gel films with 4×2 mm size gave a resistance of $2.5 \times 10^6 \Omega$ at 8.3 nm thickness and $0.61 \times 10^6 \Omega$ at a thickness of 28 nm, respectively. The resistivity is independent of film thickness, and is 6–$7 \times 10^{10} \Omega \cdot cm$. An In_2O_3-gel film of 41 nm thickness has a resistance of 24 Ω, and the corresponding resistivity is calculated as $5 \times 10^5 \Omega \cdot cm$. This value is smaller by a factor of 10^5 than that of TiO_2-gel films. This difference in resistivity is close to that between TiO_2 crystal

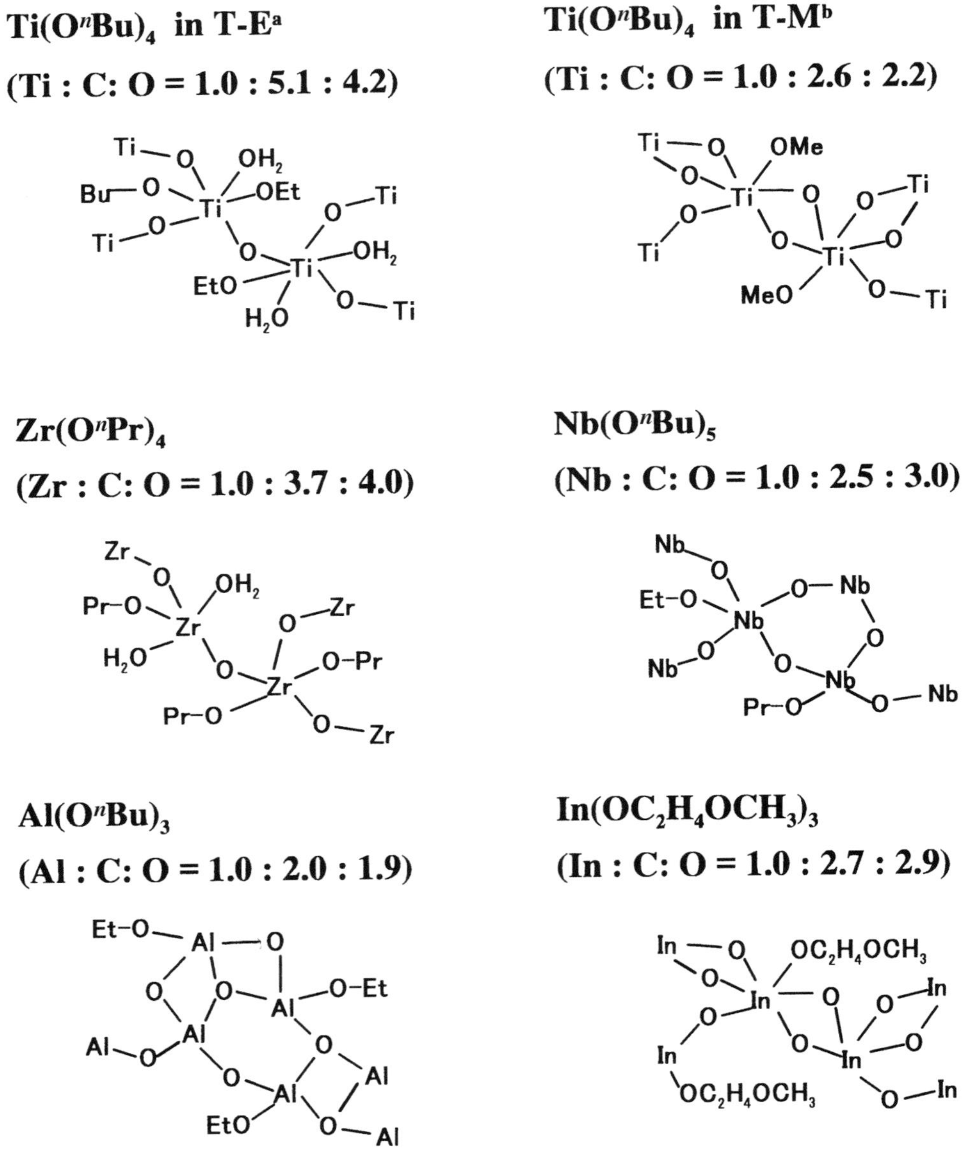

Ti(O^{n}Bu)$_4$ in T-E^a
(Ti : C: O = 1.0 : 5.1 : 4.2)

Ti(O^{n}Bu)$_4$ in T-M^b
(Ti : C: O = 1.0 : 2.6 : 2.2)

Zr(O^{n}Pr)$_4$
(Zr : C: O = 1.0 : 3.7 : 4.0)

Nb(O^{n}Bu)$_5$
(Nb : C: O = 1.0 : 2.5 : 3.0)

Al(O^{n}Bu)$_3$
(Al : C: O = 1.0 : 2.0 : 1.9)

In(OC$_2$H$_4$OCH$_3$)$_3$
(In : C: O = 1.0 : 2.7 : 2.9)

Figure 6.6. Condensation structures of metal oxide gels presumed by elemental composition estimated by XPS measurements. ^{a}T-E, toluene/ethanol 1/1; ^{b}toluene/methanol 5/1.

($\sim 10^2$ Ω·cm) and In_2O_3 crystal ($\sim 10^{-4}$ Ω·cm). Although the resistivity of the oxide gel films is much larger than that of the crystal, it is noted that there is a similar difference as in crystals. The apparent resistivity of an In_2O_3-gel film (5×10^5 Ω·cm) is 0.6 Ω/nm, in nanometer thickness. This resistance is equal to the value of a conventional ITO electrode (cross section: 2 mm $\times$ 2000 Å, length: 5 cm). Therefore, the ultrathin In_2O_3-gel film is useful as a conductive material at nanometer thickness.

6.2.3 *Organic/inorganic composite films*

The surface sol-gel process takes advantage of surface hydroxyl groups for adsorption of metal alkoxides. The surface hydroxyl groups need not be restricted to hydrolyzed metal oxide layers. We found that polyhydroxyl compounds, polymers and small organic molecules are adsorbed readily onto the surface of the oxide layer formed by the surface sol-gel process [19]. The poly-hydroxyl compounds on the surface provide free hydroxyl groups, and metal alkoxides are subsequently adsorbed. Thus, it is possible to extend the surface sol-gel process to the alternate adsorption technique of polyhydroxyl organic compounds and metal alkoxides. A typical preparative operation is shown in Fig. 6.7. A QCM resonator is immersed into metal alkoxide solutions for given periods of time. Then, it is washed in organic solvent, and intact chemisorbed alkoxides are hydrolyzed in water. The QCM frequency shift is measured after drying by flushing with nitrogen gas. Then, the resonator is immersed in a solution of a polyhydroxyl compound, rinsed in water, and the frequency is measured after drying. By repeating these procedures, an alternately layered film of organic and inorganic components is obtained. Figure 6.8 shows the QCM frequency change during the alternate adsorption of $Ti(O^nBu)_4$ (100 mM in 1:1 v/v toluene/ethanol) and poly(acrylic acid) (PAA; 10 mg/ml in water). The linear frequency decrement indicates regular growth of TiO_2-gel/PAA multilayers on the electrode. In each step, the frequency change is 68 ± 34 Hz for $Ti(O^nBu)_4$ and 87 ± 26 Hz for PAA, and thicknesses of TiO_2-gel layer and PAA layer calculated from these frequencies are 14 ± 4 Å and 13 ± 7 Å, respectively. Frequency shifts for the alternate adsorption of $Ti(O^nBu)_4$ and starch are also shown in Fig. 6.8.

Figure 6.9a shows a scanning electron micrograph of the cross section of a TiO_2-gel/PAA multilayer film. The film located on a gold electrode looks very uniform, with a constant thickness of 350 ± 20 Å. The density was estimated as 1.9 g/cm^3 from the SEM thickness and the total QCM frequency change (2508 Hz) during film growth. FT-IR spectra of this film gave peaks at 1560 cm^{-1} (Ti-carboxylate complex) and 1720 cm^{-1} (free carboxylate acid) with

I. Ichinose et al.

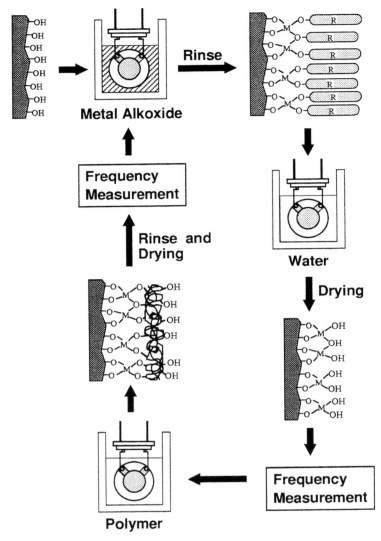

Figure 6.7. QCM measurement of the alternate adsorption process of metal alkoxide and organic hydroxyl polymer.

similar intensities [20]. Therefore, it is presumed that half of the carboxylate group of PAA is coordinated to titanium atoms of the TiO_2-gel layer. We schematically illustrate the chemical structure of the alternate film in Fig. 6.9b. Both the TiO_2-gel and PAA layers are approximately 1 nm thick, and are densely attached to each other. Compactness of the film is supported by the film density of 1.9 g/cm^3, which is larger than the densities of each component (TiO_2-gel; 1.7 g/cm^3, PAA; 1.4 g/cm^3).

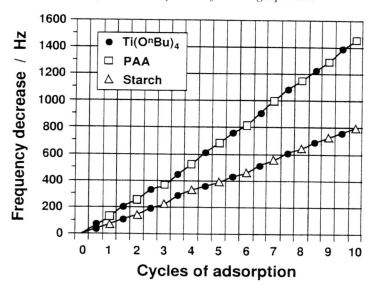

Figure 6.8. QCM frequency shift on alternate adsorption of $Ti(O^nBu)_4$ and organic polymer:$Ti(O^nBu)_4$ (100 mM in 1:1 v/v toluene/ethanol, 30 °C) (●); poly(acrylic acid) (10 mg/ml in ethanol, 30 °C) (□); starch (1 mg/ml in water, 40 °C) (△).

The alternate adsorption can be applied to other hydroxyl molecules such as poly(vinyl alcohol) (PVA) and starch (see Fig. 6.8), glucose, maltose, and porphyrin with carboxyl groups (TCPP). QCM frequency shifts and adsorption conditions for the compounds given in Chart 6.1 are summarized in Table 6.2. In the case of PVA, a uniform film was similarly obtained, but the thickness of the polymer layer was 2–3 times larger than that of PAA. A molecularly smooth film was also obtained for TCPP with four carboxyl groups, as will be described later. On the other hand, combination of $Ti(O^nBu)_4$ and sugars did not always give composite films with smooth surfaces, although a constant mass increase was observed from QCM measurement. The number of hydroxyl groups included in one molecule is a determinant factor for regular growth of composite films. For example, bi-functional compounds such as 1,10-decane-diol, hydroquinone, and terephthalic acid did not give alternate films with $Ti(O^nBu)_4$. It is probable that these compounds are easily desorbed during the rinsing process and/or adsorption process of $Ti(O^nBu)_4$. In contrast, PAA and PVA can undergo regular alternate adsorption even with other metal alkoxides $(Al(O^nBu)_3, Zr(O^nPr)_4, BaTi(OR)_x$, etc.). The combination of hydroxyl polymers and metal alkoxides is highly diverse. Undoubtedly, this technique will become very important as a preparative means of nano-composite materials.

Further examples of the alternate organic/inorganic adsorption include combinations with cyclodextrin, polyrotaxane, and dendrimer, which play an

(a)

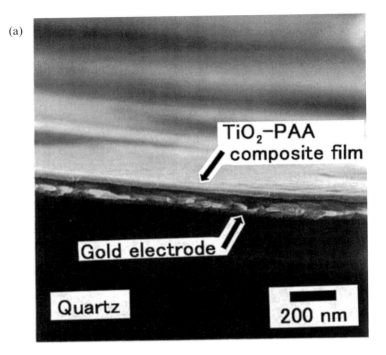

(b)

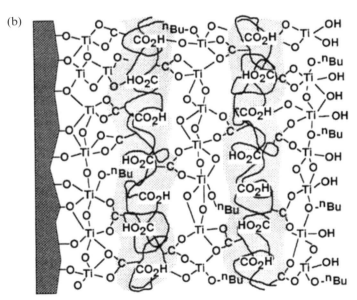

Figure 6.9. (a) Scanning electron micrograph of a TiO_2/PAA film on a Au-coated resonator. Total frequency shift: 2508 Hz. (b) Schematic illustration of a TiO_2/PAA film.

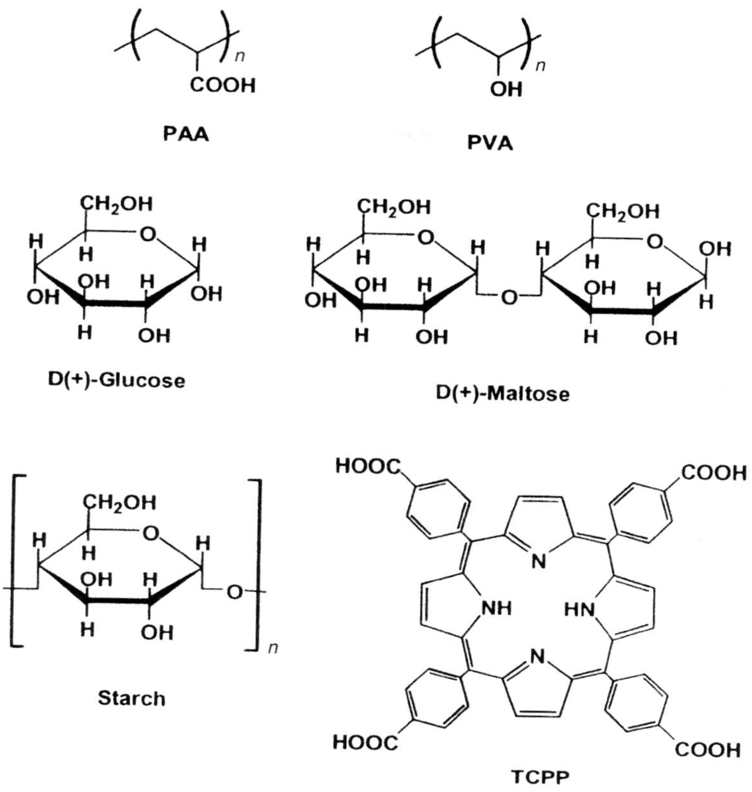

Chart 6.1.

important role in host/guest chemistry. Figure 6.10a shows QCM frequency shifts for alternate adsorption of $Ti(O^nBu)_4$ (100 mM in toluene/ethanol = 1/1) and α-cyclodextrin (10 mM in water). In this combination, regular adsorption with a frequency change of 34 ± 6 Hz for $Ti(O^nBu)_4$ and 26 ± 10 Hz for α-cyclo-dextrin was observed. A very similar adsorption behavior has been observed for β-cyclodextrin and γ-cyclodextrin. In the case of the fourth generation den-drimer (sugar ball; 0.2 mg/ml) which has 64 glucose units on the outer surface, the regular film growth was achieved by repeating two adsorption cycles for $Ti(O^nBu)_4$ against one adsorption of the sugar ball. Average frequency shifts are 27 ± 12 Hz for $Ti(O^nBu)_4$ and 375 ± 40 Hz for the sugar ball. A greater fre-quency shift (about 650 Hz) was observed for the fifth generation sugar ball. In contrast, a frequency change for a linear polyrotaxane (cyclodextrin/polyethy-lene glycol composite) was 21 ± 8 Hz, and was almost the same as that for cyclodextrins [21]. As described above, smooth organic/inorganic multilayers are not necessarily formed when sugar derivatives are used in combination with

Table 6.2. *Alternate adsorption of $Ti(O^nBu)_4$ and polyhydroxyl compounds*

metal alkoxide			polyhydroxyl compound			frequency change	
compd	concn/solvent[a]	immersion time/ temp.[b]	compd	concn/solvent[a]	immersion time/ temp.[b]	metal alkoxide	organic compd
Ti $(O^nBu)_4$	100mM/T-E	3 min/30°C	PAA	1 wt%/E	10 min/30°C	87 ± 26 Hz	68 ± 34 Hz
		3 min/30°C	PVA	0.5 wt%/W	10 min/30°C	91 ± 41 Hz	149 ± 31 Hz
		3 min/40°C	starch	0.1 wt%/W	10 min/40°C	46 ± 23 Hz	39 ± 14 Hz
		3 min/40°C	glucose	100mM/W	20 min/40°C	71 ± 26 Hz	59 ± 34 Hz
		3 min/40°C	maltose	100mM/W	20 min/40°C	69 ± 15 Hz	91 ± 41 Hz
		3 min/40°C	TCPP	1mM/E	10 min/30°C	75 ± 10 Hz	64 ± 32 Hz

Notes:
[a] T-P, toluene/propanol 1/1; E, ethanol; W, water. [b] Adsorption, rinsing, and hydrolysis were carried out at the designated temperature.

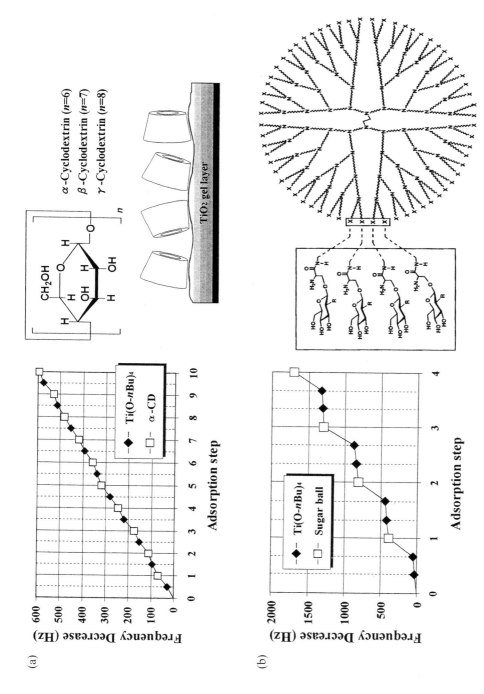

Figure 6.10. QCM frequency shifts on the alternate adsorption of $Ti(O^nBu)_4$ with an α-cyclodextrin (a) and a sugar ball (b).

TiO$_2$-gel [23]. In fact, spherical aggregates of 10–100 nm in diameter are frequently seen on the film surface. It is possible that sugar molecules that are weakly bound with TiO$_2$-gel re-assemble on the film surface to produce the spherical aggregates.

Very recently, two-dimensional alignment of metal nanoparticles was successfully conducted by using the TiO$_2$-gel surface [22]. A long-alkyl disulfide with two hydroxyl groups at both molecular termini (HO(CH$_2$)$_{11}$SS(CH$_2$)$_{11}$OH) was used as a protective agent for small gold particles with an average diameter of 4.7 nm. When a solid substrate overlaid with a TiO$_2$-gel film was immersed in an aqueous dispersion of a gold nanoparticle ([Au] = 1.5×10^{-3} mol/l), a condensely packed monolayer of the nanoparticles was obtained. Saturation of the adsorption requires 10 h, in contrast to adsorption of polymeric PVA and starch, which requires only 10 min. Au nanoparticle/TiO$_2$-gel multilayers were obtainable by alternating formation of a TiO$_2$-gel layer for 15 cycles and adsorption of Au nanoparticles. Unique catalytic and optical properties are expected for organized metal nanoparticles in oxide gel matrices.

6.2.4 *Photocurrent generation*

Photocurrent generation has been found for the alternate assembly of metal oxide gel and porphyrin–tetracarboxylic acid (TCPP), where TCPP acts as an antenna molecule for photoenergy capturing. Absorption spectral changes during the assembly process of Ti(O^{n}Bu)$_4$ and TCPP are shown in Fig. 6.11a. Absorption peaks attributed to the Soret-band of TCPP (424 nm) and to TiO$_2$-gel (near 240 nm) increase with the number of adsorption cycles, indicating regular growth of the composite film. A TiO$_2$-gel/TCPP film formed on a QCM resonator has a constant thickness of 300 ± 10 Å after 13 cycles, as confirmed by cross-sectional observation by scanning electron microscopy (Fig. 6.11b). The total QCM frequency shift is 1864 Hz. From these values, the film density was estimated as 1.7 g/cm^3, and the thickness increase for each cycle was calculated to be 23 Å. Reflection FT-IR spectra proved that nearly half of the carboxyl groups of TCPP molecules were coordinated to titanium atoms.

Figure 6.12 shows photocurrent generation from alternate assemblies of TCPP and various metal alkoxides on gold-evaporated substrates. IP5, NP5, TP5 and SP5 denote layered assemblies of (In$_2$O$_3$-gel/TCPP)$_5$, (Nb$_2$O$_5$-gel/TCPP)$_5$, (TiO$_2$-gel/TCPP)$_5$, and (SiO$_2$-gel/TCPP)$_5$, respectively. Measurements were carried out in an oxygen-saturated electrochemical cell in 0.1 M aqueous NaClO$_4$, equipped with a platinum counter electrode and a reference electrode of Ag/AgCl in saturated KCl solution. Light from a Xe lamp was

(a)

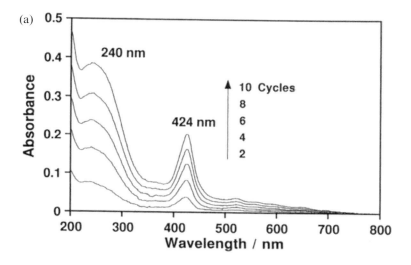

(b)

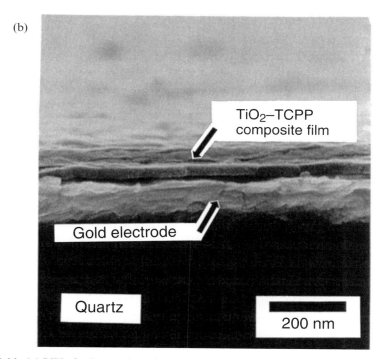

Figure 6.11. (a) UV-vis absorption changes for every two cycles of alternate adsorption of Ti(O^{n}Bu)$_4$ and TCPP. (b) Scanning electron micrograph of a TiO$_2$/TCPP film on a Au-coated resonator. Total frequency shift: 1864 Hz.

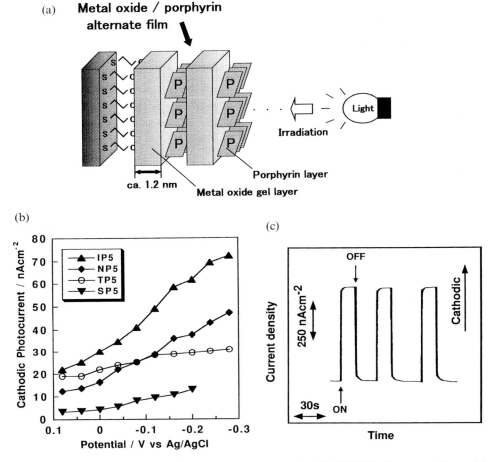

Figure 6.12. Photocurrent generation from metal oxide/TCPP alternate films. (a) Schematic illustration of a metal oxide/TCPP multilayer, (b) current/bias plots (irradiation; 0.3mW/cm² at 420 nm), (c) photocurrent from (TiO₂-gel/TCPP)₅ assembly prepared on an ITO electrode (0.3 mW/cm² at 420 nm, −0.2 V vs Ag/AgCl).

irradiated to the sample through a monochromator. QCM measurements and SEM observations combined indicate that all these films have (metal oxide/TCPP) units of 20 ± 3 Å thickness. Relative amounts of TCPP in the multilayer assemblies were determined from the absorbance of UV-vis spectra as IP5 : TP5 : NP5 : SP5 = 1.0 : 0.6 : 0.4 : 1.0. The amount of the porphyrin adsorbed on the Nb_2O_5-gel layer is small relative to that on the TiO_2-gel, and it was greater on In_2O_3-gel and on SiO_2-gel. Upon irradiating light at 420 nm (0.3 mW/cm²) to these multilayers, photocurrent was observed in a direction from gold electrode to the composite films. The photocurrent under the applied bias of $-0.2V$ increases in the order SP5 < TP5 < NP5 < IP5. Apparently, NP5 and IP5 films generate photocurrent more efficiently, and the efficiency of SP5 is low. It is clear that the photocurrent generation in these assemblies is deeply affected by the nature of the oxide-gel layer. From systematic experiments, it is found that the most efficient photocurrent generation was found for the In_2O_3-gel/TCPP film prepared on an ITO electrode. The response behavior is shown in Fig. 6.12c. Immediately after irradiating light at 420 nm (0.3 mW/cm^2) to the IP5 multilayer, cathodic photocurrent of 0.58 μA/cm^2 was generated, and it disappeared as soon as the irradiation was stopped [23]. This photocurrent value is about 15 times larger than that on In_2O_3-gel/TCPP film on a gold electrode, after normalization with absorbance of each film at 420 nm. This difference probably arises from the fact that electron injection from the ITO electrode to the In_2O_3-gel layer is more facile.

The cathode photocurrent is in proportion to the number of TCPP layers at least up to 10 cycles. The efficiency is greatly dependent on the kinds of oxide gel. These experimental observations suggest that electron transfer from the electrode to the porphyrin via the oxide gel layer is an essential mechanism of the photocurrent generation. Oxygen molecules as an electron acceptor readily diffuse in the oxide gel films of about 20 nm thickness. The electron transfer from the electrode to the porphyrin is assisted by satisfactory conductivity of the gel layer. The overall photocurrent value is considerably smaller than the conventional wet solar cell [11]. However, modification of the electrode surface by ultrathin oxide gel films will facilitate the design of novel light harvesting devices.

The surface sol-gel process based on stepwise adsorption from solution is applicable to any surface with hydroxyl groups, regardless of the size and shape of the substrate. This modification will be possible, in principle, on the inner surface of tubes and porous materials. It is also possible to design various organic/inorganic multilayer films by the selection of adsorption cycle, sequence, and chemical structure of adsorbates. These points become an advantage that is not found in the preparative techniques of ultrathin films such as conventional dry process and the LB method. Good optical and electronic properties will be expected, because composite films with nanometer precision, in which organic and inorganic components are molecularly contacted, are produced under simple and mild conditions. Polymer/metal oxide composites will also give mechanically excellent properties. Through the nanostructure control of materials, wide developments towards the chemical applications such as catalysts, separation, molecular recognition, etc., will be especially expected. In the next section, as an example, we describe molecule imprinting by means of the surface sol-gel process and the molecular recognition.

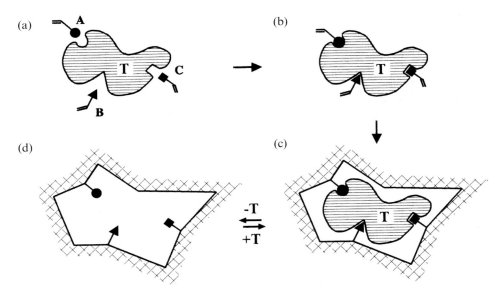

Figure 6.13. Schematic representation of the imprinting of specific cavities in cross-linked polymers.

6.3 Molecular imprinting

6.3.1 Some past examples

Recently, molecular imprinting has become recognized as a means of preparing sensors and separation materials or catalysts [24]. In this approach, specific binding sites are created by imprinting the structure of given guest molecules in three-dimensional cross-linked matrices of organic polymers (and metal oxides). In pioneering work, Dickey prepared imprinted silica gels in the presence of alkyl oranges [25]. The adsorption of alkyl oranges was enhanced for silica gels that had been formed in the presence of the particular alkyl orange as template. Since that time, various investigations have been conducted on molecular imprinting mostly by using cross-linked polymers as solid matrices. In the 1970s, Wulff and co-workers proposed the synthesis of polymers that were substrate-selective due to spatial fixation of several functional groups that non-covalently interacted with the template molecules, as illustrated in Fig. 6.13 [26]. Similar studies have been extensively carried out till the present time, and their application to the separation of complicated organic compounds and biologically active substances is the most important target [27].

In spite of the earlier study by Dickey, inorganic matrices have rarely been employed for molecular imprinting, except for the following studies. Mosbach and co-workers allowed organic silanes to polymerize on the surface of porous silica particles in aqueous solution [28]. The resulting polysiloxane copolymers acted as superior supports for high-performance liquid chromatography when imprinted with dye molecules. Morihara and others developed the 'footprint' technique, where tailor-made catalysts for trans-acylation were designed by imprinting transition-state analogues onto aluminium ion-doped silica gel [29]. Maier and co-workers similarly imprinted a transition-state analogue for transesterification on amorphous silicon dioxide [30]. Very recently, Pinel *et al.* again took up imprinting of silica gel and showed that regiospecificity toward cresol-isomers was successfully imprinted by using *o*-cresol as a template [31].

Molecular imprinting attracts practical interests, especially in the form of ultrathin films, since separation capability can be enhanced by improvement of adsorption rate. Sagiv prepared mixed monolayers of trichloro-*n*-octadecylsilane (OTS) and a modified dye as a template on glass [32]. The dye molecules alone were removed because they were not covalently bound. The stable network of silane monolayer provides the imprinted site for the dye molecule. On the other hand, Kodakari *et al.* obtained a silica overlayer on tin oxide (SnO_2) by chemical vapor deposition (CVD) using pre-adsorbed benzoate anion as a template [33]. The resulting silica overlayer acted as molecular sieve, owing to formation of an imprinted cavity. In these examples, the site of molecular imprinting has to be located at the surface and/or the crevice close to the surface, in order to secure efficient adsorption of guest molecules. This suggests that the positive use of ultrathin films should be advantageous. Organized molecular films such as Langmuir–Blodgett multilayers and surface-bound monolayers may appear to be good candidates for this purpose. However, they are not necessarily suitable, since the flexible structural modification that is required for the imprinting process is, in principle, not compatible with the ordered molecular organization.

As described above, the surface sol-gel process gives rise to oxide gel films of nano-size thickness, and is capable of incorporating various organic molecules by sequential chemisorption and activation. We discovered that such ultrathin films are very useful as matrices of molecular imprinting, and carried out extensive research [34]. The molecular imprinting can be performed in two different approaches: the pre-mixing approach and the layer-by-layer approach (Fig. 6.14). The pre-mixing approach involves complexation of template organic molecules with metal alkoxides, and the complex is solubilized in an organic solvent. This is used for organic species that are not adsorbed by themselves. The layer-by-layer approach involves alternate adsorption of template molecules and metal alkoxide, and it can be used for polar molecules that are effectively adsorbed from aqueous solution, amino acids and peptides.

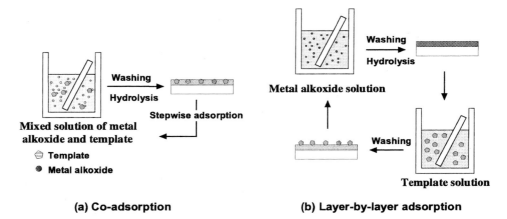

Figure 6.14. Two different approaches of molecular imprinting by the surface sol-gel process; the pre-mixing approach (a), and the layer-by-layer approach (b).

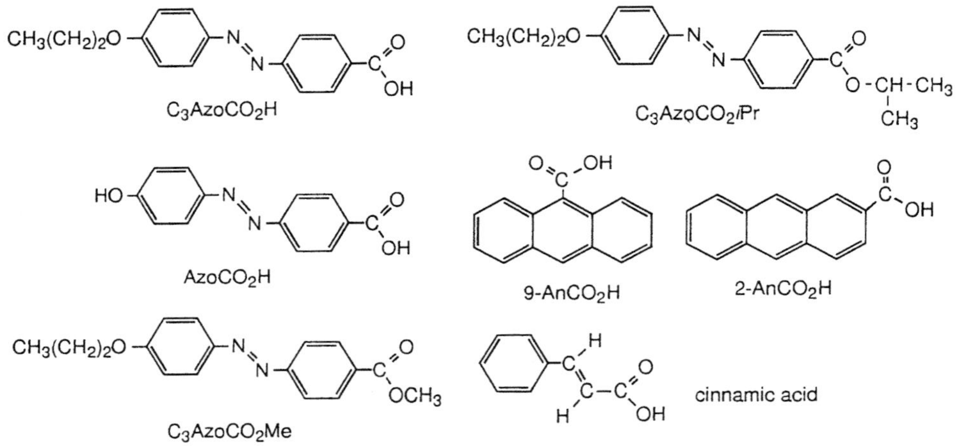

CH$_3$(CH$_2$)$_2$O —◯— N=N —◯— C(=O)OH

C$_3$AzoCO$_2$H

CH$_3$(CH$_2$)$_2$O —◯— N=N —◯— C(=O)O-CH-CH$_3$ / CH$_3$

C$_3$AzoCO$_2$iPr

HO —◯— N=N —◯— C(=O)OH

AzoCO$_2$H

9-AnCO$_2$H

2-AnCO$_2$H

CH$_3$(CH$_2$)$_2$O —◯— N=N —◯— C(=O)OCH$_3$

C$_3$AzoCO$_2$Me

cinnamic acid

Chart 6.2.

Molecule-sized cavities that are imprinted with the structure of template molecules are created by removing organic components.

6.3.2 Imprinting of aromatic carboxylic acids by the surface sol-gel process

Incorporation and removal of templates

Mixtures of Ti(O^{n}Bu)$_4$ and C$_3$AzoCO$_2$H (see Chart 6.2) were dissolved in a 2:1 toluene/ethanol (vol/vol) medium, and stirred at room temperature for more than 12 h. Ion-exchanged water was added by 2–4 times relative to the amount

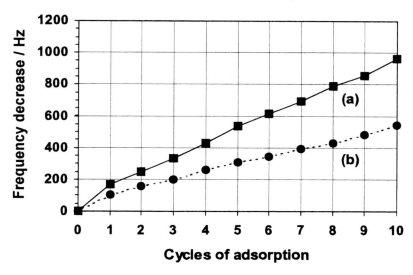

Figure 6.15. QCM frequency shifts during the adsorption of $Ti(O^nBu)_4/C_3AzoCO_2H$ complex. Stock solutions were diluted 20 times by toluene. (a) Stock solution: $Ti(O^nBu)_4$; 100 mM, C_3AzoCO_2H; 25 mM, H_2O; 275 mM, in toluene/ethanol = 2:1. (b) Stock solution: $Ti(O^nBu)_4$; 100 mM, C_3AzoCO_2H; 50 mM, H_2O; 350 mM in toluene/ethanol = 2:1. The period of adsorption is 1 min at room temperature, followed by washing for 1 min in toluene.

of $Ti(O^nBu)_4$, and the mixtures were aged for one to several hours and diluted with toluene. TiO_2-gel thin films were prepared from these mixtures by the surface sol-gel process, and the template molecule (C_3AzoCO_2H) was bound in the form of a complex with Ti in those films. The formation of the titanium–carboxylate complex could be confirmed by the presence of the carbonyl stretching bands at 1534 cm^{-1} and 1416 cm^{-1} in FT-IR reflection spectra [20]. The $v_{c=o}$ peak (1685.5 cm^{-1}) of free carboxylic acid was not found. It is clear that the structure of the Ti complex in the solution (1547 cm^{-1}, 1410 cm^{-1}) is maintained in the thin film.

A stock solution of 100 mM $Ti(O^nBu)_4$ and 25 mM (or 50 mM) C_3AzoCO_2H was diluted with toluene by 20 times and used as a dipping solution. The QCM frequency decrease due to adsorbed mass in each cycle is given in Fig. 6.15 for both the 25 mM and 50 mM template solutions. Regular film growth is confirmed by the linear frequency decrease in both cases. The averaged frequency change was 100 ± 30 Hz for one cycle, corresponding to a thickness of 1.8–2.7 nm, if the composite gel density is assumed to be 1.0–1.5 g/cm^3. When the concentration of the template was 50 mM, the frequency decrease was 54 ± 18 Hz for one cycle. The extent of the growth is greater at the lower concentration of the template molecule.

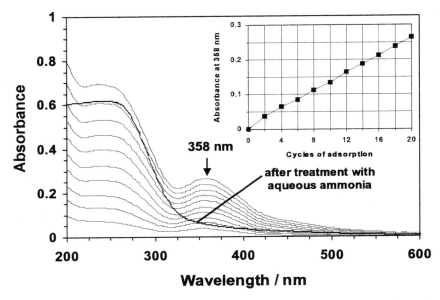

Figure 6.16. UV-vis absorption spectral change due to adsorption of Ti(O^{n}Bu)$_4$/C$_3$AzoCO$_2$H complex. The conditions are the same as those of Fig. 6.15. The inset shows the absorbance changes at 358 nm.

Incorporation of the template molecule and its removal can be confirmed by the UV-vis absorption peak of the azobenzene moiety. The absorption peaks due to the azobenzene moiety at 358 nm and the TiO$_2$-gel near 250 nm regularly increased in proportion to the adsorption cycle, which indicates that the azobenzene template is constantly incorporated into the gel film. The bound template molecule was removed by treating surface gel films with 1 wt % aqueous ammonia for 30 min. The UV absorbance at 358 nm completely disappeared upon ammonia treatment, showing complete removal of the template molecule: see Fig. 6.16.

The amount of the template molecule introduced in one adsorption cycle onto one side of the plate is determined to be 1.76 molecule/cycle·nm^2 by using the difference in absorbance before and after the template removal. This figure corresponds to the mass increase of 264 ng during 10 cycles, as estimated from the surface area of the electrode and the molecular weight of the template. It occupies 29 wt % of the total absorbed mass increase (1008 Hz, 907 ng), and is explained by the film composition of Ti$_4$O$_4$(OH)$_4$(O^{n}Bu)$_4$C$_3$AzoCO$_2$H, where the template molecule included is 31.6 wt %. This agrees well with the original mixing ratio in the stock solution. The elemental composition estimated by XPS measurement, Ti : C : O : N = 1.0 : 8.9 : 4.9 : 0.5, is consistent with the estimate that one C$_3$AzoCO$_2$H template molecule is incorporated with four titanium atoms.

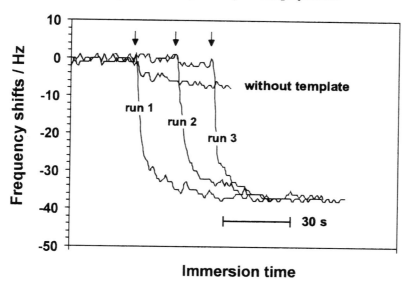

Figure 6.17. *In situ* QCM frequency decreases due to rebinding of the template molecule: 5 ml of 50 mM C_3AzoCO_2H in THF was added into 1 ml of CH_3CN at the time marked with an arrow to give a C_3AzoCO_2H concentration of 0.25 mM.

Molecular recognition

The above-mentioned imprinted films were used for guest binding experiments based on *in situ* QCM measurement in acetonitrile. The imprinted films were placed in CH_3CN, after the QCM frequency equilibrated, and guest solutions in THF were added to give a final guest concentration of 0.25 mM. A frequency decrease due to mass increase by rebinding of C_3AzoCO_2H was observed. The rebinding was saturated within 1 min, and the template molecule was again desorbed by thorough washing of the QCM electrode with 1 wt % aqueous ammonia. The frequency change by desorption–adsorption cycles is highly reproducible, giving a value of 32 ± 1 Hz, as illustrated in Fig. 6.17. In contrast, a similarly prepared TiO_2-gel film without the template shows a much smaller frequency decrease of only 2–3 Hz. It is clear that the rebinding of the C_3AzoCO_2H molecule is made possible by the preceding imprinting process and that the imprinted site remains intact after repeated guest binding experiments [34].

The identical imprinted film was used for binding of other carboxylic acid derivatives of Chart 6.2, and some of the binding data are shown in Fig. 6.18. The adsorption processes essentially finished in less than 1 min. The original template molecule gave a frequency decrease greater than those of other related carboxylic acids. The imprinting efficiency is summarized in Table 6.3. For the C_3AzoCO_2H-imprinted film, the relative binding efficiency decreases in the

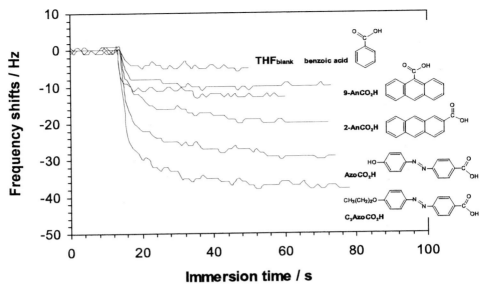

Figure 6.18. *In situ* QCM frequency decreases due to binding of a series of carboxylic acids. Conditions of *in situ* experiments are identical to those of Fig. 6.17.

order of $AzoCO_2H$, $2\text{-}AnCO_2H$, cinnamic acid, benzoic acid, $9\text{-}AnCO_2H$, and octanoic acid. The structurally closest acid, $AzoCO_2H$, gives a relative efficiency of 80–90%, and other carboxylic acids (cinnamic acid, benzoic acid and octanoic acid) are bound less efficiently. Anthracenecarboxylic acids show 30–60% binding efficiencies. The 2-isomer is a better substrate than the 9-isomer, because the former isomer appears structurally closer to the original template. Conversion of the carboxylic acid function to the corresponding ester depresses the binding efficiency: i.e. C_3AzoCO_2Me and $C_3AzoCO_2{}^iPr$.

The imprinting study was similarly conducted by using $2\text{-}AnCO_2H$ and $9\text{-}AnCo_2H$ as template, as given in Table 6.3. The binding efficiencies for these template molecules are greater than those of non-template guests in both cases. It is clear that the imprinting effect is observed for different aromatic carboxylic acids. A graphical presentation of the relative binding efficiency is given in Fig. 6.19.

It is evident from the above results that the guest selectivity is associated with molecular shape and functionality. The carboxylic acid group was an important element of recognition. FT-IR spectra indicated that the azobenzene carboxylic acid was covalently linked to titanium during the imprinting process. Unfortunately, the existence of the same linking could not be confirmed in rebinding experiments. The carboxylic acid in the guest molecule is bound

Table 6.3. *Relative binding efficiency of imprinted TiO$_2$ films for various guest molecules*

guest molecule MW	*in situ* frequency decrease[a] (Hz/pmol)			relative binding efficiency[b]		
	IF$_{C_3Azo}$	IF$_{2-An}$	IF$_{9-An}$	IF$_{C_3Azo}$	IF$_{2-An}$	IF$_{9-An}$
benzoic acid (122.12)	5/37	12/88	12/88	0.36	0.61	0.64
octanoic acid (144.21)	5/31	9/56	8/50	0.31	0.39	0.36
cinnamic acid (148.16)	8/49	13/79	17/88	0.48	0.54	0.75
2-AnCo$_2$H (222.24)	15/61	36/146	30/121	0.60	**1.00**	0.88
9-AnCO$_2$H (222.24)	3/32	26/105	34/138	0.33	0.72	**1.00**
AzoCO$_2$H (242.32)	24/88	30/111	33/123	0.88	0.76	0.89
C$_3$AzoCO$_2$H (284.31)	32/101	33/104	29/92	**1.00**	0.71	0.67

Notes:
[a]Corrected for the frequency shift due to solvent THF (5Hz). [b]The ratio of amounts of bound guest molecule and the template molecule. All QCM data are corrected by molecular weights of bound species. IF; an abbreviation of imprinted film.

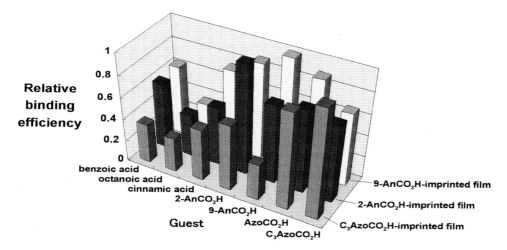

Figure 6.19. Graphical presentation of the relative binding efficiency for various guest molecules. The data are the same as those listed in Table 6.3.

either through covalent complexation or through noncovalent hydrogen bonding, as illustrated in Fig. 6.20. The aromatic moiety in the guest would be surrounded by the hydrophobic domain of TiO$_2$-gel that is composed of the titanium–oxygen network and the unhydrolyzed butoxy group. The elongated hydrophobic cavity thus formed would be connected to a polar hydrogen-bonding site where the carboxylic acid moiety is accommodated.

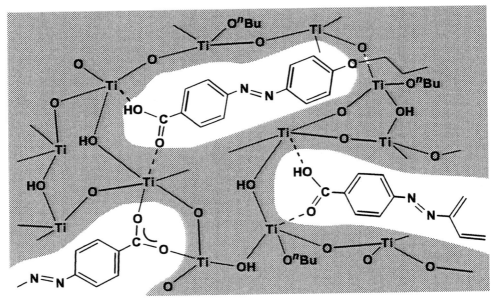

Figure 6.20. Schematic illustration of C_3AzoCO_2H imprinted film.

6.3.3 Imprinting of amino acids and peptides

Protected amino acids

The application of molecular imprinting to biologically active substances such as amino acids and proteins is extremely important. As the first trial, we examined molecular imprinting of protected amino acids in a TiO_2-gel thin film [35].

Ti(O^{n}Bu)$_4$ (100 mM) and carbobenzyloxy-L-alanine (Cbz-L-Ala, 25 mM) were mixed in toluene/ethanol and stirred at room temperature for more than 12 h. After addition of water and aging for several hours, the stock solution was diluted with toluene, and subjected to the surface sol-gel process. Uniform adsorption was observed up to 15 cycles with frequency shifts of 140–160 Hz per cycle. The template molecule, Cbz-L-Ala, was removed from the gel film by dipping in 1 wt % aqueous ammonia, as confirmed from the disappearance of characteristic peaks of the titanium–carboxylate complex and carbamate in reflection FT-IR spectra.

In situ QCM measurements were carried out for re-binding of the original template and other molecules in CH_3CN. Figure 6.21 shows selected examples of the guest binding experiment. The binding is rapid and saturated in 30–60 s. The frequency decrease and relative binding efficiency for each guest molecule in the Cbz-L-Ala-imprinted film are shown in Table 6.4. Among the peptide

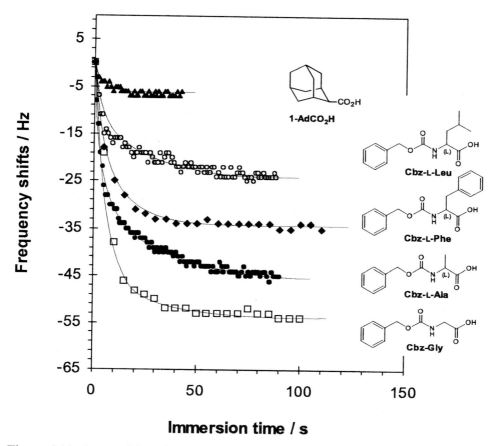

Figure 6.21. *In situ* QCM frequency decreases due to binding of a series of guest molecules in a TiO$_2$-gel film imprinted with Cbz-L-Ala.

Table 6.4. *Rebinding efficiency of Cbz-L-Ala imprinted TiO$_2$ film*

guest molecule MW	saturation $-\Delta F$ of binding (Hz/pmol)	relative binding efficiency (mol/mol)
benzoic acid (122.12)	5/37	0.21
octanoic acid (144.21)	7/44	0.25
cinnamic acid (148.16)	0/0	0
1-AdCO$_2$H (180.25)	7/35	0.20
Cbz-Gly (209.20)	53/230	1.32
Cbz-L-Ala (223.23)	43/170	**1.00**
Cbz-L-Pro (249.27)	33/120	0.69
Cbz-L-Leu (265.31)	24/81	0.47
Cbz-L-Phe (299.33)	34/100	0.59

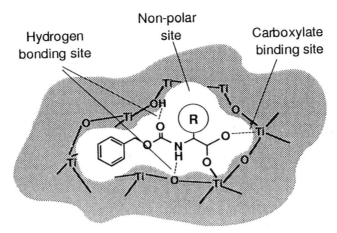

Figure 6.22. Schematic illustration of a Cbz-amino acid imprinted film. **R** denotes the side chain of Cbz-amino acid derivatives.

derivatives, Cbz-Gly showed the largest binding (53 Hz) and the extent of binding became suppressed with increasing sizes of the side chain. It is noteworthy that the original template, Cbz-L-Ala (43 ± 3 Hz), is less efficiently bound than the smaller Cbz-Gly. However, the binding efficiency is not determined solely by the size of guest molecules. For example, adamantane-1-carboxylic acid, 1-AdCO$_2$H, gave much smaller binding than all the amino acid derivatives, in spite of its smaller molecular weight. Organic carboxylic acids such as benzoic acid and cinnamic acid also give relatively low binding efficiency. All the protected amino acids are bound better than the conventional carboxylic acids. The binding data strongly suggest that the imprinting has produced a specific receptor site in the TiO$_2$-gel film. As illustrated in Fig. 6.22, this receptor cavity is probably composed of a carboxylate site, a hydrogen-bonding site, and hydrophobic site(s). The carboxylate-binding site is common to all guest molecules. The TiO$_2$-gel film contains the hydroxyl group (Ti—OH), the hydrogen-bond site and the hydrophobic site (Ti-O^{n}Bu in addition to the network of Ti—O—Ti) to accommodate non-polar moiety. These varied interaction sites act cooperatively to realize selective recognition of amino acid derivatives.

The efficiency of molecular recognition may be improved by creating additional hydrogen bonding sites in the TiO$_2$-gel. As an example, a composite gel film was prepared from trimethoxysilane with a urea unit (TMS-PU) and Ti(O^{n}BU)$_4$. TMS-PU and Cbz-β-Ala were mixed in organic solvent in the presence of water. After aging and addition of Ti(O^{n}Bu)$_4$, the stock solution was subjected to the surface sol gel process, and the template molecule was removed

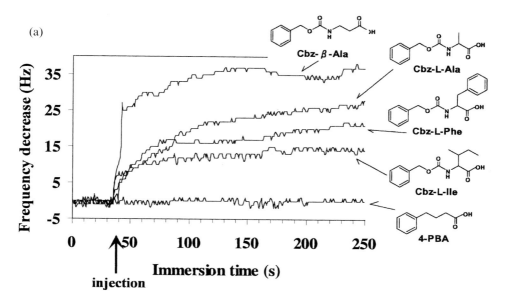

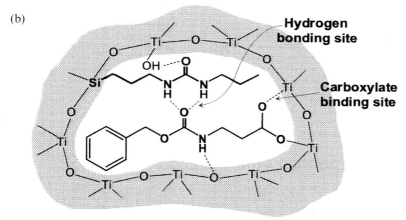

Figure 6.23. (a) *In situ* QCM frequency decreases due to binding of a series of guest molecules in a TiO$_2$-gel/TMS-PU film imprinted with Cbz-L-Ala. (b) Schematic illustration of Cbz-β-Ala imprinted film.

by aqueous ammonia. Binding experiments shown in Fig. 6.23a indicate that the template molecule, Cbz-β-Ala (20 μM), gave the greatest binding. It is interesting that the adsorption of Cbz-β-Ala was much more rapid than that of other amino acid derivatives. The absorbed mass decreased in the order of Cbz-L-Ala, Cbz-L-Phe, Cbz-L-Ile, the adsorption of 4-PBA was not observed. As illustrated in Fig. 6.23b, this gel film is probably composed of carboxylate site and urethane hydrogen bonding site.

(a)

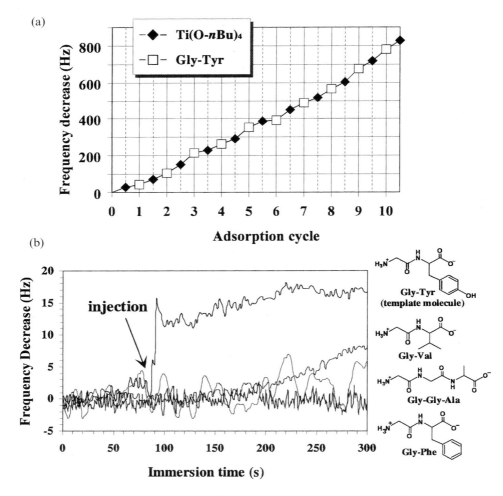

Figure 6.24. (a) QCM frequency shifts during alternate adsorption of Ti(O^{n}Bu)$_4$ (100 mM in toluene/ethanol = 2:1) and Gly-Tyr (10 mM in water). (b) *In situ* QCM frequency decreases due to binding of a series of dipeptides.

Other peptides

It is difficult to make water-soluble peptides form complexes with titanium alkoxide, because they are not soluble in organic solvents. Therefore, such templates cannot be imprinted by the complexation approach. Instead, they could be imprinted in TiO$_2$-gel films by the alternate adsorption approach with Ti(O^{n}Bu)$_4$. Figure 6.24a shows a plot of alternate layer-by-layer assembly of 100 mM titanium butoxide (3 min adsorption in toluene/ethanol) and 10 mM glycyl-L-tyrosine (Gly-L-Tyr, 10 min adsorption in water). The template molecule was removed by treatment with 10 mM aqueous sodium hydroxide, as

Table 6.5. *Rebinding data for* L- *and* D-*enantiomers in TiO$_2$-gel films imprinted with Cbz-amino acid derivatives*

imprinted film	saturation $-\Delta F$ of binding (pmol/Hz)		α^b
	guest $_{\text{L-form}}$	guest $_{\text{D-form}}$	
Cbz-L-Ala	173/43^a (65)	153/38 (60)	1.13
Cbz-D-Ala	157/39 (67)	177/44 (70)	1.13
Cbz-L-Leu	129/38 (74)	75/22 (58)	1.73
Cbz-D-Leu	81/24 (62)	139/41 (79)	1.71
Cbz-L-Phe	120/40 (69)	60/20 (49)	2.00
Cbz-D-Phe	96/32 (44)	162/54 (66)	1.68

Notes:
^{a}Frequency change except for surface adsorbed mass.
^{b}Enantioselectivity factor (α) in each imprinted film is a ratio of amounts of bound template molecule and its enantiomer (M-template/M-enantiomer, mol/mol).

confirmed from the disappearance of IR characteristic peaks of the template molecule. Similar TiO$_2$-gel films could be prepared for glycyl-L-glutamic acid (Gly-L-Glu), glycylglycine (Gly-Gly), and glycylglycylglycine (Gly-Gly-Gly). These peptide-imprinted TiO$_2$-gel films are able to recognize the peptides at low concentrations less than 10 μM in water. The Gly-L-Tyr-imprinted film adsorbs the template molecule itself most effectively, at a rapid rate, as shown in Fig. 6.24b. The absorbed mass became about a half when glycyl-L-valine (Gly-L-Val) was injected as guest. In the case of glycyl-L-phenylalanine (Gly-L-Phe) and glycylglycyl-L-alanine (Gly-Gly-L-Ala), the frequency was not stabilized, or their adsorptions did not occur at all. The TiO$_2$-gel film imprinted with Gly-L-Glu adsorbs Gly-L-Glu better than Gly-L-Tyr by a factor of two.

6.3.4 *Enantioselective recognition of amino acids*

The preceding results indicated that subtle geometrical structures of amino acids and peptides were imprinted in TiO$_2$-gel matrices. It is conceivable that the imprinting effect is not restricted to geometrical differentiations.

Imprinted TiO$_2$-gel films were prepared in a similar way by using L- or D-amino acid derivatives (Cbz-Ala, Cbz-Leu, Cbz-Phe) as templates. The template molecule is removed by treating with 0.1 wt % aqueous ammonia. The rebinding data for the L- and D-enantiomers in these imprinted films are shown in Table 6.5. The enantioselectivity factor (α) in each imprinted film is a ratio of

(a)

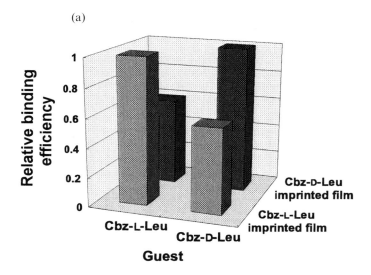

(b)

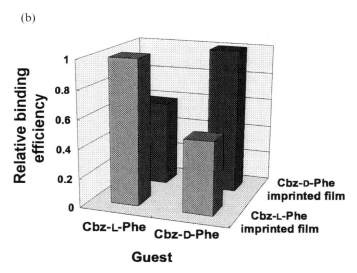

Figure 6.25. Graphical presentation of the relative binding efficiency for Cbz-amino acid derivatives. (a) Cbz-Leu imprinted film, (b) Cbz-Phe imprinted film.

the amounts of bound template molecule and its enantiomer (M-template/M-enantiomer, mol/mol). Figure 6.25 shows selected examples of relative binding efficiency [36].

In the Cbz-Ala imprinted films, the frequency change between the template and its enantiomer was 5 or less, and the enantioselectivity factor was about 1.1 in either of the L- and D-imprinted films. In contrast, the template molecule was bound much better than its enantiomer in the Cbz-Leu- or Cbz-Phe-imprinted

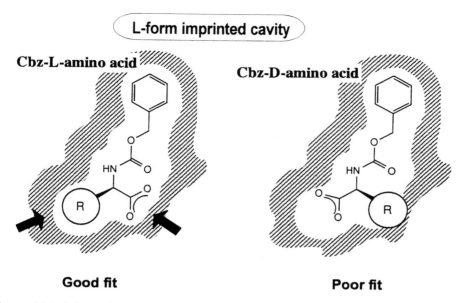

L-form imprinted cavity

Cbz-L-amino acid

Cbz-D-amino acid

Good fit

Poor fit

Figure 6.26. Schematic illustration of the enantioselective binding in a TiO_2-gel film imprinted with a Cbz-amino acid derivative of L-form.

film. The difference in the frequency change between these optical isomers was about 16–22 Hz, and the chiral separation factor amounted to 1.7–2.0. The enantioselectivity is attained only when all three substituents (–COOH, Cbz-NH, and side chain) on the α carbon are recognized. When a D-enantiomer is adsorbed in an L-imprinted film, the configuration of side-chain and carboxyl group is reversed, as shown in Fig. 6.26.

6.3.5 Enantioselective recognition of sugars

As described in Section 6.2.3, some sugar compounds such as monosaccharides, starch, and sugar balls are effectively adsorbed on the surface of TiO_2-gel films. This led to molecular imprinting of monosaccharides [37]. Although sugars are too insoluble in organic solvents for the purpose of complexation with metal alkoxides, they are solubilized upon complexing with $Ti(O^nBu)_4$. Thus, TiO_2-gel composite films could be prepared by the surface sol-gel process after $Ti(O^nBu)_4$ (200 mM) and D(+)-glucose (D(+)-Glc, 20 mM) were complexed in toluene/ethanol (2:1). The template molecule is again removed readily by treating with 1 wt % aqueous ammonia. Binding experiments were carried out in water for sugar derivatives (1 mM), and the results are given in Table 6.6. The imprinted film shows the highest adsorption for D(+)-Glc, which is the template molecule. Five other monosaccharides with equal molecular weight

Table 6.6. *Rebinding data for sugar derivatives in a TiO₂-gel film imprinted with D(+)-glucose*

guest molecule MW	saturation $-\Delta F$ of binding (Hz/pmol)	relative binding efficiency (mol/mol)	α^b
D(+)-glucose (180.16)	111 (77)^a/555	**1.00**	**1.00**
D(+)-glucose (180.16)	69 (35)/345	0.62	1.6
D(+)-mannose (180.16)	59 (25)/295	0.53	1.9
D(+)-galactose (180.16)	51 (17)/255	0.46	2.2
D(+)-fructose (180.16)	50 (16)/250	0.45	2.2
D(+)-ribose (150.13)	65/390	0.70	1.4
D(+)-maltose (342.31)	64/160	0.30	3.3

Notes:
^{a}Frequency change except for surface adsorbed mass. ^{b}Enantioselectivity factor (α):M-template/M-enantiomer, mol/mol.

and equal number of hydroxyl groups give much smaller binding than the template. They have relative binding efficiencies of 0.4–0.6 and separation factors (α) of 1.4–2.2. These binding data are related to adsorption at the interior and surface of the TiO₂-gel film. When the surface adsorbed mass (34 Hz in the case of D(+)-Glc) is excluded, the separation factor (α) is certain to increase. The selectivity is also influenced by the molecular size of guests, since the relative binding efficiency of D(+)-Mal (0.29) is low.

6.4 Concluding remarks

The metal oxides prepared by conventional baking or by the CVD method are, in general, chemically stable, crystalline materials, and show excellent mechanical, electrical, optical, and physical properties. Flexible porous gel films obtained by the surface sol-gel process are totally different. In this chapter, we described a new preparative method for ultrathin metal oxide films by stepwise adsorption of various metal alkoxides. We named this method the surface sol-gel process. Structural characterization of the gel films thus obtained, the electrical property, and formation of nano-composites with organic compounds, were also explained. The soft porous gel contains many active hydroxyl groups at the surface and interior of the film. This facilitates adsorption of organic compounds, and consequent preparation of ultrathin metal oxide/polymer nano-composite films and organization of functional small molecules. In the nano-composites, proper selection of polymer components leads to the design of new materials with unique electrical, optical, and chemical response properties. Combination with functional small molecules may contribute to the development of new catalyst and separation materials, as well as energy conversion and molecular sensing systems.

Molecular imprinting is an equally important development of the surface sol-gel process. We can conclude from these results that imprinted TiO_2-gel films are capable of precisely recognizing the size of template molecules, the nature and position of functional groups, and molecular chirality. Particularly, the fact that monosaccharide isomers in which only the configuration of hydroxyl groups differs can be recognized sensitively indicates that the structure of imprinted TiO_2-gel is precisely controlled at the atomic level. Current investigations have aimed at the construction of highly discriminating molecular recognition systems by 3D arrangement of specific best functional groups. However, the synthesis of such host compounds is not readily achieved. For example, Wulff attempted enhancement of substrate selectivity by fixing (or imprinting) functional groups that interact with the template molecule in polymer (Fig. 6.13) [38]. Appropriated spatial organization of the hydrogen-bonding unit in cross-linked polymers of divinyl monomers has been studied extensively. When a target molecule has structurally carried functional groups, these techniques are effective. However, divinyl monomers cannot produce precisely structured cavities complementary to small guest molecules; it seems to be unsuitable for molecular imprinting in which the functional group is scarce. Furthermore, varied complementary functions have to be employed, depending on the structure of guest molecules. In contrast, TiO_2-gel has a flexible network structure because of the freedom of the O–Ti–O bond angle and, in addition, TiO_2-gel can provide metal-coordination sites, hydrogen-bonding sites, charged sites and hydrophobic domains. Such structural flexibility and functional diversity are the greatest advantages of the TiO_2-gel for molecular imprinting. An additional advantage is that the titanium oxide nano-film is readily prepared on surfaces of various kinds: particle surfaces, inner-walls of capillary, and the inside of a porous support.

6.5 References

1. Kumagai, H., Matsumoto, M., Toyoda, K., Obara, M. and Suzuki, M. (1995). *Thin Solid Films*, **263**, 47; Toda, H. (1995). *Langmuir*, **11**, 3281; Desu, S. B. (1992). *Mater. Sci. Eng. B* **13**, 299.
2. Yoshimura, M., Suchanek, W. and Han, K.-S. (1999). *J. Mater. Chem.* **9**, 77.
3. Moriguchi., I., Maeda, H., Teraoka, Y. and Kawaga, S. (1995). *J. Am. Chem. Soc.* **117**, 1139.
4. Kalachev, A. A., Mathauer, K., Höhne, U., Möhwald, H. and Wegner, G. (1993). *Thin Solid Films* **228**, 307.
5. Mirley, C. L. and Koberstein, J. T. (1995). *Langmuir* **11**, 1049; Brunner, H., Vallant, T., Mayer, U. and Hoffman, H. (1996). *Langmuir* **12**, 4614.

6. Mallouk, T. E. and Gavin, J. A. (1998). *Acc. Chem. Res.* **31**, 209.
7. Kleinfeld, E. R. and Ferguson, G. S. (1994). *Science* **265**, 370.
8. Iler, R. (1966). *J. Colloid Interface Sci.* **21**, 569; Lvov, Y., Ariga, K., Onda, M., Ichinose, I. and Kunitake, T. (1997). *Langmuir* **13**, 6195.
9. Keller, S. W., Kim, H.-N. and Mallouk, T. E. (1994). *J. Am. Chem. Soc.* **116**, 8817; Lvov, Y., Ariga, K., Ichinose, I. and Kunitake, T. (1996). *Langmuir*, **12**, 3038.
10. Ingersoll, D., Kulesza, P. L. and Faulkner, L. R. (1994). *J. Electrochem. Soc.* **141**, 140; Ichinose, I., Tagawa, H., Mizuki, S., Lvov, Y. and Kunitake, T. (1998). *Langmuir* **14**, 187.
11. Bach, U., Lupo, D., Comte, P., Moser, J. E., Weissörtel, F., Salbeck, J., Spreitzer, H. and Grätzel, M. (1998) *Nature* **395**, 583.
12. Ichinose, I., Senzu, H. and Kunitake, T. (1996). *Chem. Lett.*, 831; Ichinose, I., Senzu, H. and Kunitake, T. (1997). *Chem. Mater.* **9**, 1296.
13. Kase, A., Asakaru, K., Egawa, C. and Iwasawa, Y. (1986). *Chem. Lett.*, 855; Asakura, K. and Iwasawa, Y. (1991). *J. Phys. Chem.* **95**, 1711.
14. Asakura, K. and Iwasawa, Y. (1988). *Chem. Lett.*, 633.
15. Kleinfeld, E. R. and Ferguson G. S. (1994). *Mater. Res. Soc. Symp. Proc.* **351**, 419.
16. Sauerbery, G. (1959). *Z. Phys.* **155**, 206; Ebara, Y. and Okahata, Y. (1994). *J. Am. Chem. Soc.* **116**, 11 209.
17. Bradley, D. C. (1989). *Chem. Rev.* **89**, 1317.
18. Iwasaki, Y. (2000). Master's thesis, Dept. of Chemistry and Biochemistry, Kyushu University.
19. Ichinose, I.,Kawakami, T. and Kunitake, T. (1998). *Adv. Mater.* **10**, 535.
20. Brinker, C. J. and Scherer, G. W. (1990). *Sol-Gel Science, The Physics and Chemistry of Sol-Gel Processing*. Academic Press, San Diego; Roger, C. and Hampden-Smith, M. J. (1992). *J. Mater. Chem.* **2**, 1111.
21. Sugar balls and a polyrotaxane were synthesized by Okada *et al.* in Nagoya University and by Harada *et al.* in Osaka University, respectively.
22. Yonezawa, T., Matsune, H. and Kunitake, T. (1999). *Chem. Mater.* **11**, 33.
23. Ichinose, I., Suto, M., Akiyama, T., Niidome, Y., Yamada, S. and Kunitake, T. (1998). *Polym. Prep. Jpn.*, 48.
24. Hedborg, E., Winquist, F., Lundstorm, I., Andersson, L. I. and Mosbach, K. (1993). *Sensors and Actuators A* **37–38**, 796; Kriz., D., Ramström, O. and Mosbach, K. (1997). *Anal. Chem. News & Features*, June 1, 345A; Haupt, K. and Mosbach, K. (2000). *Chem. Rev.* **100**, 2495.
25. Dickey, F. H. (1949). *Proc. Natl. Acad. Sci.* **35**, 227; Dickey, F. H. (1955). *J. Phys. Chem.* **59**, 695.
26. Wulff, G. and Sarhan, A. (1972). *Angew. Chem. Int. Ed. Eng.* **11**, 341.
27. Kempe, M. and Mosbach, K. (1995). *J. Chromatogr. A* **694**, 3; Ramström, O. and Ansell, R. J. (1998). *Chirality* **10**, 195; Yu, C. and Mosbach, K. (1997). *J. Org. Chem.* **62**, 4507.
28. Glad, M., Norrlow, O., Sellergren, B., Siegbahn, N. and Mosbach, K. (1985). *J. Chromatogr.* **347**, 11.
29. Morihara, K. Kurihara, S. and Suzuki, J. (1998). *Bull. Chem. Soc. Jpn* **61**, 3991; Morihara, K., Takiguchi, M. and Shimada, (1994). *Bull. Chem. Soc. Jpn* **67**, 1078.
30. Heilmann J. and Maier, W. F. (1994). *Angew. Chem. Int. Ed. Engl.* **33**, 471.
31. Pinel, C., Loisil, P. and Gallezot, P. (1997). *Adv. Mater.* **9**, 582.

32. Sagiv, J. (1979). *Isr. J. Chem.* **18**, 346; Sagiv, J. (1980). *J. Am. Chem. Soc.* **102**, 92.
33. Kodakari, N., Katada, N. and Niwa, M. (1995). *J. Chem. Soc. Chem. Commun.*, 623; Kodakari, N., Katada, N. and Niwa, M. (1997). *Chem. Vap. Deposition* **3**, 59.
34. Lee, S.-W., Ichinose, I. and Kunitake, T. (1998). *Langmuir* **14**, 2857.
35. Lee, S.-W., Ichinose, I. and Kunitake, T. (1998). *Chem. Lett.*, 1193.
36. Lee, S.-W., Ichinose, I. and Kunitake, T. (1999). *Spr. Symp. Proc.,Chm. Soc. Jpn* **1**, 276.
37. Lee, S.-W. and Kunitake, T. (2001). *Molecular Crystal & Liquid Crystal*, in press.
38. Wulff, G. (1995). *Angew. Chem. Int. Ed. Engl.* **34**, 1812.

7

The hierarchy of open-framework structures in metal phosphates and oxalates

SRINIVASAN NATARAJAN AND C. N. R. RAO

After providing a brief description of zeolitic structures, we discuss the hierarchy of structures of open-framework metal phosphates ranging from zero-dimensional monomeric units and one-dimensional linear chains to complex three-dimensional structures. Aspects related to the likely pathways involved in the assemblage of these fascinating structures are examined, pointing out how the formation of the complex three-dimensional structures of open-framework metal phosphates involves the transformation and assembly of smaller units. Besides the role of the four-membered monomer, the amine phosphate route to the formation of the three-dimensional structures is discussed. The last step in the formation of these structures from preformed units of the desired structure is likely to be spontaneous. Our recent studies of open-framework metal oxalates have shown the presence of a hierarchy of structures. Reactions of amine oxalates with metal ions yield members of the oxalate family with differing complexity.

7.1 Introduction

Zeolites and related aluminosilicates constitute a vital family of microporous materials with immense applications in catalysis, sorption and separation processes [1–3]. The discovery of aluminophosphates is an important landmark in the science of these materials [4]. All these materials are, in general, synthesized under hydrothermal conditions by making use of template molecules [2]. The template molecules are usually organic amines and they are involved in the formation of these framework structures in more ways than one. While it is difficult to pinpoint the exact manner in which the amines participate or direct the formation of these inorganic structures, it is generally believed that their size and shape are crucial in determining the pore structure. In recent years, a variety of open-framework structures formed by divalent metal phosphates

214

have been prepared hydrothermally in the presence of organic amines. These studies have led to a hierarchy of open-framework structures and also throw some light on the role of the amine. We will discuss these aspects at some length in this chapter and show how these complex open-framework metal phosphates evolve from a basic building unit such as a four-membered ring or a linear chain phosphate. In this context, the role of amine phosphates as likely intermediates will be examined. We will also describe some of the recent results on open-framework metal oxalates, and point out how the three-dimensional oxalate structures involve a building up process from small molecular units, just as in the phosphate. We then briefly describe the phosphate–oxalate hybrid structures. What is common to the different inorganic structures with open architectures is the presence of an organic amine, which seems to be essential for this formation. Then, all these materials seem to form by processes involving supramolecular organization and self-assembly.

7.2 Zeolite structures

Before we discuss open-framework metal phosphates, it would be useful briefly to examine zeolitic structures, since they constitute the first microporous materials. The framework structures of zeolites generally consist of AlO_4 and SiO_4 (TO_4) tetrahedra linked by the sharing of oxygen atoms. The individual TO_4 (T = Si/Al) tetrahedra in the zeolites are nearly regular, but the T–O–T bond angles about the apical oxygen atoms have values in the range ~ 125–$180°$. This is one reason why there are so many zeolitic structures. The zeolite structures are classified in terms of the secondary building units (SBUs), consisting of four or more tetrahedra (four-membered rings) [5]. In Fig. 7.1, we present two simple zeolitic structures built-up of four- and six-membered SBUs. Many other SBUs such as five-rings, eight-rings and more complex units such as double four-rings (cubes) and double six-rings (hexagonal prisms) are also known. In addition to the SBUs, zeolitic structures are understood in terms of the simpler sheet structures generated by the linkages between these SBUs [6]. Thus, many of the zeolitic structures can be constructed from two-dimensional three-connected nets (layers). This is illustrated in the case of gmelenite in Fig. 7.2. The fourth connection of the TO_4 tetrahedra (T = Si, Al) is directed either up or down and is used to join these two-dimensional three-connected nets to form three-dimensional four-connected nets [6]. This method of constructing zeolite structures is useful for pentasil zeolites, which include ferrierite, mordenite, ZSM-5 and zeolite-β. The use of the nets is helpful to visualize structures containing stacking disorder as well. For example, the formation of intergrowths in erionite and offretite [7] can be thought of as arising from the non-uniform stacking

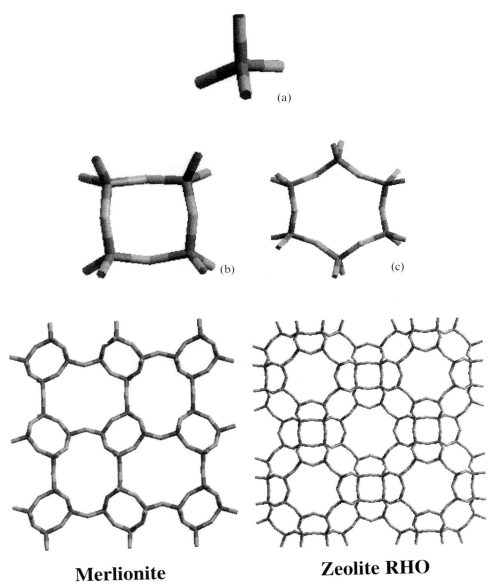

Merlionite   **Zeolite RHO**

Figure 7.1. Figure showing the assembly of zeolitic structures from fundamental building units: (a) the primary tetrahedral unit, (b) four-membered ring, (c) six-membered ring.

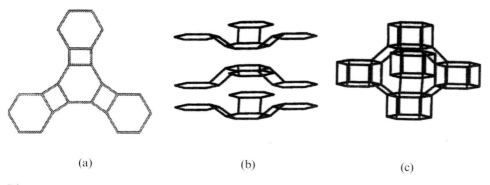

Figure 7.2. The formation of gmelenite (c) from a stack (b) of two-dimensional three-connected nets (a).

sequence of the two-dimensional three-connected nets common in both structures.

Synthetic zeolites have been known since the 1940s, through the seminal work of Barrer [8] and Milton [9]. The synthesis employs alkali or organic cations in the starting gels to produce aluminosilicate structures with variable Si/Al ratios, under hydrothermal conditions at a basic pH of ~ 10. Hydrothermal crystallizations are multi-component heterogeneous reactions involving several processes including equilibrium reactions and nucleation and growth. These processes are interdependent and change with time and temperature. Careful studies have provided insight into the possible mechanism of formation of zeolites under the synthetic conditions [10]. One of the early mechanistic considerations for zeolitic crystallization is the application of Ostwald's law of successive transformations, which states that in competing species, the least stable phase crystallizes first until the most stable product forms. Ostwald's law is generally obeyed in the formation of zeolites with high Al content. Recent efforts have focussed on the mechanism of formation of zeolites, in particular the role of the organic cations, and have shed light on some aspects related to the synthesis of zeolites. Although we do not have a complete understanding, one assumes that the synthesis follows the following stages.

Reactants → Reactant gel → Precursors → Nucleation → Crystal growth

There are many curious features of zeolite synthesis. Thus, different framework structures are formed by the same amine, and the same framework structure is also formed by the use of different amines [11]. Such observations have prompted investigations into the role of the amines in the formation of these structures. The role of the organic amine in the synthesis of zeolites can be classified into three types: 'templating', structure-directing and space-filling. Here, 'templating' refers to the formation of a unique structure, which reflects the geometrical and electronic structure of the template. Structure-direction describes the process where a specific organic amine preferentially leads to the synthesis of a structure by suitably influencing factors such as pH, solubility and electrostatic interactions. Space-filling is a process in which the organic amine excludes water and enhances the interactions in the organic-framework composite and thereby increases the thermodynamic stability.

There are other mechanistic considerations pertinent to the understanding of zeolite formation. These pertain to the solution-mediated transport and the solid hydrogel transformation. The solution-mediated process involves the dissolution of the reagents in the solution phases followed by the transport of the dissolved silicate species by diffusion to the nucleation sites where phase growth occurs. The solid hydrogel transformation is the reorganization of a solid phase from an initially amorphous state to one with long-range order. A schematic description of these processes is presented in Fig. 7.3. We shall not discuss zeolites in this contribution, our primary purpose being to provide a coherent picture of open-framework metal phosphates and related systems. Details of these processes are discussed at length in many of the reviews on the subject [12].

7.3 Open-framework metal phosphates

The structures of the variety of open-framework metal phosphates prepared and characterized to-date reveal the existence of the following hierarchy: one-dimensional linear chains possessing corner-shared four-membered rings, one-dimensional ladders possessing edge-shared four-membered rings, two-dimensional layers and three-dimensional structures. Of these, the linear chain and the ladder structures are the simplest and contain four-membered rings. One may, therefore, consider the four-membered ring to be the basic building unit of the open-framework metal phosphates. It is noteworthy that the two-dimensional layers and the three-dimensional structures are the most commonly observed in metal phosphates, the lower-dimensional ladder and chain structures being rare.

7.3.1 One-dimensional structures

The one-dimensional framework solids constitute a very small fraction of the gamut of structures known in phosphate-based materials. The existence of one-dimensional structures, however, has been known for some time in the case of aluminum, gallium and zinc phosphates [13–16]. The structures consist of polymeric chains of $[MP_2O_8]^{n-}$ units supported by the structure-directing

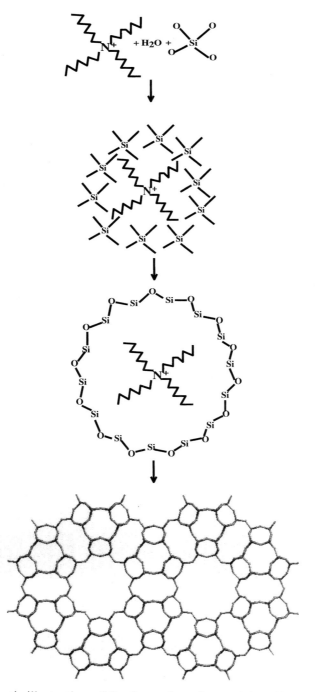

Figure 7.3. Schematic illustration of the formation of a model zeolite. Adapted from the work of Davis and Lobo [10].

amines, which are variably protonated depending upon the synthesis conditions. As mentioned earlier, the one-dimensional materials can have two types of structural arrangement; linear chain and ladder. The linear chain is the simplest possible structure amongst the open-framework structures. We show the linear structure of a zinc phosphate in Fig. 7.4a. The most commonly found one-dimensional structure is the edge-shared ladder, shown in Fig. 7.4b. A simple edge-shared four-membered ladder structure consisting of just four-membered rings made from $[M_2P_2O_4]$ units has been isolated in only one Sn(II) phosphate [17], and most of the other known ladder structures contain phosphate groups hanging from the metal. Corner-shared linear chain structures, on the other hand, are extremely rare. Although few aluminum phosphate linear chain structures were known [13], a zinc phosphate with a corner-shared linear chain structure has been isolated only recently [18]. The framework structure of the chain zinc phosphate in Fig. 7.4a consists of a corner linkage between the Zn and P via the oxygens. Unlike the ladder, the chain zinc phosphate structure does not contain hanging phosphate units.

In addition to the above one-dimensional structures, a *strip*-like cobalt phosphate structure has been isolated. The strip consists of two corner-shared chains fused together forming the architecture shown in Fig. 7.5a–c [18]. The one-dimensional linear chains are connected via a three-coordinated oxygen atom, giving rise to such an arrangement. This type of structure could be a key intermediate in the formation of framework solids from the basic one-dimensional chains.

One can visualize the formation of the two-dimensional structures from the ladders. The ladders can undergo rotation, hydrolysis and condensation to form the layers, as shown schematically in Fig. 7.6. Such layered Zn and Co phosphate structures have been isolated and characterized [19–21].

7.3.2 Two-dimensional structures

The two-dimensional structures are extended networks formed by the linking of the metal–oxygen polyhedra and the phosphate tetrahedra. These are sheet structures and often resemble those of naturally occurring clay minerals. The sheets are usually anionic and the protonated (cationic) amine molecules, located between the two sheets, render the framework neutral. The two-dimensional structures are intermediates between the one-dimensional chains and the three-dimensional structures, and the literature on phosphate networks contains descriptions of several layered materials, owing to the wide compositional diversity exhibited by them [22–24]. The layered materials are of interest because they act as precursors for the three-dimensional structures.

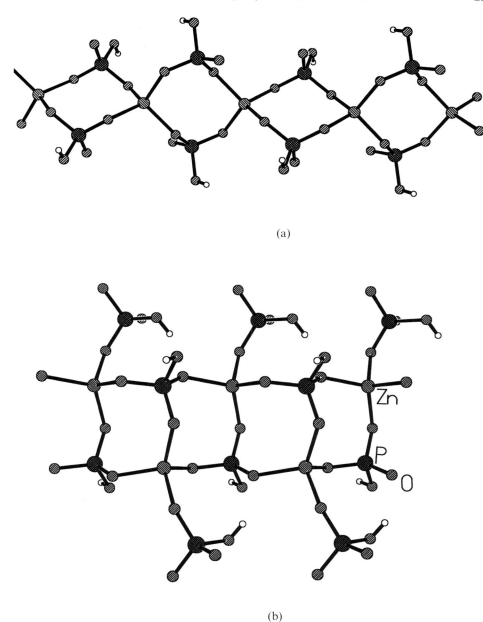

(a)

(b)

Figure 7.4. (a) The simple linear chain structure formed by corner-shared four-membered rings. (b) The ladder structure with edge-shared four-membered rings.

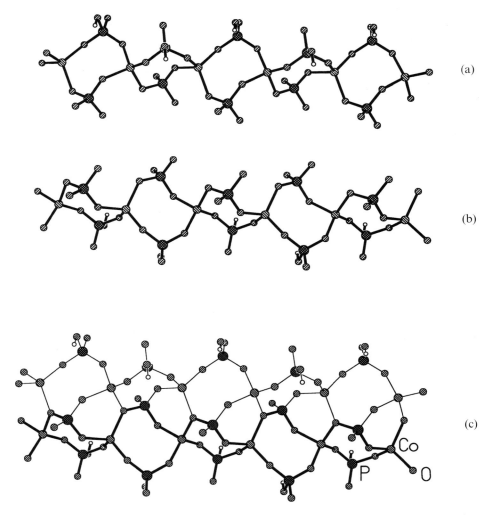

Figure 7.5. The formation of a *strip*-like cobalt phosphate (c) by the fusion of two corner-shared four-membered rings (a and b).

In the aluminophosphate family, three main structural classes with different metal to phosphorus ratios are known. The layered aluminophosphates normally have P/Al ratios of 4:3 and 3:2 and a new series with P/Al ratio of 2:1 has been reported [25]. The templates play an important role in the formation of these materials and there have been some attempts to rationalize the formation of these solids [25]. The two-dimensional architecture is generally formed by strictly alternating PO_4 and AlO_4 tetrahedra. Of the four oxygen atoms coordinated to each tetrahedral phosphate, three are shared with adjacent Al atoms,

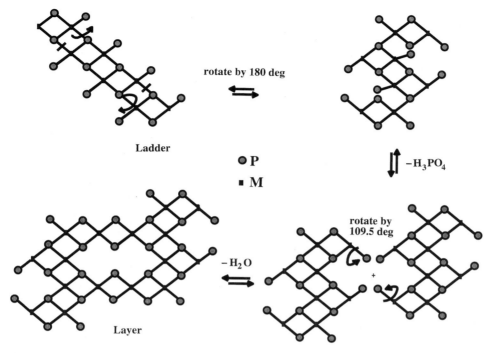

Figure 7.6. Schematic illustration of the formation of a layered architecture from the edge-shared four-membered ladder structure.

leaving a terminal oxygen atom. Charge-balance is achieved by the protonation of the PO_4 units forming HPO_4 or H_2PO_4 species.

A different scenario is encountered in the case of layered open-framework compounds of divalent elements. Owing to differences in the charges, the phosphate units in the layer structures are generally protonated [19–21]. In what follows, we shall discuss few representative examples of the layered Sn^{II}, Co^{II} and Zn phosphates.

The layered tin(II) phosphate, $[NH_3(CH_2)_3NH_3]^{2+}2[SnPO_4]^-$, is based on a network of strictly alternating SnO_3 and PO_4 units [26]. The SnO_3 and PO_4 units form infinite layers with the framework formula $[SnPO_4]^-$. Charge neutrality is achieved by the incorporation of the organic amine in its diprotonated form; there are 0.5 $[NH_3(CH_2)_3NH_3]^{2+}$ ions per framework formula unit. The Sn atoms are coordinated to three oxygens and occupy the vertex of a trigonal pyramid; the lone-pair presumably occupies the fourth vertex of the tetrahedron. The connectivity between the SnO_3 and PO_4 moieties forms the layered architecture with four- and eight-membered apertures within the layer along the *a* axis as shown in Fig. 7.7. Each four-membered ring is surrounded by four

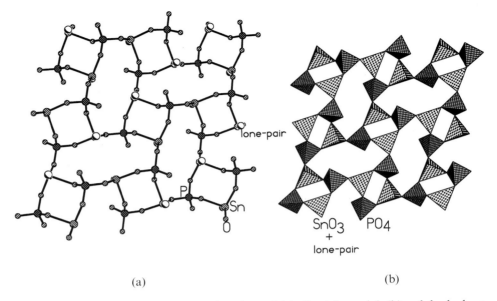

(a)   (b)

Figure 7.7. A single layer of tin(II) phosphate: (a) ball-stick model, (b) polyhedral representation. Note that the lone-pair electrons are needed to complete the tetrahedra for Sn(II) (Vaidhyanathan and Natarajan [26]).

eight-membered rings, and each eight-membered ring is connected to four four-membered rings. The structure-directing amine molecule, DAP, is located in between the layers (Fig. 7.8). The individual layers are held together by the doubly protonated DAP molecules (Fig. 7.8) through hydrogen bonding. The lone-pair of electrons associated with Sn(II) plays an important role in the structures of these materials. The stereoactive lone-pair manifests itself in the lattice by creating open-space between the two layers in the tin(II) phosphates. The lone-pair of electrons of Sn(II) points perpendicular to the plane of the layers.

A typical layered cobalt phosphate, $[C_4N_2H_2]_{1.5}[Co_2(H_2PO_4)_2(PO_4)H_2O]$, shown in Fig. 7.9, consists of vertex linked CoO_4, $PO_3(OH)$ and PO_4 tetrahedra forming layers, which are held together by hydrogen-bond interactions [18]. The connectivity between CoO_4, HPO_4 and PO_4 forms four-membered rings, which are joined to give rise to a layered topology based on a two-dimensional network of bifurcated 12-membered rings. The 12-membered ring consists of 12 T atoms (T = Co or P) formed by six cobalt and six phosphorus atoms which strictly alternate. These layers are arranged along the bc plane. The four-membered rings form a zig-zag ladder that is connected by HPO_4 moieties forming the 12-membered aperture. The di-protonated piperazinium cation occupies the space between the layers.

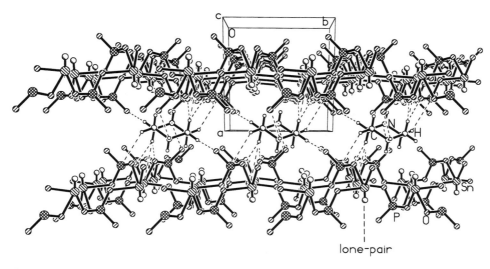

Figure 7.8. Layer arrangement of the tin(II) phosphate. Note that the lone-pair of electrons points into the inter-lamellar space (Vaidhyanathan and Natarajan [26]).

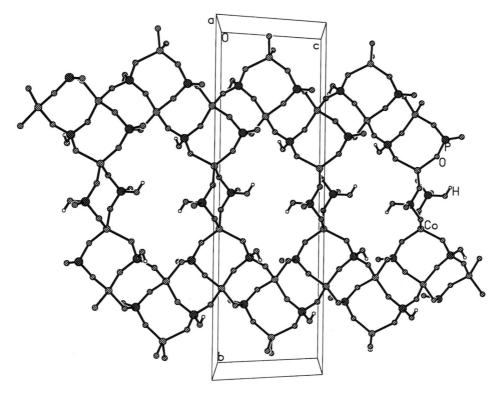

Figure 7.9. A layered cobalt phosphate. Note that the zig-zag ladders are connected by two phosphate groups forming 12-membered apertures within the layer (Rao *et al.* [18]).

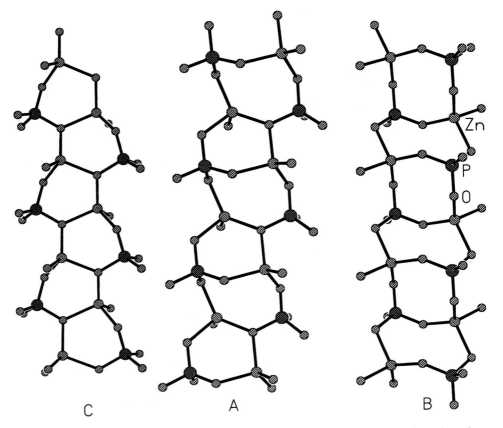

Figure 7.10. Different types of one-dimensional chain observed in the zinc phosphate family (Neeraj *et al.* [27]).

The synthesis and structure of an unusual layered zinc phosphate, $[NH_3(CH_2)_2NH(CH_2)_2NH_3]^{2+}2[Zn_2PO_4(HPO_4)]^-$, possessing layers with steps has been described recently [27]. This structure has an unusual step-like feature; the connectivity between ZnO_4 and PO_4 tetrahedron gives rise to layers. The structure-directing agent, diethylenetriamine (DETA), is doubly protonated and occupies spaces between the layers. The connectivity between the ZnO_4 and PO_4 units is such that it produces three distinct types of chain, labelled A, B and C in Fig. 7.10. The A type chain consists of alternate stacking of three- and four-membered rings [involving Zn(1) and Zn(2)] while the B type chain is made up of only four-membered rings [involving only Zn(2) and P(2)]. The C type chain is made up of only three-membered rings [involving only Zn(1) and P(1)] (Fig. 7.10). The chains are connected to one another forming the layer shown in Fig. 7.11. This is the first instance where a continuous three-member ladder (chain) is present in an open-framework material. The presence of the

Figure 7.11. A single layer of the zinc phosphate. The connectivity between the one-dimensional chains (Fig. 10) gives rise to a step within the layer (Neeraj *et al.* [27]).

exclusive three-membered chains of ladders creates some strain in the layer, which is otherwise nearly planar, and causes the formation of a step and a ladder-like feature in the material. Furthermore, we find an ABAC repeating unit along the *bc* plane giving rise to the repeat ring sequence of 3343344433 along the *c* axis. The framework is also characterized by the presence of infinite Zn−O−Zn chains. The Zn−O−Zn linkage is accompanied by the trigonal coordination of the bridging oxygen atoms, the third coordination being always to a P atom. The infinite one-dimensional Zn−O−Zn chain is formed only by the Zn(1) atoms (forming the three-membered ladders), but isolated short chain Zn−O−Zn linkages involving both Zn(1) and Zn(2) atoms are also present in the structure. An interesting aspect of this layered zinc phosphate is the connectivity between the Zn atoms within each layer. From the Zn sub-network presented in Fig. 7.12, it can be seen that the arrangement resembles a *fish backbone*. Such a feature has been seen for the first time in this material. The Zn(1) forms the vertebrae onto which Zn(2) atoms are grafted completing the *fish backbone* type arrangement. This type of connectivity may be attributed to the presence of the three-membered chains and the three-coordinated oxygen atoms.

7.3.3 Three-dimensional structures

The three-dimensional structures are characterized by the presence of channels and voids where the structure-directing amine molecules usually reside. In many of the aluminophosphates it has been possible to remove the amine

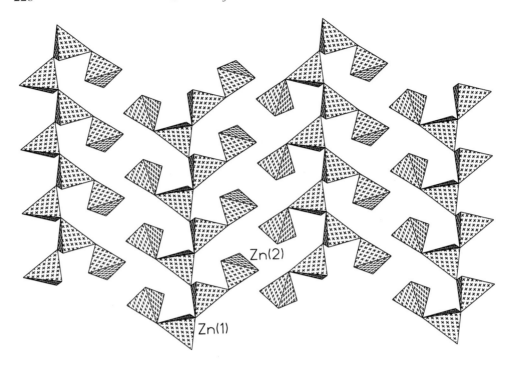

Figure 7.12. The Zn–O–Zn sub-network of the zinc phosphate. The connectivity
resembles a *fish backbone* (Neeraj *et al.* [27]).

molecules by post-synthesis treatments such as calcination and leaching, to
render them porous. A large number of such three-dimensional structures have
been reported in the literature [4]. It has also been possible to synthesize AlPO
structures analogous to the naturally occurring or synthetic aluminosilicates.
Most of the three-dimensional AlPOs have Al/P ratio of 1:1.

The divalent metal phosphates, on the other hand, present an interesting situation wherein the total charge on the metal phosphate $(2^+ + 5^+ = 7^+)$ is identical to that of an aluminosilicate $(3^+ + 4^+ = 7^+)$. A large number of divalent
metal phosphates have been isolated in the past few years and some of them are
identical to the zeolites. The three-dimensional framework phosphates have
been reviewed in detail by Cheetham *et al.* [28].

We now examine a few select examples of three-dimensional metal
phosphate structures. A novel iron phosphate, $[(C_4N_3H_{16})(C_4N_3H_{15})]$
$[Fe_5F_4(H_2PO_4)(HPO_4)_3(PO_4)_3]H_2O$, was prepared recently by employing hydrothermal methods in the presence of diethylenetriamine (DETA) [29]. The
three-dimensional structure can be considered as made from layers along

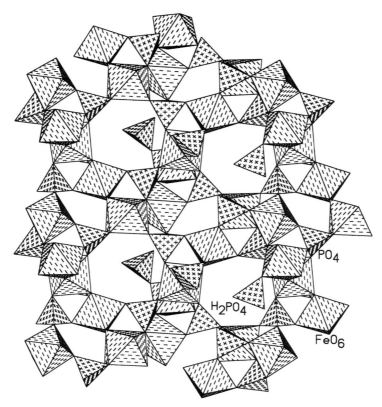

PO$_4$

H$_2$PO$_4$

FeO$_6$

Figure 7.13. The layer-type arrangement seen in an iron phosphate. The H$_2$PO$_4$ units project into the eight-membered pores within the layers (Choudhury *et al.* [29]).

the [*001*] direction, consisting of a network of FeO$_6$, FeO$_5$F, FeO$_4$F$_2$ octahedra and PO$_4$ tetrahedra (Fig. 7.13). The framework has the formula [Fe$_5$F$_4$(H$_2$PO$_4$)(HPO$_4$)$_3$(PO$_4$)$_3$]$^{5-}$. Charge neutrality is achieved by the presence of the organic structure-directing amine (DETA) in its protonated form. There are two molecules of DETA in the unit cell, one that is triply protonated and the other that is doubly protonated. The layers are connected to each other via phosphate, completing the three-dimensional architecture (Fig. 7.14). This connectivity creates large elliptical voids bound by 24 T atoms (T = Fe, P) forming one-dimensional channels along the [*010*] direction, in which the DETA and water molecules reside. The width of the channels is 15.3 × 4.5 Å (longest and shortest atom–atom contact distances, not including van der Waals radii). To our knowledge, this is the first synthetic open-framework iron phosphate material with such large voids. The layers themselves contain pores bound by eight T atoms (T = Fe, P) and the H$_2$PO$_4$ groups protrude into this

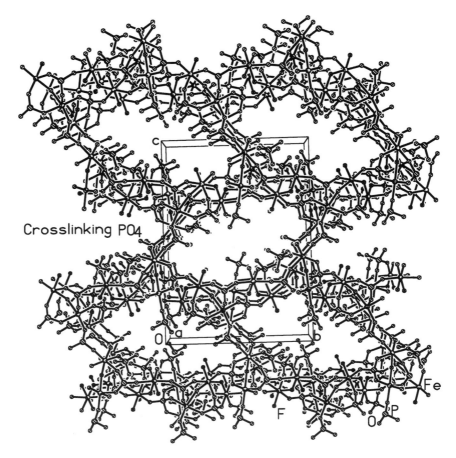

Figure 7.14. Figure showing the large elliptical tunnels. The layer-type arrangement (Fig. 7.13) gets crosslinked by phosphate tetrahedra (Choudhury *et al.* [29]).

opening rendering the pores inaccessible. Along the [*100*] direction, the structure has another narrow channel bound by 16 T atoms and the protruding H_2PO_4 and HPO_4 moieties occupy this channel.

Using the same structure-directing agent, DETA, two members of the zinc phosphate family have also been isolated [30, 31]. The zinc phosphate, $[NH_3(CH_2)_2NH_2(CH_2)_2NH_3][Zn_4(PO_4)_3(HPO_4)]$, *DETA-ZnPO-1*, was synthesized hydrothermally using diethylenetriamine (DETA) as the structure-directing agent. The structure is built from the networking of ZnO_4, PO_4 and HPO_4 tetrahedral units. The most interesting aspect of this Zn phosphate is that it crystallizes in a polar space group $P2_1$. The entire framework of *DETA-ZnPO-1*, can be considered to be built from the networking of three-, four-, six- and eight-membered rings. The three- and four-membered rings are connected together, edge-wise, forming one-dimensional helical columns along the *b* axis as shown in Fig. 7.15. The figure shows how these columns are interconnected via the HPO$_4$ group forming an eight-membered channel system along the *a* axis. This eight-membered channel along the *a* axis is connected to other eight-membered channels along the *b* axis, forming a helical interconnected one-dimensional channel system. The amine and water molecules are situated in these channels. Figure 7.16 shows the connectivity between the ZnO$_4$ and PO$_4$ moieties that creates the other eight-membered channel system along the *b* axis. Thus, *DETA-ZnPO-1* possesses an interpenetrating eight-membered channel system.

In *DETA-ZnPO-2*, DETA bonds directly to the Zn center and act as a ligand in addition to being the structure-directing agent [31]. Thus, [NH(CH$_2$)$_2$NH$_2$(CH$_2$)$_2$NH$_3$][Zn$_5$(PO$_4$)$_4$], (*DETA-ZnPO-2*), consists of a network of ZnO$_4$, PO$_4$ and ZnO$_3$N moieties, connected by Zn$-$O$-$P bonds. The framework has the formula [Zn$_5$(PO$_4$)$_4$]$^{2-}$ and charge neutrality is achieved by the incorporation of the doubly protonated amine molecule. The polyhedral connectivity between the ZnO$_4$ and PO$_4$ units leads to the formation of infinite Zn$-$O$-$Zn chains and three-membered rings via the three-coordinated oxygen atoms. The structure also possesses four-membered Zn$_2$P$_2$O$_4$ units. The three- and four-membered rings, in *DETA-ZnPO-2*, share the edges forming a one-dimensional chain (Fig. 7.17). The individual ribbons (columns of one-dimensional chains) are joined together by ZnO$_3$N tetrahedra giving rise to a 10-membered channel system along the *a* axis. Such a linkage between one-dimensional chains by the ZnO$_3$N moiety requires the amine molecule to be in the middle of the 10-membered channel (10.36 $\times$ 4.62 Å; oxygen to oxygen contact distance excluding the van der Waals radii).

7.3.4 Mechanistic pathways

From the above discussions it is clear that the amine has an important role in the formation of open-framework metal phosphates. Recent studies from this laboratory have shown that in the family of zinc phosphates, almost all the members in the hierarchy of open-framework structures can be synthesized using the same structure-directing agent, triethylenetetramine (TETA) [32]. In Fig. 7.18 we show all the structures obtained with TETA along with the relative concentrations of the amine and the phosphoric acid employed in the synthesis. We see that the one-dimensional (ladder) and the two-dimensional (layer) structures are obtained when the relative concentration of phosphoric acid is high with the ratio of concentrations of the acid and the amine in the range 3:1–4:1. This is in contrast to the three-dimensional structures where the ratio

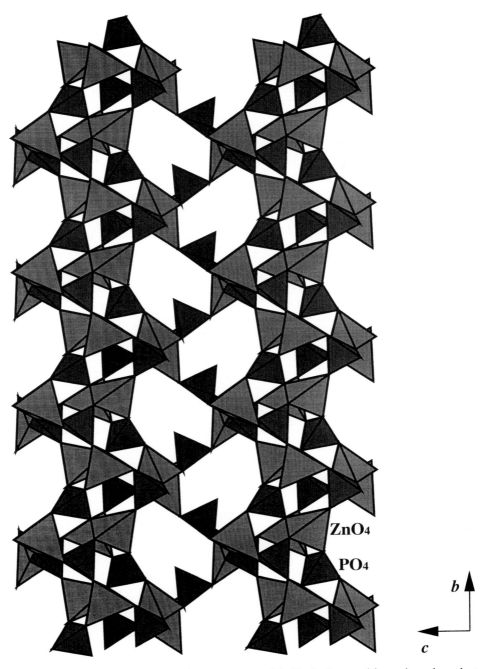

Figure 7.15. Figure showing the inter-connected helical observed in a zinc phosphate (Neeraj *et al.* [30]).

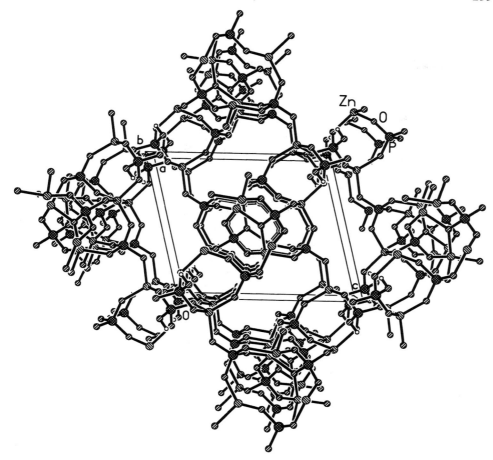

Figure 7.16. Top view of the helical channels of the zinc phosphate (Neeraj *et al.* [30]).

is in the range 2–1:1. The ligation of the amine to the metal occurs when this ratio is 1:2 or when the amine concentration is very high; low concentration of the metal ions with respect to the amine would also favor ligation. An examination of the available literature also supports these findings. Thus, chain and layer zinc phosphates are obtained with ethylenediamine (*en*) [33], 1,3-diamino-2-hydroxy propane (DAHP) [33], 1,3-diaminopropane (DAP) [19] and diethylenetriamine (DETA) [27], where the phosphoric acid to amine ratio is high (3–4:1); the ratio is close to 1:1 in the case of a three-dimensional structure. Metal ligation by the amine reported in zinc phosphates with diazabicyclo[2,2,2]octane (DABCO) [34] and DAP [35] occurs when the relative amine concentration is high (acid:amine ~ 1:3–4).

The structure of the open-framework phosphate obtained appears to

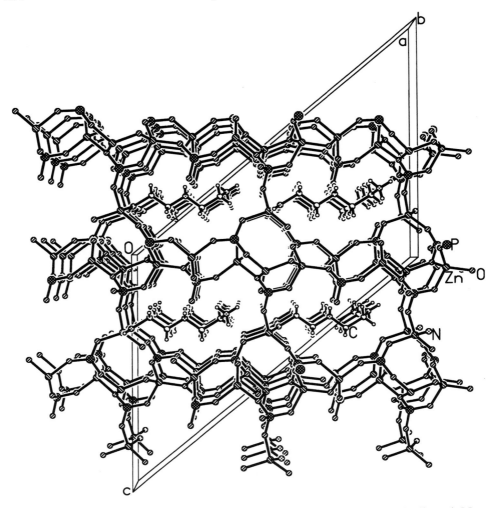

Figure 7.17. Structure of the zinc phosphate with the amine acting as the ligand. Note that the amine molecules are arranged such as to form one-dimensional 10-membered channels (Neeraj *et al.* [31]).

depend on the relative concentrations of the amine and phosphoric acid, high acid concentrations generally favoring low-dimensional structures. High amine concentration is necessary to obtain structures where the amine acts as a ligand. While these observations are supported by recent synthetic reports in the literature with *en*, DAP, DABCO and other amines, it would be important to carefully investigate this problem by determining the pK_a of the various amino groups in the polyamine employed and monitoring the pH values before, after and during the reaction.

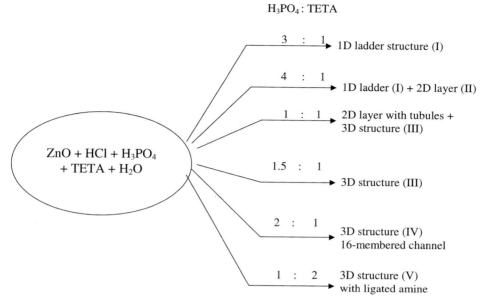

Figure 7.18. Various open-framework zinc phosphate structures formed from a single amine, TETA, by the variation of the synthetic conditions (Choudhury *et al.* [32]).

Based on the known aluminophosphates, Oliver *et al.* [36] proposed a mechanism of formation of the hierarchy of structures. According to these workers, the linear chain, the edge-shared ladder structures with essentially four-membered rings with $M_2P_2O_4$ composition (M = metal), layers with four-, six- and other higher membered apertures and finally the more complex three-dimensional structures constitute the hierarchy. An examination of Fig. 7.19 indicates that the basic unit in a supramolecularly organized framework solid is a four-membered ring. Pure four-membered ring structures (zero-dimensional monomers) have been isolated recently in a tin(II) [37] and zinc phosphates [38]. The isolated four-membered units are held in place by hydrogen bonding with the amine molecules. The four-membered unit, probably, constitutes the most basic building block of the open-framework metal phosphate and is likely to play a crucial role in poly-condensation into higher membered units with more complex structures. The aluminum phosphate structures are made of M_2P_2 rings connected via the oxygens, which are corner- or edge-shared. One-dimensional chains are formulated to transform under experimental conditions to allow the formation of other interconnected chains through hydrolysis, condensation and self-assembly pathways. The new chains can be crystallized from the system, or condense further to form more complex porous layer or open-framework phosphate based structures. However, this assumption is still

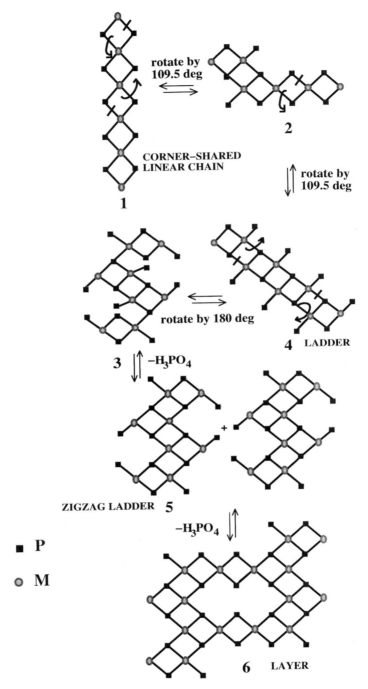

Figure 7.19. A schematic showing the formation of layer structures starting from a chain phosphate.

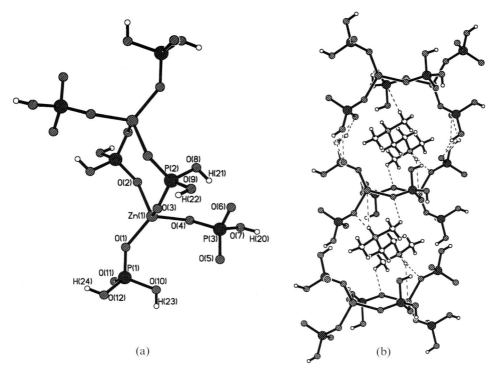

(a)   (b)

Figure 7.20. (a) Zinc phosphate monomer with the HPO$_4$ and H$_2$PO$_4$ units hanging from the Zn center. (b) Channel-like arrangement formed by hydrogen-bonded inter-actions between the monomer units. The amine molecules occupy the center of such channels (Neeraj *et al.* [38]).

under trial and more such structures are needed to arrive at a confirmation for the above model.

A four-membered ring monomer and its transformation to a layer structure

Recently, a zinc phosphate monomer of the composition [C$_6$N$_2$H$_{18}$] [Zn(HPO$_4$)(H$_2$PO$_4$)$_2$], was isolated by reacting ZnO, HCl, H$_3$PO$_4$ and water at room-temperature in the presence of NNN'N'-tetramethyl ethylenediamine [38]. The structure of the monomer consists of four-membered rings formed by ZnO$_4$ and PO$_2$(OH)$_2$ tetrahedra. The PO$_3$(OH) and PO$_2$(OH)$_2$ moieties hang from the Zn center as shown in Fig. 7.20a and are stabilized by extensive intra-molecular multi-point hydrogen bonding involving the phosphate units as well as the doubly protonated amine molecule, forming a sheet-like structure (Fig. 7.20b).

On heating in water at 50 °C for 2 days, the monomer transforms into a layered structure, [C$_6$N$_2$H$_{18}$][Zn$_3$(H$_2$O)$_4$(HPO$_4$)$_4$]. The structure of the layered zinc phosphate comprises a network of ZnO_4, $ZnO_2(H_2O)_4$ and $PO_3(OH)$ moieties, in which the vertices are shared. The connectivity between these units gives rise to macroanionic layers with bifurcated eight-membered apertures within each layer, as shown in Fig. 7.21a. One of the zinc atoms, Zn(2), links only with P(1), and the remaining Zn(2)−O linkages are terminal water molecules which point into the eight-membered aperture within the layers. As is typical of layered structures, the charge compensating cationic amine molecules are situated in between the anionic inorganic layers, as shown in Fig. 7.21b. The doubly protonated amine molecule interacts with the framework via multi-point hydrogen bonding and is partly responsible for the observation of such an architecture.

It is clear from above that the structures of the monomer and the layer are somewhat related. It is conceivable that a building unit of the type seen in the monomer can give rise to many framework structures, but to date only one structure with a layered architecture has been isolated. In Fig. 7.22, a schematic of the plausible pathway of the transformations of the monomer into other open-framework structures is presented. Thus, the monomer (shown as **1** in Fig. 7.22) can assemble and transform into a corner-shared linear chain, **3**, via encapsulation of a metal ion between the hanging phosphate groups of two monomeric units followed by condensation. The corner-shared chain can easily get converted into an edge-shared ladder, **5**, by acid hydrolysis of the Zn−O−P bond followed by a bond rotation, which can subsequently form the layer, **8**. The formation of the layer structure can be visualized from the monomer by the loss of a phosphate from **1** to form **9**, which after losing another phosphate can condense to form a layer structure, **11** (via **10**). The layer **11** further reacts with Zn^{2+} ions and adds on a $ZnO_2(H_2O)_4$ unit to give the layer architecture. Various steps can be written as follows:

$$2[Zn(HPO_4)(H_2PO_4)_2] \xrightarrow{-H_2PO_4} [Zn_2(HPO_4)_2(H_2PO_4)_3] \xrightarrow[+Zn]{-H_2PO_4} [Zn_3(HPO_4)_4]$$

Clearly, the monomer containing the four-membered ring emerges as the primary building block of open-framework metal phosphates. The formation of the layer structure may require free Zn^{2+} ions in solution, and it is possible that they are produced during the reaction by the dissociation of the monomer (**1**).

Role of amine phosphates as intermediates

It is becoming apparent that the variety of structures within a family of metal phosphates is on the rise. While the variety of structures in a given metal phosphate system, often obtained with the same amine, may arise because of the

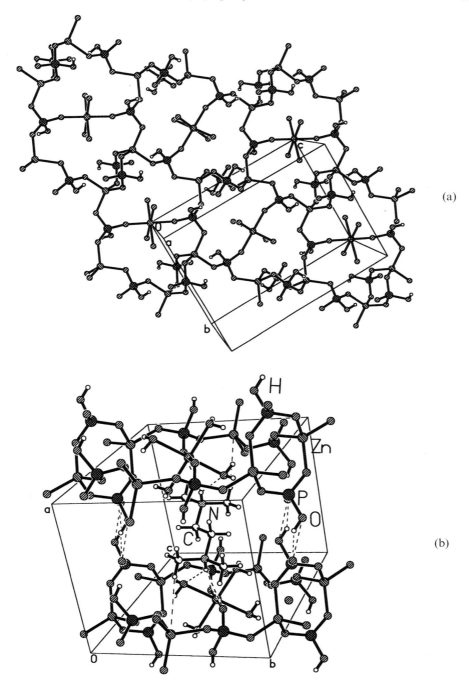

(a)

(b)

Figure 7.21. (a) Layered zinc phosphate obtained from the monomer. (b) The layer arrangement with the amine molecules in between the layers (Neeraj *et al.* [38]).

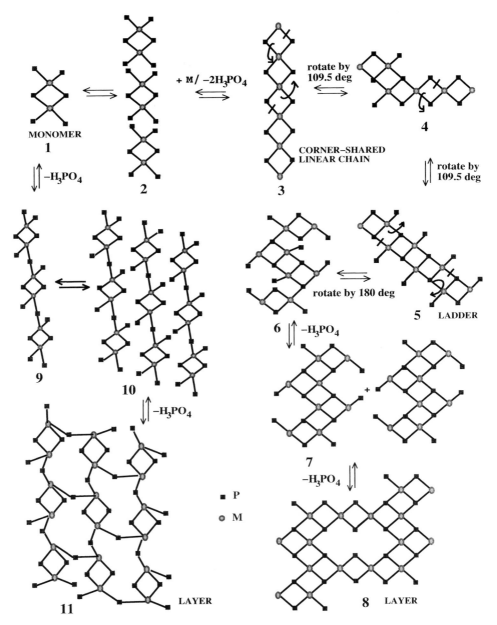

Figure 7.22. A plausible mechanism for the formation of the layer structure from the monomer. Note that the steps involve hydrolysis, condensation, etc.

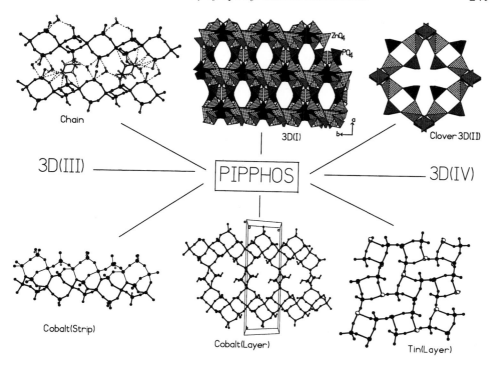

Figure 7.23. Various open-framework structures that are obtained by the reaction of a single amine phosphate, PIP (see text), with Zn^{II} ions (Rao *et al.* [18]).

small energy differences amongst them or owing to the kinetic control of the reactions, the exact role of the amines is not fully understood. During the hydrothermal synthesis of the metal phosphates, amine phosphates are often formed as additional products. It has been demonstrated recently that the reaction of amine phosphates with metal ions under hydrothermal conditions, in the absence of additional phosphoric acid, gives rise to open-framework metal phosphates [39]. Thus, a single amine phosphate – piperazine phosphate (PIP) – on reaction with Zn^{II} ions, gives rise to a plethora of open-framework metal phosphates depending upon the composition and other related reaction parameters [18], as shown in Fig. 7.23. PIP reacts with Co^{2+} ions and gives rise to two interesting phosphates that are the direct result of the condensation of the linear chains and ladders. The reaction of PIP with Sn^{2+} ions yields a simple layered structure formed by the fusion of two ladders. The formation of such a variety of open-framework metal phosphate architectures starting from one amine phosphate suggests the importance of the amine phosphate route in the synthesis of open-framework structures, and the possible role of the amine phosphates in the mechanism of formation of these compounds. What is especially interesting is that some of the amine phosphates react with metal ions under mild conditions (even at room temperatures) to yield open-framework structures. These findings from this laboratory indicate the seminal role of the amine phosphates in the formation of framework structures.

The close similarity between the structures of the amine phosphate and the final product of the reaction of the amine phosphate with metal (Zn^{2+}) is shown in Fig. 7.24. The amine phosphate consists of a hydrogen-bonded network with water molecules and resembles the loosely hydrogen-bonded structures involved in the synthesis of aluminosilicates. The mechanism of formation of the initial ladder phosphate can be understood in terms of the displacement of the water molecules from the amine phosphate by the Zn^{2+} ions.

It is to be noted that, in spite of the advances made, there is no control over the structure of the products obtained, because of the comparable energies of the various structures. The amine phosphate route, however, gives us the hope that rational synthesis of the open-framework metal phosphates may become possible in the not too distant future. If the amine phosphate plays a crucial role in the formation of open-framework metal phosphates, it should be possible to obtain a metal phosphate with a structure even simpler than the linear chain, such as the monomeric four-membered ring metal phosphate. It is noteworthy that the zinc phosphate monomer described earlier, comprising just a four-membered ring, has been obtained by the reaction of an amine phosphate with Zn^{2+} ions under mild conditions.

The process of building up

It is interesting to ponder the formation of complex open-framework structures from the primary or secondary building units. The mode of formation of open-framework structures from a linear chain involves corner-shared four-membered rings [36, 38, 40]. It appears, however, that the primary reaction involves the formation of the four-membered ring. The synthesis of a metal phosphate monomer comprising only a four-membered ring and its subsequent condensation into a layered open-framework structure therefore assumes importance in our understanding of the possible pathways involved in the formation of open architectures. It would be worthwhile to investigate the condensation and/or transformation of complex building units, such as the linear chain and ladder structures. The interaction between chains and ladders with monomeric four-membered rings would also be of interest. The transformation of the linear chain into the ladder or layer structures probably involves hydrolysis, rotation and condensation, and it is likely that once such SBUs (chains and ladders) are formed, the rest of the transformations are very facile (nearly spontaneous). *In situ* synchrotron X-ray diffraction studies, in the case

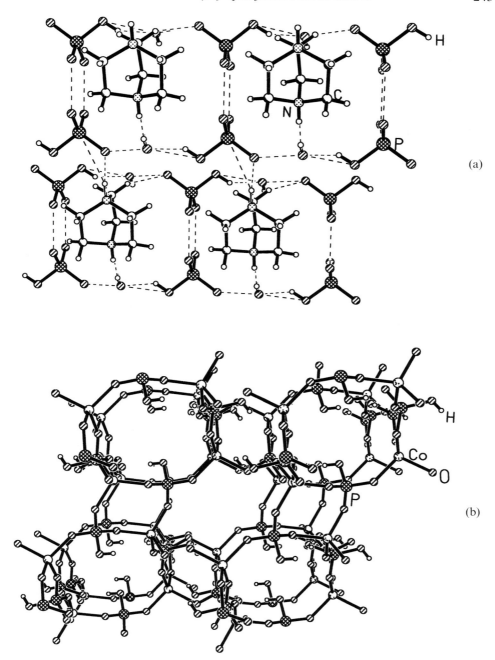

(a)

(b)

Figure 7.24. (a) The structure of an amine phosphate. (b) The structure of the final open-framework metal phosphate. Note that the structure of the metal phosphate can be generated simply by replacing the water molecules by the metal atoms.

of gallium phosphates, support the spontaneous nature of the transformation
of the preformed precursor units into the open-framework structure [41]. *In
situ* NMR studies appear to indicate that a four-membered ring phosphate is
first formed in the hydrothermal reaction, and then undergoes further transformation [42].

An examination of the literature shows the three-dimensional open-
framework structures to be most common; while several ladder structures are
known, there are only a few linear chain structures reported hitherto. The four-
membered ring monomeric phosphate is equally rare. Thus, the frequency of
occurrence of the different structures seems to be related to the nature of the
open-framework structure. Since the low-dimensional structures as well as the
four-membered ring are likely to be precursors of the complex three-
dimensional structures, it is conceivable that they are difficult to isolate, considering that further assemblage of these units probably occurs spontaneously. It
is also to be noted that the four-membered ring unit is isolated only when a
complex amine is employed, but the linear chain is formed with a regular
amine, such as piperazine. The isolation of the lower-dimensional structures
may in some way be related to the nature of the amine as well. There is reason
to believe a tri- or a tetra-amine might give rise to more open-framework structures than a simple mono- or di-amine.

7.4 Open-framework oxalates

Open-framework materials based on oxalates have been isolated in the presence of structure-directing amines. Although honeycomb architectures of
transition metal oxalates has been known for some time, it is only in the past
two years that open architectures of oxalates have been synthesized in the presence of amines. The first such materials synthesized are tin(II) oxalates [43, 44],
followed by oxalates of zinc [45]. Bi-dentate carboxylates of metals are,
however, known to form open-framework structures [46, 47].

In the tin(II) oxalate, $2[C(NH_2)_3][Sn_4(C_2O_4)_5]$, synthesized hydrothermally
using guanidine as the structure-directing agent [43], the structure consists of
macro-anionic sheets of formula $[Sn_2(C_2O_4)_{2.5}]^-$. The charge compensation is
achieved by the incorporation of protonated guanidine molecules. Of the two
distinct tin atoms, Sn(1) is four-coordinated (forming a distorted square-
pyramidal arrangement) and Sn(2) is six-coordinated (forming a pentagonal
bi-pyramidal arrangement). As can be seen from Fig. 7.25, the square-
pyramidal and the pseudo-pentagonal bi-pyramidal coordinations are
obtained taking the lone-pair of electrons into consideration. It is interesting
that, of the six oxygens coordinated to the Sn atoms in $2[C(NH_2)_3][Sn_4(C_2O_4)_5]$,

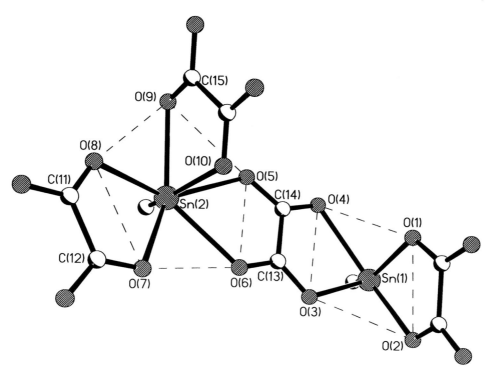

Figure 7.25. Different co-ordination environments for the Sn atoms in an open-frame-work tin(II) oxalate (Natarajan *et al.* [44]).

five oxygens lie in a plane forming a pentagon with the sixth oxygen along with the lone-pair of electrons forming two vertices, above and below the plane of the pentagon (Fig. 7.25). This resembles the classic 14-electron species observed in interhalogen compounds such as $[IF_6]^-$. The other Sn atom forms a square pyramidal arrangement with the lone-pair of electrons again forming the vertex. The Sn atoms are arranged such that lone-pairs of the neighboring Sn atoms point in opposite directions. The linking of the SnO_4 and SnO_6 units with the C_2O_4 moieties results in a unique saw-tooth layered network, as shown in Fig. 7.26. The sheets contain 20-membered corrugated rings (Fig. 7.27a), of width $\sim 8 \times 15$ Å, wherein two guanidinium cations and two water molecules reside (Fig. 7.27b). The guanidine molecules are present within the plane of the layer and are held together through hydrogen bonding interactions with the oxygens of the framework. This is further confirmed by the short O . . . H distances (1.972–2.277 Å) and the bond angles larger than 150° (the ideal angle for a linear/planar interaction is 180°). Similar interactions involving guanidinium cations have been observed in layered phosphates [48]. The presence of the amine molecules within the plane of the pores in the tin(II) oxalate also

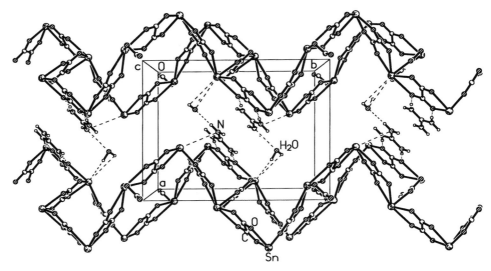

Figure 7.26. A tin(II) oxalate showing the saw-tooth arrangement between two adjacent layers. The amine and water molecules occupy the inter-lamellar space (Natarajan *et al.* [44]).

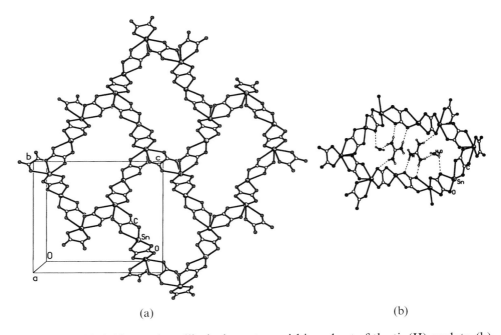

(a)   (b)

Figure 7.27. (a) A 20-member elliptical aperture within a sheet of the tin(II) oxalate. (b) A single 20-member elliptical aperture with the amine and water molecules. Two amine molecules are positioned within the 20-membered aperture (see text) (Natarajan *et al.* [44]).

suggests that the large apertures are a consequence of the occupation of the guest molecules within the apertures and are necessary to provide the structural stability. Removal of the guest molecules generally leads to the collapse of the framework structure.

A zinc oxalate, $2[C_3H_7NH_3][Zn_2(C_2O_4)_3]$, with a 20-membered channel and with an open-framework structure has been synthesized recently using propylamine as the structure-directing agent [45]. The structure of the zinc oxalate consists of a network of ZnO_6 octahedra and oxalate units; three oxalate units connect with each Zn atom to give rise to a three-dimensional connectivity. Of the three oxalate units, two connect via an *in-plane* linkage and the third one is cross-linked to the Zn atom in an *out-of-plane* manner, resulting in an interrupted honeycomb structure with a 20-member elliptical ring, as shown in Fig. 7.28. This type of connectivity between the Zn and oxalate units is unique and does not appear to have been encountered earlier. The elliptical pores in the zinc oxalate were formed by the linkage between 10 Zn and 10 oxalate units lying in the same plane and the other oxalate units connect the elliptical pores such that two such rings are perpendicular to each other. Although the largest pore opening in this zinc oxalate is a 20-membered one it appears smaller in projection. Thus, along the *b* axis, the structure has the appearance of a 12-membered square-channel system made from six Zn and six oxalate units ($\sim 8.5 \times 8.2$ Å; longest atom–atom contact distance not including van der Waals radii) (Fig. 7.29). The 20-member aperture has the dimensions $\sim 17.2 \times 6.9$ Å. Along the *c* axis it appears to have another 12-membered channel made from eight Zn and four oxalate units of 8.3×6.1 Å diameter (Fig. 7.30). The organic structure-directing agent, *n*-propylamine, which is highly disordered, occupies these channels along with water molecules. There is considerable hydrogen-bond interaction between the guest molecules and the host structure.

The two novel structures described above are closely related. In the layered Sn(II) oxalate, the 20-membered aperture results from linkages between four- and six-coordinated Sn(II) atoms and the oxalate units. There is three-dimensional connectivity in the zinc oxalate, and yet there are certain similarities between its structure and that of the Sn(II) oxalate. An examination of the connectivity patterns between the oxalates and M^{2+} ions (M = Zn or Sn) in both solids reveals that the zinc oxalate can be derived from the tin oxalate structure by the replacement of the four-coordinated Sn(II) atoms with a hexacoordinated Zn atom having two in-plane connectivities and one out-of-plane connectivity with the oxalate units as shown in Fig. 7.31. The out-of-plane connectivity is responsible for the three-dimensional nature of the structure in the zinc oxalate (Figs. 7.29 and 7.30).

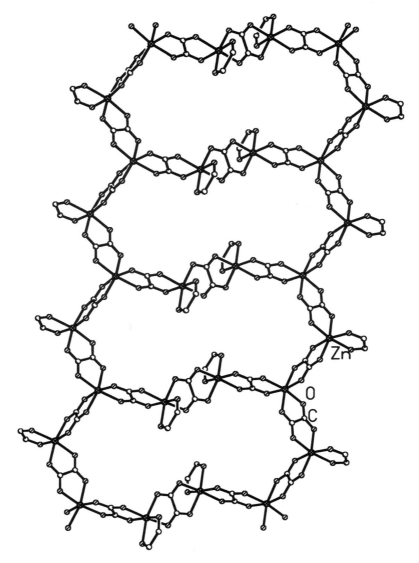

Figure 7.28. Zn oxalate showing 20-membered elliptical channels (Vaidhyanathan *et al.* [45]).

7.4.1 *Amine oxalates as intermediates*

Because the reaction of the amine phosphates with metal ions yields a variety of metal phosphates with different structures [18, 39], it is conceivable that the amine carboxylates also play a similar role in the formation of metal carboxylates with open architectures. Reports on the synthesis and characterization of

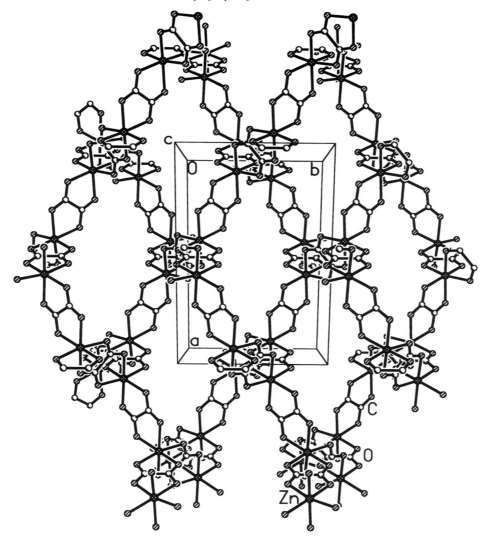

Figure 7.29. Zn oxalate showing the 12-membered one-dimensional channels (Vaidhyanathan *et al.* [45]).

amine oxalates are, however, very few. The reaction of amine oxalates and Zn^{II} leads to the formation of zinc oxalates with one-, two- and three-dimensional architectures, in addition to the simple monomeric and dimeric compounds. Furthermore, the various structures are mutually related. The amine oxalates were reacted with Zn^{II} ions under hydrothermal conditions, which resulted in the formation of a zinc oxalate monomer with guanidine, a dimer with piperazine, a chain with DABCO, a layered structure with honeycomb

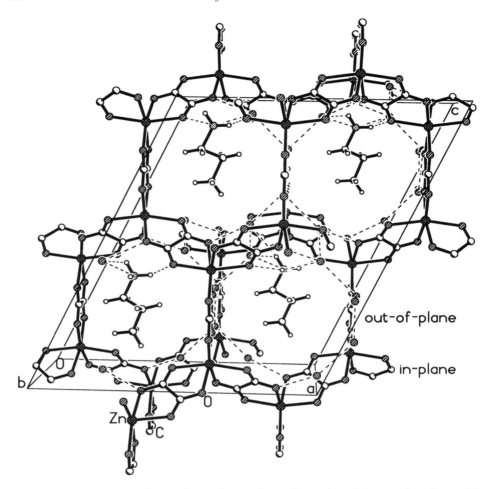

Figure 7.30. One-dimensional channels in a three-dimensional zinc oxalate formed by the linkages involving *in-plane* and *out-of-plane* linkages (Vaidhyanathan *et al.* [45]).

architecture with piperazine and a three-dimensional channel structure with propylamine [49].

The structure of the zinc oxalate monomer consists of two oxalate units directly linked to Zn atoms, which are also bonded to two water molecules. The monomeric zinc oxalate units are held by strong hydrogen bonds with the monoprotonated guanidine (Fig. 7.32a). In the zinc oxalate dimer, two Zn atoms are connected by one oxalate unit, and possess two terminal oxalates. The dimeric zinc oxalate is hydrogen bonded to the diprotonated amine, piperazine (Fig. 7.32b). In the chain zinc oxalate, the connectivity between Zn and the oxalate units is such as to form a one-dimensional chain with each Zn

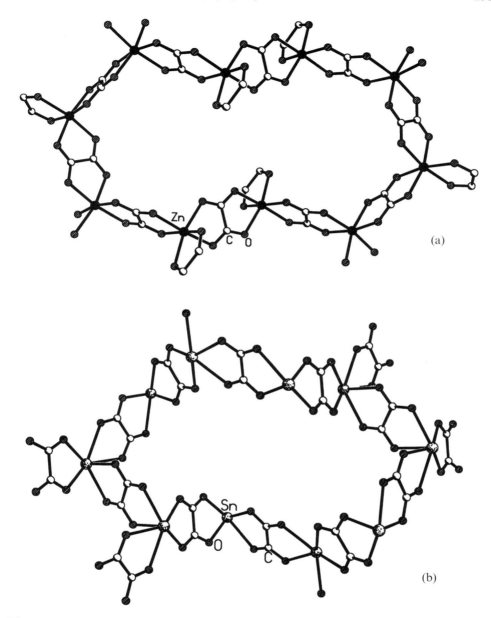

Figure 7.31. (a) A single 20-membered aperture in a Zn oxalate. (b) 20-membered aperture in the tin(II) oxalate. The four-coordinated Sn atom is replaced with a Zn atom with *in-plane* and *out-of-plane* connectivity giving rise to three-dimensional connectivity in the zinc oxalate (Vaidhyanathan *et al.* [45]).

S. Natarajan and C. N. R. Rao

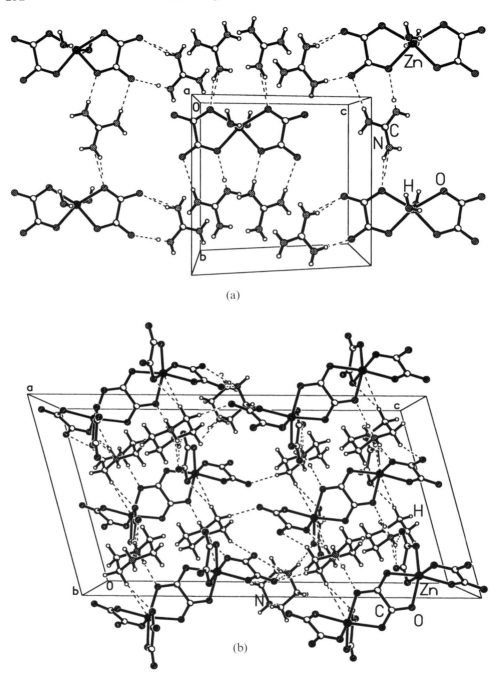

Figure 7.32. (a) Structure of the monomeric zinc oxalate. Dotted lines represent hydrogen-bond interactions. (b) Structure of the zinc oxalate dimer. The dimer and the amine alternate in a plane. Water molecules are omitted for clarity. Dotted lines are hydrogen-bond interactions (Vaidhyanathan *et al.* [49]).

possessing a terminal oxalate unit. The diprotonated DABCO is situated in between these chains and interacts with the terminal oxalate via hydrogen bonds (Fig. 7.33a). Zinc oxalate dihydrate, $[Zn(C_2O_4)(H_2O)_2]$, with a chain architecture has been known for some time [50], but the chain structure in Fig. 7.33a is the first example of a one-dimensional zinc oxalate synthesized with the structure-directing amine. The layer structure of the zinc oxalate involving the honeycomb motif has been observed in many of the divalent metal oxalates. The diprotonated piperazine sits in the middle of a 12-membered ring (six Zn and six oxalate units) as shown in Fig. 7.33b. The zinc oxalate with propylamine gives rise to a three-dimensional structure, consisting of a 20-membered aperture which on projection forms 12-membered square channels. The amine sits in the middle of the channels.

What is significant is that zinc oxalates with one-, two- and three-dimensional architectures, in addition to the monomeric and dimeric oxalates, could be synthesized by using amine oxalates. In Fig. 7.34, the various types of structure obtained are presented to demonstrate the similarities and relationships. As can be seen, the structure of the dimer can be derived from that of the monomer, the chain from the dimer, and the layer from the chain. Just as the four-membered ring monomeric phosphate unit plays a crucial role in the building of framework phosphates [18, 38], it is possible that the monomeric and dimeric oxalates are involved in the construction of the extended oxalate framework structures.

7.5 Open-framework phosphate-oxalates

A new class of mixed open-framework materials consisting of both the phosphate and the oxalate units has been synthesized in the presence of structure-directing agents [51–54]. The solids thus prepared usually contain sheets made on metal phosphates, which are pillared by the oxalate units. In addition to the interesting architectures exhibited by them, the phosphate-oxalates also show remarkable adsorptive and other properties. In the case of the transition metal phosphate-oxalates, interesting magnetic properties have also been observed. We will discuss some select cases of the phosphate-oxalates of iron in the following.

The iron phosphate, $[NH_3(CH_2)_2NH_3]_{1.5}[(Fe_3PO_4(HPO_4)_3(C_2O_4)_{1.5}].xH_2O$ ($x = 1.5$–2.0), has been shown to possess large unidimensional channels [51]. This material consists of a network of FeO_6, PO_4 and HPO_4 polyhedra forming a neutral inorganic layer of formula $Fe_3P_4O_{13}(OH)_3$ with the FeO_6 octahedra and PO_4 tetrahedra strictly alternating within the layer. The layers are porous, encompassing a circular 12-membered ring, as shown in Fig. 7.35a. The

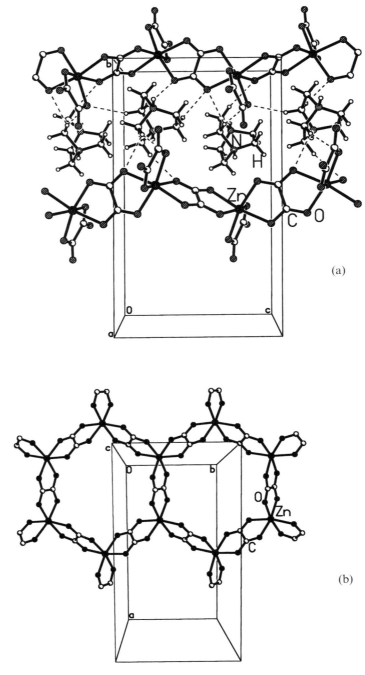

(a)

(b)

Figure 7.33. (a) Linear chain zinc oxalate. The oxalate chains are separated by the amine. (b) Layered zinc oxalate showing the honeycomb architecture. Amine and water molecules are not shown (Vaidhyanathan *et al.* [49]).

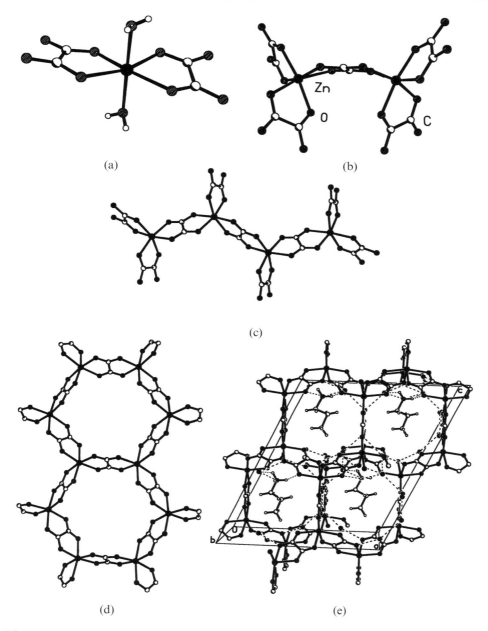

(a)

(b)

(c)

(d)

(e)

Figure 7.34. The hierarchy of zinc oxalate structures: (a) monomer, (b) dimer, (c) one-dimensional chain, (d) two-dimensional layer and (e) three-dimensional structure. Note the close relationships between them (Vaidhyanathan *et al.* [49]).

12-membered rings are surrounded by a series of four-membered rings, of which there are two distinct types. In one set of these rings, the phosphorus atoms are part of the wall of the 12-ring, whilst in the other, a phosphoryl group caps a six-membered ring and alternates above and below the plane of the 12-ring (P(1)), as shown in Fig. 7.35b. The layers are held in position by the oxalate units bonded to the iron centers. Thus, the oxalate moieties act as pillars holding the layers apart, and together.

The inorganic layer in the iron phosphate-oxalate is neutral and the negative charge required to neutralize the positive charge of the di-protonated amine is provided by the oxalate pillars. This is in contrast to the situation that generally obtains in open-framework layered phosphates where the layers are usually anionic [19–24]. Another remarkable feature of the structure is that within each iron phosphate sheet there is a supermesh of apertures of c. 5 Å free diameter (Fig. 7.35a). The oxalate and the organic amine units are positioned so that a unidimensional channel is created through the structure, in a direction perpendicular to the sheets. This results in a solid with unidimensional pores similar to those in aluminosilicate zeolites [55], certain detemplated aluminophosphates [4] and in the supramolecularly assembled organic channel structure formed between trithiocyanuric acid and 4,4′-bipyridyl [56]. The inorganic part of the framework in the iron phosphate-oxalate, $Fe_3P_4O_{13}(OH)_3$, is essentially identical to the macroanion $[Al_3P_4O_{16}]^{3-}$ present in many layered aluminophosphates [22, 23]. An alternative way to describe the topology of the two-dimensional net in this iron phosphate is by removing the capping PO_4 group (Fig. 7.35b). We then see a 4.6.12 net (*viz.* 82a in the nomenclature of Smith *et al.* [5]) commonly observed in aluminosilicate zeolites. The water molecules occupy the empty channels formed by the network and the amine molecules are positioned towards the edge of the 12-membered pore (Fig. 7.35b).

In addition to the iron phosphate-oxalates described above, where iron phosphate layers are pillared by oxalate units, it has become possible to obtain a truly hybrid iron phosphate-oxalate [54]. The new type of iron phosphate-oxalates possess oxalates, wherein the oxalate unit, besides acting as a bridge between the metal phosphate layers, is also part of the layer system. The presence of oxalates performing two types of function in these hybrid materials is noteworthy. Such a dual role of the oxalate unit is similar to the zinc oxalate structure described earlier. Thus, the oxalate units in iron phosphate-oxalates act like the phosphate unit in metal phosphates, and as the oxalate unit in metal oxalates. We discuss one special case where such a dual functionality of the oxalate units has been observed. The structure of the iron phosphate oxalate, $[C_3N_2H_{12}][(Fe_2(HPO_4)_2(C_2O_4)_{1.5}]_2$, consists of layers of formula $[(Fe_2(HPO_4)_2(C_2O_4)]$, linked by another oxalate unit completing the anionic

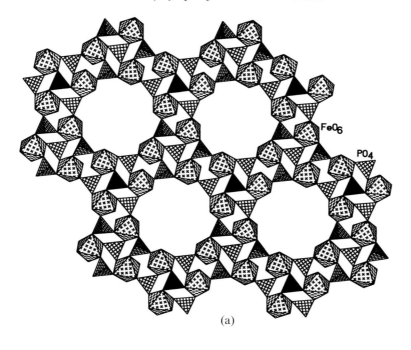

(a)

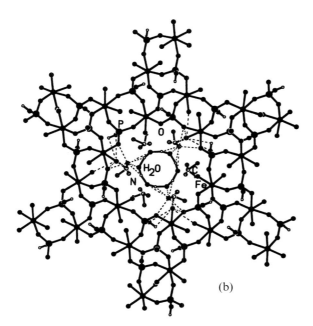

(b)

Figure 7.35. (a) A 12-membered pore opening in the iron phosphate-oxalate. (b) One 12-membered pore with the amine and water molecules (Choudhury *et al.* [51]).

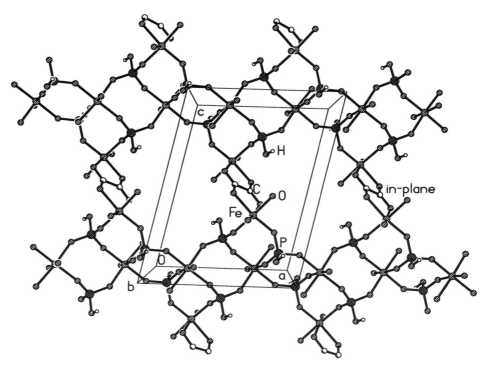

Figure 7.36. A single layer of the iron phosphate-oxalate. The layers are formed by the linkages involving the oxalate units forming truly hybrid layers (Choudhury *et al.* [54]).

framework [54]. Charge neutrality is achieved by the incorporation of organic amines in its di-protonated form. Thus, there are $0.5[C_3N_2H_{12}]^{2+}$ molecules per framework formula. The structure comprises a network of FeO_6, PO_4 and C_2O_4 moieties with each iron bound to six oxygens, which are, in turn, bound to carbon and phosphorus atoms completing the network. Of the two iron atoms, one is connected to two oxalate and phosphate units respectively and the other is bonded with one oxalate and four phosphate units. The FeO_6 and PO_4 units are connected to each other in such a way that they form zig-zag one-dimensional ladder-like chains which are linked via an oxalate unit forming an oxalate-phosphate layer as shown in Fig. 7.36. This is indeed a unique inorganic–organic hybrid layer network. The hybrid layers are linked via another oxalate unit acting like a bridge as shown in Fig. 7.37. Thus, two different types of oxalate unit occur in this material; one *in-plane* (with respect to the layer) and the other *out-of-plane*. To our knowledge, this is the first example of the existence of two different types of oxalate moiety in such a material. The linkages between the oxalates and the oxalate-phosphate layers create channels of

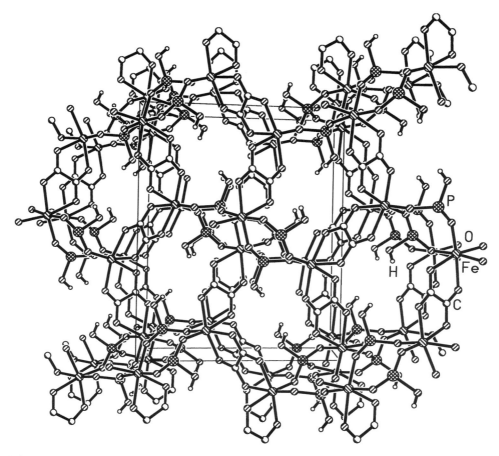

Figure 7.37. One-dimensional channels in the hybrid iron phosphate-oxalate (Choudhury *et al.* [54]).

width 5.2×4.5 Å (shortest atom–atom contact distance not including the van der Waals radii) along the *a* axis (Fig. 7.37).

As mentioned above, the iron phosphate-oxalate layers in this material are cross-linked by the (*out-of-plane*) oxalate units as in most of the phosphate-oxalates. It is interesting that similar dual functionality has also been observed in the zinc oxalate described earlier. The zinc oxalate also contains both the *in-plane* and *out-of-plane* oxalate linkages to create three-dimensional connectivity, and possessing channels [45]. In Fig. 7.38, we show the structure of this material to illustrate the presence of the oxalates within the layers as well as a bridge between the layers. This dual functionality of the oxalate units, in the Zn oxalate, gives rise to an elliptical aperture made by the linkages between 10 Zn and 10 oxalate units within the same plane, with the other oxalate unit

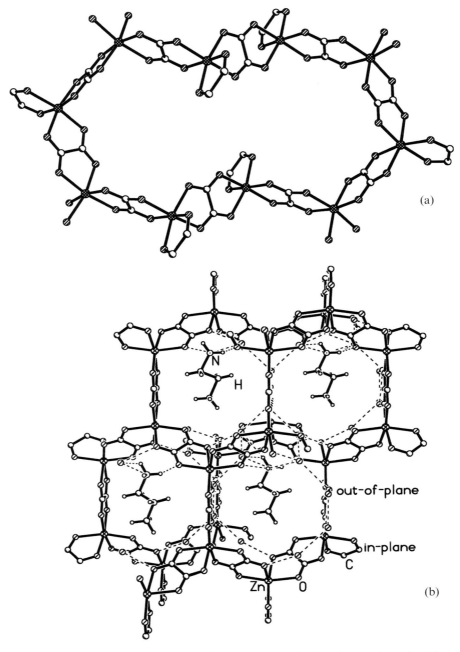

Figure 7.38. (a) A single 20-membered aperture seen in the zinc oxalate. (b) Channels in the three-dimensional zinc oxalate. Note that the *out-of-plane* connectivity is responsible for the three-dimensional nature of the structure (Choudhury *et al.* [54]).

connecting the elliptical pores such that two such rings are perpendicular to each other (Fig. 7.38b).

7.6 Concluding remarks

Research on inorganic open-framework structures, in particular those of metal phosphates, over the past decade has enabled the discovery of many novel materials possessing channels and other features of potential technological application. It is significant that one is in a somewhat better position today to design these structures based on the understanding of the processes involved in their formation. We are, however, far from being able rationally to design materials with the desired dimensionality or porosity. There is still much to be done to unravel fully the role of the amine and the process(es) involved in the assembly of complex three-dimensional structures from simpler units. One feature that has become apparent is that the formation of open-framework structures is not slow and step-wise throughout, but is likely to involve the spontaneous assembly of preformed units. The formation of the precursor building units of the desired structure (size and shape) therefore assumes significance. Careful studies published recently have shown that both zero-dimensional (monomeric) and one-dimensional (ladder) zinc phosphates transform to higher dimensional layer and channel structures under simple reaction conditions [57]. These processes are yet to be investigated. *In situ* X-ray diffraction and NMR studies will certainly be of great use in this regard. There is still considerable scope to explore newer structures possessing novel properties, such as ferromagnetic channels. It would be of great value if one can find ways to remove the amines present in the channels or between the layers in the open-framework phosphates. Recent work on the open-framework metal oxalates has shown not only that their formation involves assemblage from smaller units but also the possibility of synthesizing other inorganic structures with useful properties and interesting structures.

7.7 References

1. Dyer, A. (1988). *An Introduction to Zeolite Molecular Sieves*. Wiley, Chichester.
2. Barrer, R. M. (1982). *Hydrothermal Chemistry of Zeolites*. Academic Press, London.
3. Zosztack, R. (1989). *Molecular Sieves, Principles of Synthesis and Identification*.Van Nostrand Reinhold Catalysis Series, NY.
4. Lok, B. M., Messina, C. A., Patton, R. L., Gajek, R. T., Cannan, T.R. and Flanigen, E. M. (1982). *J. Am. Chem. Soc.* **106**, 6092.
5. Smith, J. V. (1977). *Am. Miner.* **62**, 703; Smith, J. V. (1988). *Chem. Rev.* **88**, 149.
6. O'Keeffe, M. and Hyde, B. G. (1980). *Phil. Trans. R. Soc. Lond.* A **295**, 553.

7.   Gard, A. J. and Tait, J. M. (1971). *Am. Chem. Soc. Adv. Chem. Ser.* **101**, 230.

8.   Barrer, R. M. (1948). *J. Chem. Soc.* 127; Barrer, R. M. and Denny, P. J. (1961). *J. Chem. Soc.* 971.

9.   Milton, R. M. (1959). *US Patent*, 2882423.

10. Davis, M. E. and Lobo, R. F. (1992). *Chem. Mater.* **4**, 756.

11. Kokotailo, G. T., Lawton, S. L., Olson, D. H. and Meier, W. M. (1978). *Nature*, **272**, 437; Flanigen, E. M., Bennett, J. M., Crose, R. W., Cohen, J. P., Patton, R. L., Kirchner, R.M. and Smith, J. V. (1978). *Nature* **271**, 512; Olson, D. H., Kokotailo, G. T., Lawton, S. L. and Meier, W. M. (1981). *J. Phys. Chem.* **85**, 2238.

12. Thomas, J. M. (1990). *Phil. Trans. R. Soc. Lond.* A **333**, 173; Davis, M. E. (1997). *Chem. Eur. J.* **3**, 1745; Morris, R. E. and Weigel, S. J. (1997). *Chem. Soc. Rev.* **26**, 309; Francis, R. J. and O'Hare, D. (1998). *J. Chem. Soc., Dalton Trans.*, 3133.

13. Jones, R. H., Thomas, J. M., Xu, R., Huo, Q., Xu, Y., Cheetham, A. K. and Bieber, D. (1990). *J. Chem. Soc., Chem. Commun.*, 1170; Gao, Q., Chen, J., Li, S., Xu, R., Thomas, J. M., Light, M. and Hursthouse, M. B. (1996). *J. Solid State Chem.* **127**, 145.

14. Chippindale, A. M., Bond, A. D., Law, A. D. and Cowley, A. R. (1998). *J. Solid State Chem.* **136**, 227; Loiseau, T., Serpaggi, F. and Ferey, G. (1997). *J. Chem. Soc., Chem. Commun.*, 1093.

15. Williams, I. D., Yu, J., Gao, Q., Chen, J. and Xu, R. (1997). *J. Chem. Soc., Chem. Commun.*, 1273; Chippindale, A. M. and Turner, C. (1997). *J. Solid State Chem.* **128**, 318.

16. Zima, V. and Lii, K.-H. (1998). *J. Chem. Soc., Dalton Trans.*, 4109; Cavellec, M., Riou, D., Greneche, J.-M. and Ferey, G. (1997). *Inorg. Chem.* **36**, 2187; Cowley, A. R. and Chippindale, A. M. (1999). *J. Chem. Soc., Dalton Trans.*, 2147, and references therein.

17. Ayyappan, S., Bu, X., Cheetham, A. K., Natarajan, S. and Rao, C. N. R. (1998). *J. Chem. Soc., Chem. Commun.*, 2181.

18. Rao, C. N. R., Natarajan, S. and Neeraj, S. (2000). *J. Am. Chem. Soc.* **122**, 2810.

19. Harrison, W. T. A., Bircsak, Z., Hannooman, L. and Zhang, Z. (1998). *J. Solid State Chem.* **136**, 93; Harrison, W. T. A., Phillips, M. L. F., Clegg, W. and Teat, S. J. (1999). *J. Solid State Chem.* **148**, 433.

20. Bu, X., Feng, P. and Stucky, G. D. (1997). *J. Solid State Chem.* **131**, 387; DeBord, J. R. D., Haushalter, R. C. and Zubieta, J. (1996). *J. Solid State Chem.* **125**, 270; Kongshaug, K. O., Fjellvag, H. and Lillerud, K.P. (1999). *Chem. Mater.* **11**, 2872; Escobal, J., Pizarro, J. L., Mesa, J. L., Lezama, L., Olazcuaga, R., Arriortua, M. I. and Rojo, T. (2000). *Chem. Mater.* **12**, 376.

21. Cavellec, M., Riou, D. and Ferey, G. (1994). *J. Solid State Chem.* **112**, 441; Cavellec, M., Riou, D. and Ferey, G. (1995). *Acta Crystallogr.* **C51**, 2242; DeBord, J. R. D., Reiff, W. M., Haushalter, R. C. and Zubieta, J. (1996). *J. Solid State Chem.* **125**, 186; Riou-Cavellec, M., Greneche, J.-M., Riou, D. and Ferey, G. (1998). *Chem. Mater.* **10**, 2434.

22. Thomas, J. M., Jones, R. H., Xu, R., Chen, J., Chippindale, A. M., Natarajan, S. and Cheetham, A. K. (1992). *J. Chem. Soc., Chem. Commun.*, 929; Jones, R. H., Chippindale, A. M., Natarajan, S. and Thomas, J. M. (1994). *J. Chem. Soc., Chem. Commun.*, 565; Williams, I. D., Gao, Q., Chen, J., Ngai, L.-Y., Lin, Z. and Xu, R. (1996). *J. Chem. Soc., Chem. Commun.*, 1781; Yao, Y.-W.,

Natarajan, S., Chen, J.-S. and Pang, W.-Q. (1999). *J. Solid State Chem.* **146**, 458.

23. Jones, R. H., Thomas, J. M., Xu, R., Huo, Q., Cheetham, A. K. and Powell, A.V. (1991). *J. Chem. Soc., Chem. Commun.*, 1266; Yu, J., Li, J., Sugiyama, K., Togashi, N., Terasaki, O., Hiraga, K., Zhou, B., Qiu, S. and Xu, R. (1999). *Chem. Mater.* **11**, 1727; Li, J., Yu, J., Yan, W., Xu, Y., Xu, W., Qiu, S. and Xu, R. (1999). *Chem. Mater.* **11**, 2600, and references therein.

24. Gray, M. J., Jasper, J. D., Wilkinson, A. P. and Hanson, J. C. (1997). *Chem. Mater.* **9**, 976; Bruce, D. A., Wilkinson, A. P., White, M. G. and Bertrand, J. A. (1996). *J. Solid State Chem.* **125**, 228; Renaudin, J. and Fereu, G. (1995). *J. Solid State Chem.* **125**, 197; Yu, J. and Williams, I. D. (1998). *J. Solid State Chem.* **136**, 141; Leech, M. A., Cowley, A. R., Prout, K. and Chippindale, A. M. (1998). *Chem. Mater.* **10**, 451; Bircsak, Z. and Harrison, W. T. A. (1998). *Chem. Mater.* **10**, 3016; Cheng, S., Tzeng, J.-N. and Hsu, B.-Y. (1997). *Chem. Mater.* **9**, 1788; Gao, Q., Chen, J., Xu, R. and Yue, Y. (1997). *Chem. Mater.* **9**, 457.

25. Chen, J.-S., Pang, W.-Q. and Xu, R.-R. (1999). *Topics in Catalysis* **9**, 93, and references therein.

26. Vaidhyanathan, R. and Natarajan, S. (1999). *J. Mater. Chem.* **9**, 1807.

27. Neeraj, S., Natarajan, S. and Rao, C. N. R. (1999). *Chem. Mater.* **11**, 1390.

28. Cheetham, A. K., Ferey, G. and Loiseau, T. (1999). *Angew. Chem. Int. Ed.* **38**, 3268.

29. Choudhury, A., Natarajan, S. and Rao, C. N. R. (1999). *J. Chem. Soc., Chem. Commun.*, 1305.

30. Neeraj, S., Natarajan, S. and Rao, C. N. R. (1999). *J. Chem. Soc., Chem. Commun.*, 165.

31. Neeraj, S., Natarajan, S. and Rao, C. N. R. (1999). *New. J. Chem.* **23**, 303.

32. Choudhury, A., Natarajan, S. and Rao, C. N. R. (2000). *Inorg. Chem.* **39**, 4295.

33. Chidambaram, D., Neeraj, S., Natarajan, S. and Rao, C. N. R. (1999). *J. Solid State Chem.* **147**, 154.

34. Harrison, W. T. A., Nenoff, T. M., Eddy, M. M., Martin, T. M. and Stucky, G. D. (1992). *J. Mater. Chem.* **2**, 1127.

35. Vaidhyanathan, R., Natarajan, S. and Rao, C. N. R. (1999). *J. Mater. Chem.* **9**, 2789.

36. Oliver, S., Kuperman, A., Lough, A. and Ozin, G. A. (1996). *Chem. Mater.* **8**, 2391; Oliver, S., Kuperman, A. and Ozin, G. A. (1998). *Angew. Chem. Int. Ed.* **37**, 46, and references therein.

37. Ayyappan, S., Cheetham, A. K., Natarajan, S. and Rao, C. N. R. (1998). *J. Solid State Chem.* **139**, 207.

38. Neeraj, S., Natarajan, S. and Rao, C. N. R. (2000). *J. Solid State Chem.* **150**, 417.

39. Neeraj, S., Natarajan, S. and Rao, C. N. R. (1999). *Angew. Chem. Int. Ed.* **38**, 3480.

40. Ferey, G. (1998). *C.R. Acad. Sci. Paris, Ser. II*, 1.

41. Francis, R. J., O'Brien, S., Fogg, A. M., Halasayamani, P. S., O'Hare, D. Loiseau, T. and Ferey, G. (1999). *J. Am. Chem. Soc.* **121**, 1002.

42. Toullele, F., Personal communication.

43. Ayyappan, S., Cheetham, A. K., Natarajan, S. and Rao, C. N. R. (1998). *Chem. Mater.* **10**, 3746.

44. Natarajan, S., Vaidhyanathan, R., Rao, C. N. R., Ayyappan, S. and Cheetham, A. K. (1999). *Chem. Mater.* **11**, 1633.
45. Vaidhyanathan, R., Natarajan, S., Cheetham, A. K. and Rao, C.N.R. (1999). *Chem. Mater.* **11**, 3636.
46. Harrison, W. T. A. and Phillips, M. L. F. (1997). *Chem. Mater.* **9**, 1837.
47. Reineke, M., Eddaoudi, M., Fehr, M., Kelly, D. and Yaghi, O. M. (1999). *J. Am. Chem. Soc.* **121**, 1651, and references therein; Yaghi, O. M. and Li, H. (1995). *J. Am. Chem. Soc.* **117**, 10401; Yaghi, O. M. and Li, G. (1995). *Angew. Chem. Int. Ed.* **34**, 207; Li, H., Davis, C. E., Groy, T. L., Kelley, D. G. and Yaghi, O. M. (1998). *J. Am. Chem. Soc.* **120**, 2186; Chui, S. S.-Y., Lo, S. M.-F., Charmant, J. P. H., Orpen, A.G. and Williams, I. D. (1999). *Science* **283**, 1148.
48. Serpaggi, F. and Ferey, G. (1999). *Microporous and Mesoporous Mater.* **32**, 311; Livage, C., Egger, C., Nogues, M. and Ferey, G. (1998). *J. Mater Chem.* **8**, 2743; Livage, C., Egger, C. and Ferey, G. (1999). *Chem. Mater.* **11**, 1546; Riou-Cavellec, M., Sanselme, M. and Ferey, G. (2000). *J. Mater. Chem.* **10**, 745, and references therein.
49. Vaidhyanathan, R., Natarajan, S. and Rao, C.N.R. (2000). *J. Chem. Soc., Dalton Trans.*, 699.
50. Deyrieux, C., Bero, R. and Penelous, A. (1973). *Bull. Soc. Chim. Fr.* 25.
51. Choudhury, A., Natarajan, S. and Rao, C. N. R. (1999). *Chem. Mater.* **11**, 2316.
52. Choudhury, A. and Natarajan, S. (1999). *J. Mater. Chem.* **9**, 3113; Choudhury, A. and Natarajan, S. (2000). *Solid State Sciences* **2**, 89.
53. Choudhury, A., Natarajan, S. and Rao, C. N. R. (1999). *J. Solid State Chem.* **146**, 538.
54. Choudhury, A., Natarajan, S. and Rao, C. N. R. (2000). *Chem. Eur. J.* **6**, 1168.
55. *Atlas of Zeolite Structure Types* (W. M. Meier and D. H. Olson, Eds.) (1996). Butterworth-Heineman, London.
56. Pedireddi, V. R., Chatterjee, S., Ranganathan, A. and Rao, C. N. R. (1997). *J. Am. Chem. Soc.* **119**, 10867.
57. Rao, C. N. R., Natarajan, S., Choudhury, A., Neeraj, S. and Ayi, A. A. (2001). *Accts. Chem. Res.* **34**, 80; Ayi, A. A., Choudhury, A., Natarajan, S., Neeraj, S. and Rao, C. N. R. (2001) *J. Mater. Chem.* **11**, 1181; Choudhury, A., Neeraj, S., Natarajan, S. and Rao, C. N. R. (2001) *J. Mater. Chem.* **11**, 1537.

8

Mesoscale self-assembly of metal nanocrystals into ordered arrays and giant clusters

G. U. KULKARNI, P. JOHN THOMAS AND C. N. R. RAO

Nanocrystals of semiconductors as well as of metals covered by alkanethiols organize themselves in two-dimensional arrays. We discuss such arrays of metal nanocrystals at length, with our focus on the dependence of the structure and stability of the arrays on the particle diameter and the distance between the particles. Three-dimensional superstructures of metal nanocrystals obtained by the use of alkanedithiols are examined. These ordered two- and three-dimensional structures of thiolized metal nanocrystals are good examples of mesoscale self-assembly. The association of metal nanocrystals to give rise to giant clusters with magic nuclearity provides an even more graphic demonstration of mesoscale self-assembly.

8.1 Introduction

Mesoscale self-assembly of objects of nanometric dimensions is a topic attracting wide attention currently. It is becoming recognized that such self-assembly can occur through a variety of weak forces. Cooperative assemblies of ligated metal [1] and semiconductor nanocrystals [2], as well as of colloidal polymer spheres seem to occur through the mediation of electrostatic and capillary forces [3, 4]. The ability to engineer such assemblies extends the reach of current lithographic techniques and holds promise for a new generation of electronics of the nanoworld. In this context, synthesis and programmed assembly of metal nanocrystals assumes significance [1].

A metal nanocrystal is a tiny chunk of the bulk, measuring a few nanometers with a finite number of metal atoms in it. The forces that govern the nanocrystal structure, however, are different in many ways. Surface tension, for example, plays an important role [5] because, in a nanocrystal, a large fraction of atoms are present at the surface. Consequently, the structure of a metal nanocrystal may exhibit features unknown in the bulk such as a five-fold symmetry [6].

G. U. Kulkarni et al.

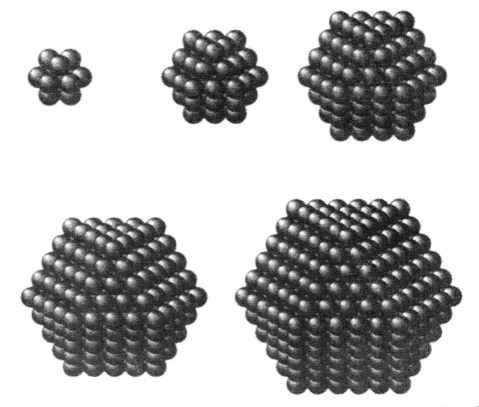

Figure 8.1. Metal nanocrystals in closed-shell configurations with magic number of atoms.

Nanocrystals with a specific number of atoms (nuclearity) are bequeathed special stability [7]. For nanocrystals of cubic close-packed metals, the numbers 13, 55, 147, 309 and 561 stand for magic nuclearities corresponding to the closure of 1, 2, 3, 4 and 5 shells respectively. A schematic illustration of magic nuclearity nanocrystals is shown in Fig. 8.1. In Fig. 8.2, we show scanning tunneling and transmission electron micrographs of polymer-protected Pd_{561} nanocrystals.

An added dimension to this area of research is the size-dependent properties of the metal and semiconductor nanocrystals owing to quantum confinement of the electronic states [8, 9]. Metal nanocrystals with volume less than 1 nm^3 have been shown to exhibit a definitive bandgap (a few tens of millielectron-volts) at room temperature [10]. The charging energies of polymer-coated nanocrystals of Au and Pd are in the range of a few hundreds of millielectron-volts following a scaling law with respect to the inverse of the particle diameter [11]. An assembly of nanocrystals, therefore, can exhibit novel properties as a result of the two-tier interaction. Two-dimensional arrays of nanocrystals of

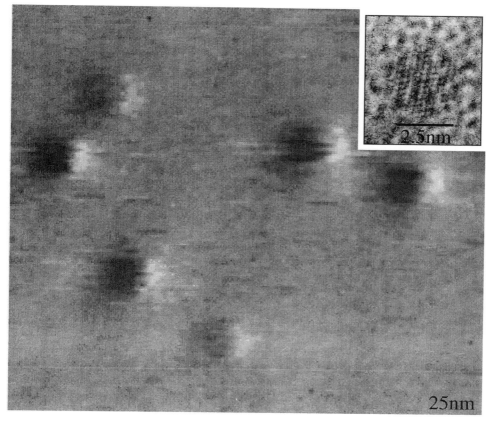

Figure 8.2. STM image of polymer coated Pd$_{561}$ nanocrystals. The nanocrystals are seen as fluffy balls against the plane background of the graphite substrate. The inset shows a high-resolution electron micrograph (HRTEM) of an individual nanocrystal. We see the characteristic 11 [111] fringes in the icosahedral shape measuring 2.5 nm. The diameter estimated from STM is ~ 3.4 nm, the difference being due to the ligand shell.

semiconductors (e.g. II–VI) and metals have been prepared successfully [1, 2, 12, 13]. In this article, we discuss the structure and stability of mesoscopic organizations of nanocrystals obtained by using alkanethiols and polyvinyl-pyrrolidone as surfactants. It is shown that nanocrystals covered with thiols form ordered arrays whereas those protected with the polymer self-aggregate into giant clusters (Fig. 8.3). Interestingly, the giant clusters exhibit magic nuclearity with respect to the number of nanocrystals.

8.2 Ordered assemblies

Surfactant molecules that self-assemble on metal surfaces have proved to be the best means of obtaining ordered arrays of nanocrystals [12]. The way in which

G. U. Kulkarni et al.

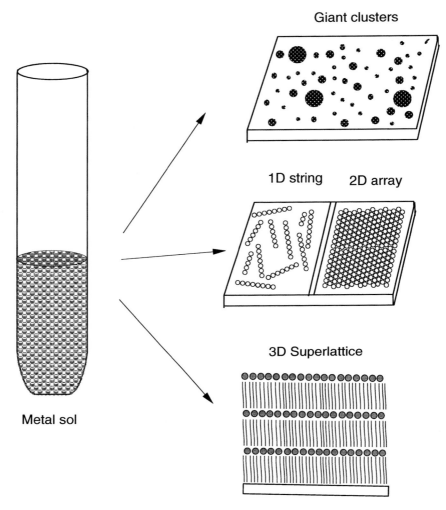

Figure 8.3. Schematic illustration of the various metal nanocrystal organizations

the nanocrystals organize themselves depends critically on the metal core diameter, the nature of the ligand, the substrate and even the dispersive medium used [14]. Thiolized nanocrystals readily arrange into two-dimensional arrays on removal of the solvent [1]. By using suitable methods, they can also be put into one-dimensional organization in the form of strings or assembled in a stepwise fashion in a three-dimensional superlattice (see Fig. 8.3). We shall first examine the nature of the two-dimensional assemblies of metal nanocrystals.

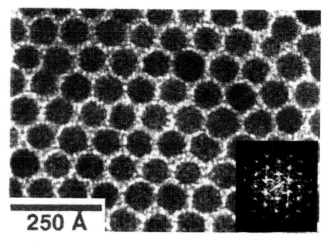

Figure 8.4. Transmission electron micrograph showing hexagonal close-packed Ag nanocrystals (diameter 7 nm) obtained by evaporating a chloroform dispersion on a carbon substrate. The average interparticle distance is 1.5 nm. The inset shows the two-dimensional power spectrum of the image (reproduced with permission from reference [18]).

8.2.1 *Two-dimensional arrays*

Gold organosols using alkane thiols as surfactants were first prepared by Schiffrin and co-workers [15] by phase transferring gold ions and carrying out reduction in the presence of the thiols. Several workers have adopted this procedure to obtain thiolized metal nanocrystals [16–20]. Whetten *et al.* [17] have centrifuged and separated out fractions containing particles of different mean sizes, to prepare two-dimensional arrays of size-selected Au nanocrystals. A two-dimensional array of dodecanethiol covered Ag nanocrystals obtained by Fitzmaurice and co-workers [18] is shown in Fig. 8.4. In another interesting experiment, Harfenist *et al.* [19] steered a mass-selected Ag cluster beam through a toluene solution of thiol and derivatized the vacuum-prepared particles. The Ag nanoparticles prepared this way were stable in air and formed extended two-dimensional arrays. Ordered two-dimensional lattices containing thiolized Au particles of two different sizes have been reported recently by Kiely *et al.* [20] (Fig. 8.5). Long-chain fatty acids have also been used for ligating metal nanocrystals. Colloidal dispersion of Co nanocrystals capped with fatty acids were found to self-assemble to yield hexagonally ordered arrays similar to those obtained with alkanethiols [21, 22]. Similarly, Ag nanocrystals capped with fatty acids of appropriate lengths yield cubic or hexagonal close-packed structures [23, 24].

A novel method of thiol-derivatizing hydrosols of metals has been recently developed by Sarathy *et al.* [25, 26]. The procedure involves mixing a hydrosol

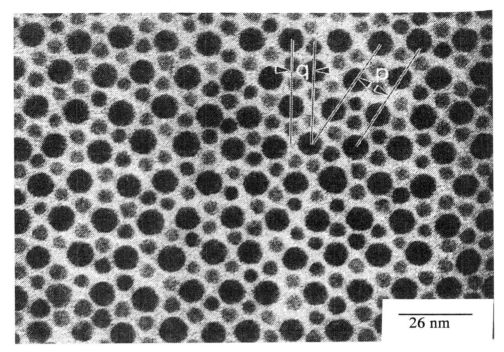

Figure 8.5. A bimodal hexagonal array of Au nanocrystals. The radius ratio of the nanocrystals is 0.58 (reproduced with permission from reference [20]).

containing metal particles of the desired size distribution with a toluene solution of an alkane thiol (butanethiol or higher members). The immiscible liquid layers thus obtained are stirred vigorously with HCl or $NaBH_4$ when the metal particles in the bottom aqueous layer gush to the upper hydrocarbon layer containing the thiol and become thiol-derivatized in this process. The completion of the derivatization is marked by a vivid interchange of the colors from the aqueous layer to the hydrocarbon layer, as demonstrated in Fig. 8.6. The advantage of this method is that well-characterized metal particles can be easily thiol-derivatized in a nonaqueous medium. A variety of hydrosols of Au, Ag and Pt have been thiolized by this procedure [25, 26].

Colloidal Au nanocrystals of different mean sizes obtained by reducing $HAuCl_4$ with partially hydrolized tetrakis-hydroxymethylphosphoniumchloride, were thiolized by the procedure described above and organized on a carbon substrate. A transmission electron micrograph of an array of Au nanocrystals with a mean diameter of 4.2 nm is shown in Fig. 8.7, as an example. The metal nanocrystals form a close-packed structure similar to the one shown in Fig. 8.4, extending over tens of nanometers with a regular spacing of ~ 1 nm between

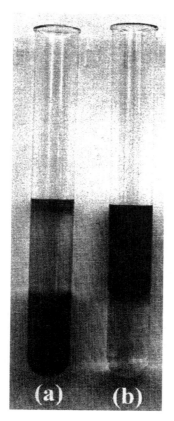

Figure 8.6. Thiol-derivatization of a hydrosol: immiscible layers of (a) a gold hydrosol (at the bottom) and a toluene solution containing dodecanethiol (on top); (b) thiol-derivatised Au sol in the toluene layer (on top) and the clean aqueous solution at the bottom (reproduced with permission from reference [26]).

them [25]. The X-ray diffraction pattern of this array exhibits a low-angle peak corresponding to a *d*-spacing of 5.0 nm (Fig. 8.7). The center to center distance between the nanocrystals deduced from the d-spacings is somewhat smaller than that expected from the dimensions of the metal cores and the thiol, suggesting some overlap of the alkane chains of the thiols on the neighboring particles. A TEM image of an array of thiol-derivatized hexagonal Pt nanocrystals (~8 nm diameter) is shown in Fig. 8.8a. These were first prepared by using an inverse micellar method, and thiol-derivatized without altering the shape or size. The hexagonal nanocrystals form close-packed structures resembling a honeycomb pattern. The wiggly pattern in the TEM image is from the second layer of the honeycomb structure sitting directly over the first layer in an AB-type packing. The STM image in Fig. 8.8b shows three layers of the assembly. We have successfully obtained two-dimensional arrays of magic nuclearity

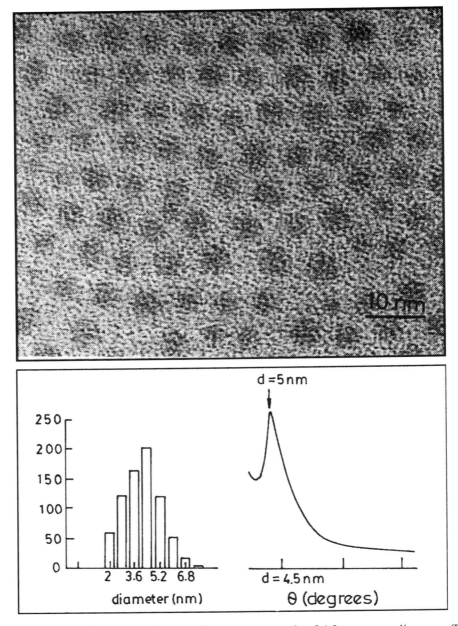

Figure 8.7. Two-dimensional arrays of Au nanocrystals of 4.2 nm mean diameter. The nanocrystals were obtained by thiol-derivatizing the hydrosol. A histogram indicating particle size distribution is given (below left). The XRD pattern from this array is also shown (below right) (reproduced with permission from reference [25]).

nanocrystals. We could prepare near mono-dispersed nanocrystals of Pd with diameters of 2.5 and 3.2 nm closely corresponding to nuclearities of 561 and 1415 respectively by employing the method described by Teranishi *et al.* [27]. The Pd_{561} nanocrystals were obtained by reducing H_2PdCl_4 with 20% ethanol–water mixture in the presence of PVP. The Pd_{1415} were obtained by changing the composition of ethanol–water to 25%. In Fig. 8.9, we show TEM micrographs obtained with such Pd_{561} and Pd_{1415} nanocrystals derivatized using octanethiol and dodecanethiol, respectively. Organized arrays of these nanocrystals extending over several microns are seen in the images. With the change of thiol from octane to dodecane, the packing of nanocrystals looks somewhat different (compare Figs. 8.9a and 8.9c with 8.9b and 8.9d, respectively). The self-assembly of the thiol molecules on the faceted nanocrystal surface perhaps determines the structure of the organization [18]. The Pd_{561} clusters coated with octanethiol (Fig. 8.9a) exhibit a pattern where a nanocrystal at the center of a hexagon is more prominent compared with the surrounding ones. The Pd_{1415} clusters coated with dodecanethiol (Fig. 8.9c) also organize in a similar fashion. These are essentially bilayered structures [28] where the nanocrystals in the top layer occupy two-fold saddle sites to form a hexagon that is rotated by 30° with respect to the hexagon below. On the other hand, the Pd_{561}–dodecanethiol (Fig. 8.9b) and Pd_{1415}-octanethiol (Fig. 8.9c) form regular hexagonal arrays. The novelty of this method is that nanocrystals of choice can be thiolized and organized. Schmidt *et al.* [29] have recently reported the formation of ordered two-dimensional monolayers of Au_{55} nanocrystals on a polymer film (Fig. 8.10).

The relative stabilities of the two-dimensional arrays of thiolized Pd nanocrystals has been investigated in some detail recently [30]. The effect of the thiol chain length on nanocrystal organization is illustrated in Fig. 8.11 for Pd nanocrystals with mean diameter of 4.5 nm. The butanethiol-coated nanocrystals form agglomerates, 20–40 nm in size, with little internal order (Fig. 8.11a). In many regions, the particles do not show distinct boundaries. The mean diameter estimated from the isolated particles (~ 4.6 nm) is similar to that of the pristine sol. Longer chain thiols, on the other hand, give rise to close-packed arrays (Figs. 8.11b–d) spread over micron-sized areas similar to those shown in Fig. 8.9. The contrasting behavior of butanethiol (Fig. 8.11a) is clearly as a result of its short chain length ($l = 0.8$ nm) and not related to the concentration of the sol, because similar conditions of preparation were used in all the cases. The role of the thiol chain length becomes more evident when we closely examine the organizations in Figs. 8.11b–d. The 4.5 nm nanocrystals covered with octanethiol form a honeycomb lattice (Fig. 8.11b), which is relatively less dense compared with the lattices in Fig. 8.9. We observe short-range order limited to

G. U. Kulkarni et al.

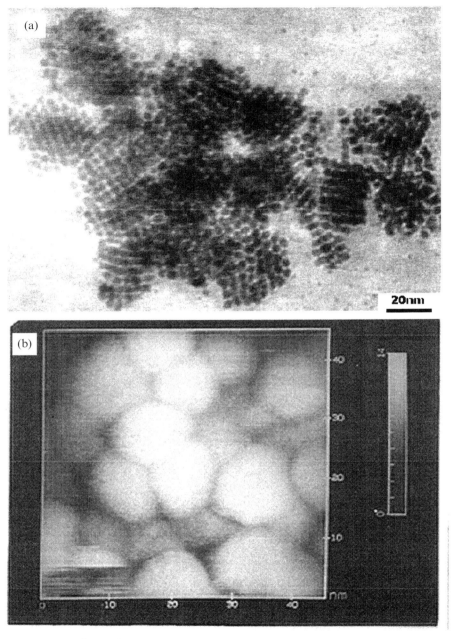

Figure 8.8. Two-dimensional array of Pt nanocrystals. (a) TEM image of thiol-derivatized Pt nanocrystals. Hexagonal particles form honeycomb-like arrays. (b) STM image showing the hexagonal stacking of the top three layers (reproduced with permission from reference [25]).

three or four hexagons of nanocrystals. The packing is facilitated in many places by a curved arrangement of the nanocrystals. On the other hand, the 4.5 nm sol derivatized with dodecanethiol and hexadecanethiol crystallizes in close-packed structures as shown in Figs. 8.11c and 8.11d, respectively. The nanocrystals are, however, not uniform in size. The mean sizes are similar to the pristine sol, but the distributions have considerably widened ($\sigma \sim 20\%$ and 18%, respectively). Unlike the near-magic nuclearity particles, the 4.5 nm particles exhibit wider size distribution in the hydrosol and are more susceptible to change in size and shape after thiolization. Despite the variations in shape and size, the nanocrystals organize into close-packed arrays as shown in Figs. 8.11c and 8.11d. These organizations are different from the opal structures observed by Ohara *et al.* [31] and resemble more closely the jammed arrangements of polydispersed spheroids [32].

An examination of the extent of thiol interdigation reveals some interesting aspects of the two-dimensional organizations [30]. Assuming a fully extended all-*trans* conformation of the thiol molecule inclined at 30° on the particle surface, the projected chain lengths (l) are ~ 0.8, 1.2, 1.7 and 2.1 nm for the butane-, octane-, dodecane- and hexadecanethiol respectively. The mean value for Pd_{561}-dodecanethiol, $<c>$, is ~ 4.1 nm (see inset in Fig. 8.9b) while the expected c value (diameter plus twice the chain length) is ~ 5.9 nm. A reduction in the c value of 1.8 nm amounts to 53% interdigation of the thiol molecule in this system. On the contrary, the octanethiol molecules on Pd_{1415} (Fig. 8.9c) exhibit 25% interdigation. The histograms shown in Figs. 8.9a and 8.9d refer to the second neighbors. The interdigation of the thiol molecules has been estimated to be 48% and 31%, respectively.

The results obtained above clearly demonstrate that the nature of nanocrystal organization depends on both the particle diameter, d, and the thiol chain length, l. Besides those mentioned above, nanocrystals of diameters 1.8, 4.5 and 6 nm coated with the various thiols were also studied. A stability diagram in terms of d and l (Fig. 8.12) was derived based on this study. In Fig. 8.12, the bright area in the middle is the most favorable d–l regime, corresponding to extended close-packed organizations of nanocrystals such as those illustrated in Figs. 8.9 and 8.11. The d/l values in this area are in the range 1.5–3.8. The dark shaded area in Fig. 8.12 includes the d/l regime giving rise to various short-range aggregations, formed when the particles are small and the chain length is large, or vice versa. As an example, we show a TEM image of Pd(6 nm)–octanethiol ($d/l \sim 5.0$) assembly in Fig. 8.13 that exhibits no long-range order. However, the nanocrystals are isolated from one another, unlike those in Fig. 8.11a. This organization resembles a collapsed monolayer of metal particles observed in the LB method on applying pressure [33].

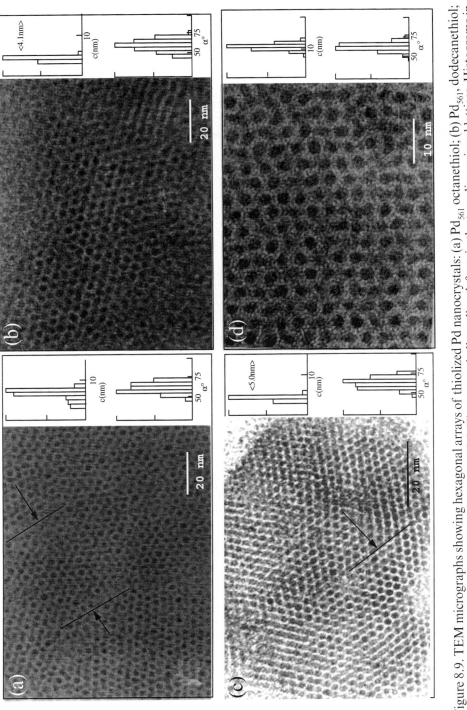

Figure 8.9. TEM micrographs showing hexagonal arrays of thiolized Pd nanocrystals: (a) Pd_{561} octanethiol; (b) Pd_{561} dodecanethiol; (c) Pd_{1415} octanethiol; and (d) Pd_{1415} dodecanethiol. The arrows indicate line defects in the two-dimensional lattices. Histograms in the insets show distributions (in percentage) in the nearest neighbor distance, c, and the angle subtended by a pair of nearest neighbors, α. Average values of c are given in the insets. The distribution in α is centered around 60° and in the case of dodecanethiol-assisted organizations, it is somewhat narrower (unpublished results of the authors). (Reproduced with permission from

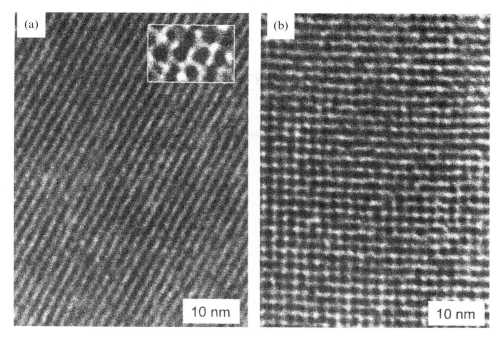

Figure 8.10. TEM images of Au_{55} monolayers showing a hexagonal (a) and a cubic (b) structure. The monolayers were prepared on a polyethyleneimine functionalized carbon grid. The magnified inset in (a) shows single clusters in the hexagonal form (reproduced with permission from reference [29]).

We have sought to understand the nature and stability of the two-dimensional arrays of thiolized Pd nanocrystals in terms of the particle diameter, d, and the chain length, l [30]. It has been suggested that the d/l ratio is a relevant parameter [33, 34]. As shown above, the penetration of the thiol shell is different depending on d and l, making the investigation more intricate. There is a general notion that ligated metal nanocrystals in close-packed arrangements are driven by entropic forces like those in hard sphere organizations. This notion has met with some success, for instance in explaining the bimodal ensembles of Au nanocrystals [20]. A simple hard sphere model, however, fails to account for the annular ring formation by thiolized metal nanocrystals [35]. Based on a study of the effect of the solvent polarity on the self-assembly of ligated nanocrystals, Korgel *et al.* [14, 18] proposed a soft sphere model taking the interparticle interaction into consideration. Accordingly, a ligated nanocrystal allows for penetration of the ligand shell up to its hard sphere limit. In this model, the total potential energy, E, is considered to be a result of two types of forces between the nanocrystals

$$E = E_{\text{steric}} + E_{\text{vdW}} \tag{8.1}$$

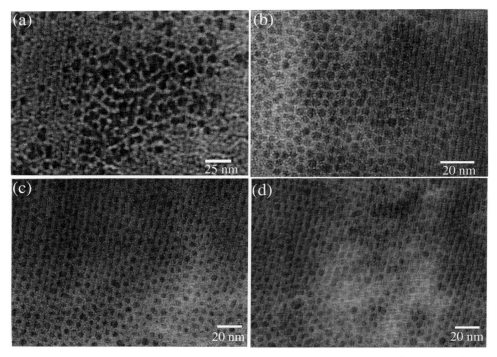

Figure 8.11. TEM micrographs showing Pd nanocrystals of mean diameter 4.5 nm organized into two-dimensional lattices by using different thiols: (a) butanethiol, (b) octanethiol, (c) dodecanethiol, and (d) hexadecanethiol. (Reproduced with permission from reference [30].)

The van der Waals interaction due to the polarization of the metal cores constitutes the attractive term and the steric interaction between the thiol molecules on the two surfaces forms the repulsive term

$$E_{\text{vdW}} = \frac{A}{12} \left\{ \frac{d^2}{\tau^2 - d^2} + \frac{d^2}{\tau^2} + 2 \ln\left(\frac{\tau^2 - d^2}{\tau^2} \right) \right\} \tag{8.2}$$

$$E_{\text{steric}} = \frac{50 d l^2}{(\tau - d)\pi\sigma_a^3} kT e^{-\pi(\tau - d)} \tag{8.3}$$

where τ is the interparticle distance. The Hamaker constant, A, for Pd nanocrystals in toluene has been estimated to be 1.95 eV [36]. The calculated diameter of the area occupied by the thiol molecule (σ_a) on the particle surface is $\sim$4.3 Å [18]. The total energy is attractive over a range of interparticle distances, the magnitude increasing as the distance falls. There could be a range of interparticle distances where the attractive energy from the van der Waals term exceeds the repulsive energy due to the steric factor, giving rise to net stabilization of the two-particle system. This is illustrated in Fig. 8.14 in the case of

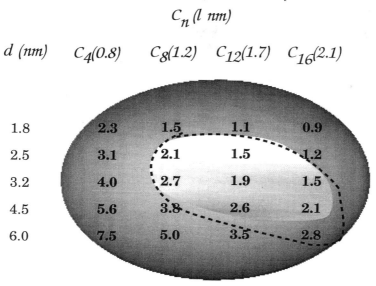

$$C_n (l \; nm)$$

d (nm)	$C_4(0.8)$	$C_8(1.2)$	$C_{12}(1.7)$	$C_{16}(2.1)$
1.8	2.3	1.5	1.1	0.9
2.5	3.1	2.1	1.5	1.2
3.2	4.0	2.7	1.9	1.5
4.5	5.6	3.8	2.6	2.1
6.0	7.5	5.0	3.5	2.8

Figure 8.12. The d–l phase diagram for Pd nanocrystals thiolized with different alkane-thiols. The mean diameter, d, was obtained from the TEM measurements on as-prepared sols. The length of the thiol, l, is estimated by assuming an all-*trans* conformation of the alkane chain. The thiol is indicated by the number of carbon atoms, C_n. The bright area in the middle encompasses systems that form close-packed organizations of nanocrystals. The surrounding darker area includes disordered or low-order arrangements of nanocrystals. The area enclosed by the dashed line is derived from calculations from the soft sphere model. (Reproduced with permission from reference [30].)

4.5 nm Pd particles. Stabilization energies of 17 and 2 meV were obtained from the calculation for particles coated with octanethiol and dodecanethiol respectively.

In Fig. 8.15 we depict the stabilization energies for d/l values corresponding to the experimentally investigated systems [30]. With shorter thiol chain lengths or larger metal cores ($d/l > 3$), we observe steep potential energy wells in the range of tens of millielectronvolts, possibly implying agglomeration of the particles. For $d/l < 1.5$, the two-particle system exhibits a shallow minimum with negligible stabilization, corresponding to a situation where long thiol chains shield the attractive interactions between the metal cores. An organization of this kind is influenced more by the directional property of the thiol chain resulting in lower order structures. For d/l values in the range ~ 2.0–3.0, the stabilization energies have moderate values (we show this regime by the dashed line in Fig. 8.12) and are comparable to the thermal energy of the nanocrystals at room temperature. This energy is perhaps just enough to bring the metal particles in the medium closer till the interaction turns repulsive. This is similar to

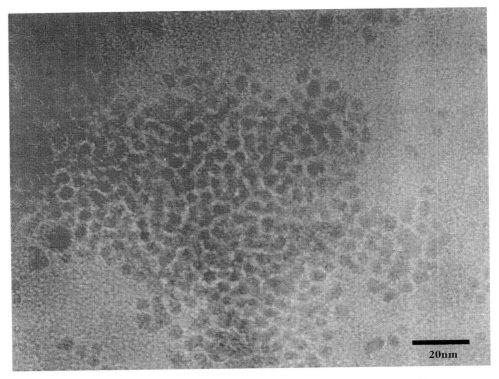

Figure 8.13. TEM micrograph showing disordered arrangement of Pd (6 nm) nano-crystals coated with octanethiol ($d/l \sim 5.0$). (Reproduced with permission from reference [30].)

the hard spheres, where the interaction energy is negligible till the spheres come in contact and the repulsion is then asymptotic [37]. Thiolized nanocrystals in the hard sphere limit are guided mainly by the entropy leading to the honeycomb organization.

In Fig. 8.16, we present the separation distances obtained in the hard sphere limit ($r_{E=0} = c_{E=0} - d$) for the different nanoparticles and thiols in the form of a grid against d/l. For a given thiol, the separation distance increases as the nanocrystal diameter decreases, more steeply when the thiol chain is longer. These results are compared with those from our experimental values discussed earlier. The experimental points generally lie close to the grid in Fig. 8.16, with some deviations on either side. It is to be noted that our calculations underestimate the distance in the case of octanethiol, the deviations being 44%, 13% and 25% for 2.5, 3.2 and 4.5 nm particles, respectively. Similarly, the values deviate by as much as 50% for the longer thiols. In other words, the thickness of the thiol shell, l', estimated from the experimental separation distances ($2l' = c - d$), is considerably smaller in many cases compared with the projected lengths (l).

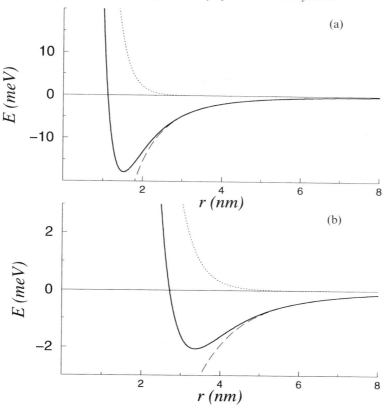

Figure 8.14. Variation of the two components and the total potential energy versus the separation distance between two nanocrystals of 4.5 nm diameter coated with (a) octanethiol and (b) dodecanethiol. (Reproduced with permission from reference [30].)

Accordingly, the ratio d/l' for the experimental crystalline lattices falls in the range 2.5–4.0 as compared with the d/l range of 1.5–3.8. Clearly, the extent of interdigation of thiol molecules plays a major role in attributing hardness to the ligated nanocrystal, which in turn decides the nature of the two-dimensional organization.

We now briefly examine organizations in systems where the dimensions of the metal core and of the thiol reach have extreme values, as for 1.8 nm particles coated with butanethiol (Fig. 8.17) [30]. The d/l value in this case is 2.3 and the stabilization energy is relatively high, (~ 10 meV, see Fig. 8.15). We see from Fig. 8.17 that the nanocrystals are engaged locally in small aggregates to form dimers, trimers or occasionally even tetramers. The aggregates mingle with one another to give rise to what could be efficient packing under the given constraints. Clearly, the butanethiol molecule, despite being attached to a small

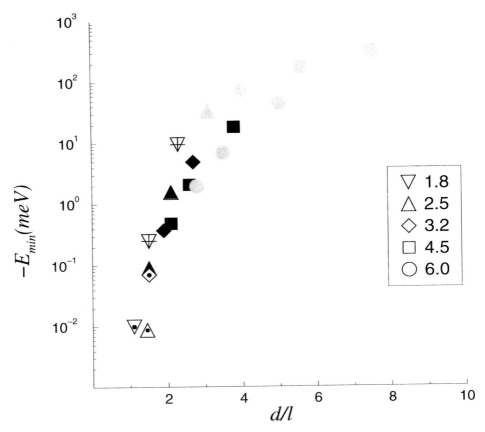

Figure 8.15. Variation of the stabilization energy against d/l for various nanocrystal–thiol systems. Symbols: filled, crystalline organization; plus signs, dimers and trimers; open, centered, one-dimensional string; grey, disordered organization. (Reproduced with permission from reference [30].)

nanocrystal (1.8 nm) fails to prevent agglomeration of the metal cores. In extreme cases such as this, the d/l criterion alone would not suffice for the formation of a nanocrystal array. Another interesting arrangement is seen in the case of the Pd_{1415} (3.2 nm) clusters covered by hexadecanethiol (d/l, 1.5) whose stabilization energy is very small (see Fig. 8.15). The metal nanocrystals are not sufficiently big to attract hexagons of partners but instead form a loose structure filled with one-dimensional strings, as shown in Fig. 8.18. Each string is identified with a series of nearly equally spaced ($c \sim 6.3$ nm, $\sigma^2 \sim 14\%$) nanocrystals, corresponding to 26% of interdigation of the hexadecanethiol molecules. The length of the strings varies between 15 and 65 nm (3 and 12 particles, respectively) with the mean at 35 nm (6 particles). A vector sum of the strings

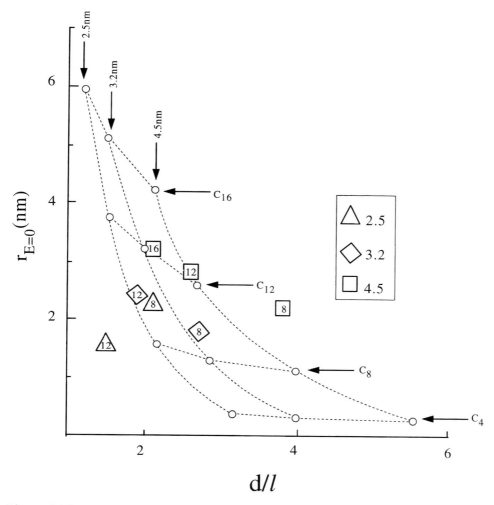

Figure 8.16. Separation distance at the zero crossover of the total potential ($r_{E=0}$) against d/l for various nanocrystal–thiol systems. The separation distances in close-packed organizations of nanocrystals obtained from TEM are also shown. (Reproduced with permission from reference [30].)

gives a small residue (6 nm), implying that the arrangement is indeed random. This type of arrangement is observed, although less prominently, in 1.8 and 2.5 nm particles covered with hexadecanethiol.

The fact that the interaction energy of nanocrystals can be continually varied by changing the interparticle distance was exploited by Heath and co-workers [38, 39]. They prepared a monolayer of Ag (3 nm) nanocrystals at the air–water interface in an LB trough and varied the interparticle distance by applying pressure. This study led to the observation of a reversible

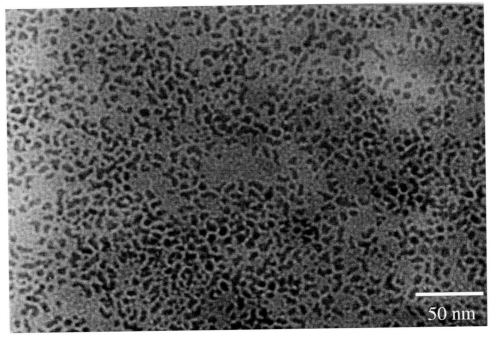

Figure 8.17. TEM micrograph showing aggregates of 1.8 nm particles covered with butanethiol. (Reproduced with permission from reference [30].)

Mott–Hubbard metal–insulator transition in the nanocrystal ensemble wherein the Coulomb gap closes at a critical distance between the particles.

8.2.2 One-dimensional arrangements

Hornayak and co-workers [40] used the ordered channels of porous alumina as templates to obtain linear arrangements of Au nanocrystals. By varying the pore size, the diameter of the nanowire could be controlled. A linear arrangement has also been obtained by coordinating Au particles (~ 1.4 nm) stabilized with phosphine ligands to single-stranded DNA oligonucleotide of the desired length and specific sequence [41]. Organization of particles in a one-dimensional lattice has met with limited success. Heath and co-workers [42] have fabricated wires of Ag nanocrystals by compressing a dispersion of Ag (4.5 nm) nanocrystals in toluene (Fig. 8.19). The wires were one nanocrystal thick, a few nanocrystals wide and extended in length from 20 to 300 nm. The interwire separation distance as well as the alignment of the wires could be controlled by compressing the film of nanocrystals.

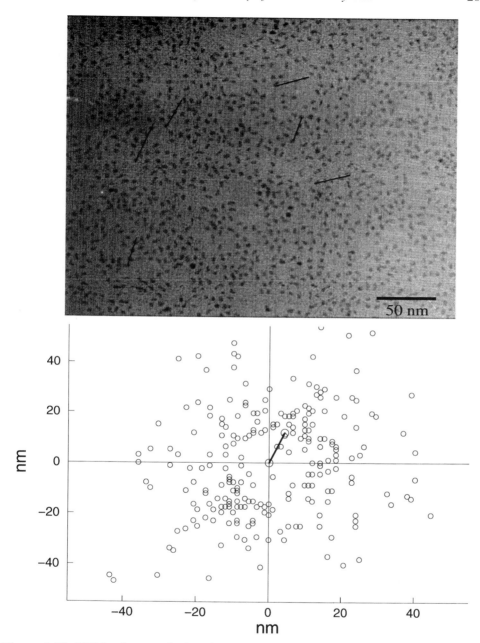

Figure 8.18. TEM micrograph showing one-dimensional strings of 3.2 nm particles held by hexadecanethiol. A few of the strings are shown by straight lines. The plot below shows the vectors depicting the length and direction of the strings with the center of the picture as the origin. The result obtained by adding the vectors is indicated as a thick line. (Reproduced with permission from reference [30].)

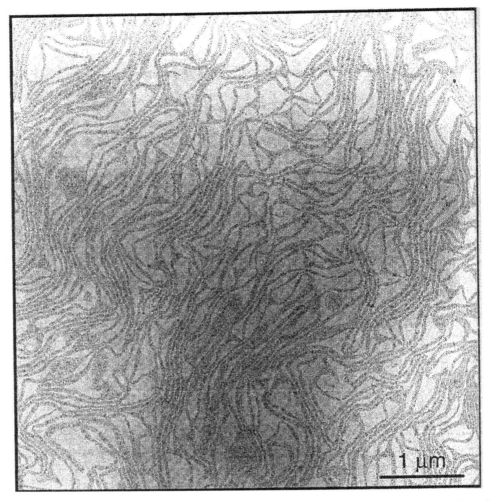

Figure 8.19. TEM image of a continuous stratum structure of a compressed LB film of Ag nanocrystals. One-dimensional strings of Ag nanocrystals are clearly seen (reproduced with permission from reference [42]).

8.2.3 Three-dimensional superlattices

Multilayer assemblies using monothiols are generally fragile. Multilayer deposition of nanocrystals is best achieved by the sequential adsorption of dithiol molecules and metal nanocrystals of the desired size, by dipping the substrate into the respective solutions with intermediate steps involving washing with toluene and drying. By using this procedure, several monometal, bimetal and metal-semiconductor superlattices have been prepared [43]. As many as five depositions of nanocrystalline arrays could be accomplished by this method.

(a)

(b)

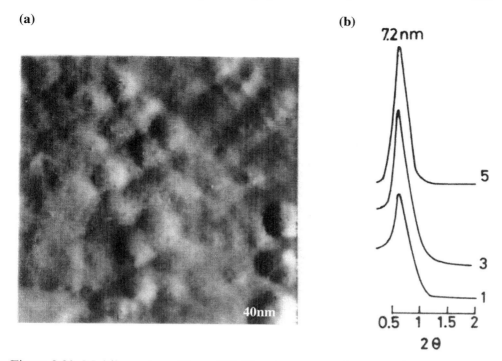

Figure 8.20. Multilayer deposition of Pt (5 nm) nanocrystals on a polycrystalline Au substrate. (a) STM image obtained after the second depostion. (b) X-ray diffraction pattern of the arrays after the first, third and fifth depositions (reproduced with permission from reference [43]).

After each deposition, the structure was characterized by STM, X-ray diffraction and by XPS. STM images showed the presence of regular arrays of nanoparticles extending over 300 nm, corresponding to the size of a typical flat terrace on the substrate (Fig. 8.20a). The images also revealed a nearly regular spacing of 2 nm between the nanoparticles. X-ray diffraction patterns recorded after successive depositions of layers (Fig. 8.20b) exhibited low-angle reflections with the *d*-spacings reflecting the particle diameter and the inter-particle distance.

Core-level XP spectra of superlattices of Pt nanocrystals ($\sim$ 5nm diameter) up to five layers deposited on an Au substrate are shown in Fig. 8.21. We see that the intensity of the Pt (4f) feature increases with the number of depositions, accompanied by a decrease in the Au (4f) intensity as the substrate becomes increasingly shadowed due to the limited escape depth of the photoelectrons. The intensities of the C (1s) and S (2p) levels of the dithiol (at 285.0 and 163.6 eV, respectively) also increase with the increasing number of depositions. A plot of the metal coverage versus the number of depositions gave a

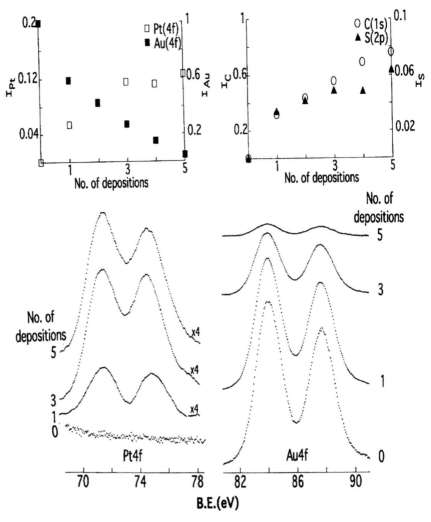

Figure 8.21. X-ray photoelectron spectra in the Pt(4f) and Au(4f) regions for the 5 nm Pt/Au system. The insets show variations in the Pt(4f) and Au(4f) intensities as well as C(1s) and S(2p) intensities with the number of depositions (reproduced with permission from reference [43]).

slope close to unity until the third deposition and increased thereafter suggesting that layer-by-layer deposition of nanoparticles had been accomplished [43]. Brust *et al.* [44] have reported the formation of multilayers of Au nanoparticles using dithiols. These workers have confirmed the layer-by-layer deposition of particle arrays by employing UV-vis spectroscopy and ellipsometry. Such assemblies can also be made with ionic spacer molecules [45, 46]. Three-dimensional superlattices involving nanocrystals of different metals (e.g. Pt,

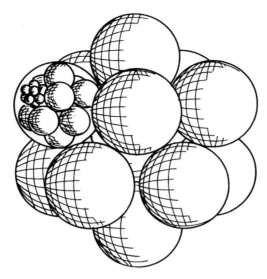

Figure 8.22. Self-similarity: schematic illustration of the formation of a cluster of metal nanocrystals (super cluster) and a cluster of superclusters. The size effect operating in nanocrystals could be invariant to scaling.

Au) and of metals and semi-conductors (e.g. Au, CdS) have also been prepared and characterized.

8.3 Giant clusters of magic nuclearity

Self-similarity in metal nanocrystal organization would manifest itself in the form of a giant cluster whose shape and size are direct consequences of the nanocrystals themselves [47]. The invariance of the shell effects in metal nano-crystals with scaling is shown schematically in Fig. 8.22. Thus, Pd_{561} nanocrys-tals would be expected to self-aggregate into a giant cluster of the type $(Pd_{561})_{561}$ under suitable conditions. Formation of such clusters was observed in the mass spectra of magic nuclearity Au_{55} nanocrystals. Secondary ion mass spectrometry indicated the presence of species with large m/z values and these were attributed to $(Au_{13})_{55}$ giant clusters [48]. The giant clusters so obtained have, however, not been isolated or imaged. We have made use of Pd nanocrys-tals with nuclearities close to 561 for this study, since the monodispersed nature of the nanocrystals is important in assisting the self-aggregation process. When the PVP-covered Pd_{561} nanocrystals were allowed to stand in an aqueous medium, the particles aggregated to form giant clusters [49]. In Fig. 8.23, we show a TEM image revealing the formation of the giant clusters. There are

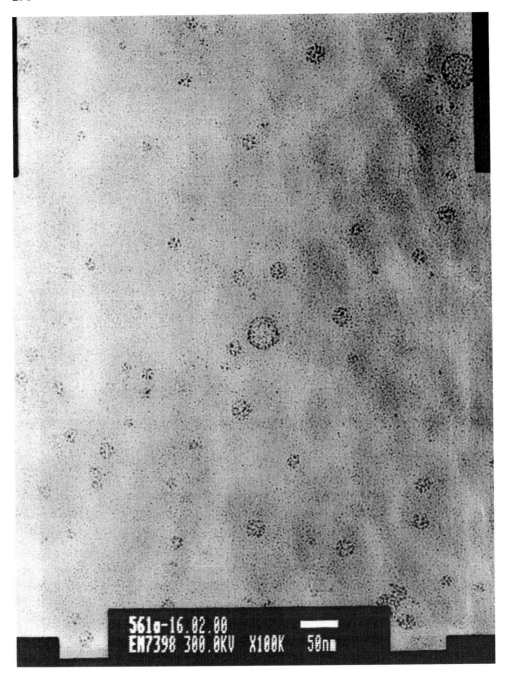

561a-16.02.00
EM7398 300.0KV X100K 50nm

Figure 8.23. TEM micrograph showing the giant clusters comprising Pd_{561} nanocrystals. The sample for TEM was prepared by the slow evaporation of a $PVP–Pd_{561}$ hydrosol. (Reproduced with permission from reference [49].)

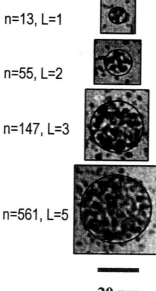

n=13, L=1

n=55, L=2

n=147, L=3

n=561, L=5

20 nm

Figure 8.24. Giant clusters enclosed in circles whose diameters correspond to magic numbers. The *n* and *L* values indicate the number of nanocrystals and closed shells respectively. (Reproduced with permission from reference [49].)

regions where the nanocrystals are densely packed in the form of aggregates. Importantly, the giant aggregates exhibit discrete sizes with diameters of 9.6, 15.6, 21.6 and 33.8 nm. In order to calculate the number of nanocrystals in such a giant cluster, we first estimated the effective volume of a nanocrystal by measuring the shortest distance between non-aggregated particles. The mean value of this distance is 4.1 nm ($\sigma < 10\%$). This value is somewhat higher than the diameter estimated from STM (see Fig. 8.2) by ~ 0.5 nm. It is amazing that the estimated volume ratios show strong preference to the magic numbers. Thus, the nuclearity of the 9.6 nm giant cluster is 13, corresponding to the closure of the first shell of nanocrystals. Similarly, the 15.6 nm giant cluster consists of 55 nanocrystals in two closed shells. Giant clusters of nanocrystals with nuclearities of 147 and 561 have also been observed. The giant clusters enclosed in the respective magic diameters are depicted in Fig. 8.24. We find excellent agreement between the experimental diameters and those calculated from the effective volume ratios. Images from the scanning electron microscope revealed the spherical shape of the giant clusters. We notice tiny spheres of ~ 10, 15 and 20 nm, corresponding to the magic nuclearity giant clusters. The spherical nature of the giant clusters was also confirmed by recording TEM images at various tilt angles. The individual Pd_{561} nanocrystals involved in the formation of giant clusters exhibit the characteristic lattice image, indicating thereby that the giant clusters truly consist of an assembly of distinguishable nanocrystals.

The giant clusters could be reproducibly formed starting from Pd_{561} nanocrystals in water, ethanol and ethanol–water mixtures and from sols with very different concentrations of the nanocrystals. It is possible that the formation of the giant clusters is facilitated by the polymer shell that encases them. Unlike Pd nanocrystals coated with alkanethiols, which self-assemble to form ordered arrays, the polymer shell effectively magnifies the facets of the metallic core, thereby aiding a giant assembly of the nanocrystals. The surface properties of the polymer-coated nanocrystals are clearly more favorable in that the interparticle interaction becomes sufficiently attractive.

8.4 Conclusions

Thiolized metal nanocrystals self-assemble to give rise to hexagonal arrays. The nature of the organization depends on the particle diameter, d, and the alkane chain length, l. Experimental observations on Pd nanocrystals relating the stability of the two-dimensional arrays to the d/l ratios find support from empirical calculations based on a soft sphere model. Nanocrystals with stabilization energies comparable to the thermal energy (a few millielectronvolts) give rise to close-packed hexagonal crystalline arrays whose d/l values are in the range 1.5–3.8. Large Pd nanocrystals ($d/l > 3$) with high stabilization energies (> 10 meV) form collapsed structures while those attached to longer chain thiols ($d/l \leq 1.5$) are associated with little stabilization, and exhibit low-order structures. The interparticle distances estimated from the soft sphere model deviate to some extent from the experimental values owing to the interdigation of thiol molecules chemisorbed on the curved nanocrystal surfaces. The extent of interdigation of thiol shells and the size distribution of particles influence the quality of the organization. A three-dimensional arrangement of the nanocrystals, on the other hand, has been carried out on a thiolized metal substrate by depositing layers of nanocrystals separated by dithiol spacers. Superlattices of Au and Pt nanocrystals as well as heterostructures involving CdS particles have been obtained in this fashion. Another example of the mesoscopic self-assembly of metal nanocrystals is the formation of giant clusters. PVP-protected Pd_{561} nanocrystals have been found to aggregate into giant clusters on removal of the solvent. The experimental observation that these clusters contain magic number of nanocrystals makes the study very interesting.

8.5 References

1. Rao, C. N. R., Kulkarni, G. U., Thomas, P. John and Edwards, P. P. (2000). *Chem. Soc. Rev.* **29**, 27.
2. Murray, C.B., Kagan, C. R. and Bawendi, M. G. (1995). *Science* **270**, 1335.
3. Terfort, A., Bowden, N. and Whitesides, G. M. (1997). *Nature* **386**, 162.
4. Mirkin, C. A., Letsinger, R. L., Mucic, R.C. and Storhoff, J. F. (1996). *Nature* **382**, 607.
5. Kirkland, A. I., Jefferson, D. E., Duff, D. G., Edwards, P. P., Gameson, I., Johnson, B. F. U. and Smith, D. J. (1993). *Proc. R. Soc. Lond. A* **440**, 589.
6. Bovin, J. O. and Malm, J. O. (1991). *Z. Phys. D* **19**, 293.
7. Martin, T. P., Bergmann, T., Göhlich, H. and Lange, T. (1991). *J. Phys. Chem.* **95**, 6421.
8. Edwards, P. P., Johnston, R. L. and Rao, C. N. R. (1999). In *Metal Clusters in Chemistry* (P. Braunstein, G. Oro and P. R. Raithby, Eds.). Wiley-Vch.
9. Alivisatos, A. P. (1996). *J. Phys. Chem.* **B100**, 13226.
10. Vinod, C. P., Kulkarni, G. U. and Rao, C. N. R. (1998). *Chem. Phys. Lett.* **289**, 329.
11. Thomas, P. John, Kulkarni, G. U. and Rao, C. N. R. (2000). *Chem. Phys. Lett.* (in press).
12. Collier, C. P., Vossmeyer, T. and Heath, J. R. (1998). *Annu. Rev. Phys. Chem.* **49**, 371.
13. Li, M., Schnablegger, H. and Mann, S. (1999). *Nature* **402**, 393.
14. Korgel, B. A. and Fitzmaurice, D. (1998). *Phys. Rev. Lett.* **80**, 3531.
15. Brust, M., Walker, M., Bethell, D., Schiffrin, J. D. and Whyman, R. (1994). *J. Chem. Soc., Chem. Commun.*, 801.
16. Sandhyarani, N., Reshmi, M. R., Unnikrishnan, R., Vidyasagar, K., Ma, S., Antony, M. P., Selvam, G. P., Visalakshi, V., Chandrakumar, N., Pandian, K., Tao, Y. T and Pradeep, T. (2000). *Chem. Mater.* **12**, 104.
17. Whetten, R. L., Khoury, J. T., Alvarez, M. M., Murthy, S., Vezmar, I., Wang, Z., Stephens, P. W., Clevend, Ch. L., Luedtke, W. D. and Landman, U. (1996). *Adv. Mater.* **8**, 428.
18. Korgel, B. A., Fullam, S., Connolly, S. and Fitzmaurice, D. (1998). *J. Phys. Chem.* **B102**, 8379.
19. Harfenist, S. A., Wang, Z. L., Whetten, R. L., Vezmar, I. and Alvarez, M. M. (1997). *Adv. Mater.* **9**, 817.
20. Kiely, C. J., Fink, J., Brust, M., Bethell, D. and Schiffrin, D. J. (1998). *Nature* **396**, 444.
21. Sun, S. and Murray, C. B. (1999). *J. Appl. Phys.* **85**, 4325.
22. Petit, C., Taleb, A. and Pileni, M. P. (1999). *J. Phys. Chem.* **B103**, 1805.
23. Pileni, M. P. (1998). *New. J. Chem.* 693.
24. Abe, K., Hanada, T., Yoshida, Y., Tanigaki, N., Takiguchi, H., Nagasawa, H., Nakamoto, M., Yamaguchi, T. and Yase, K. (1998). *Thin Solid Films* **327–329**, 524.
25. Sarathy, K. V., Raina, G., Yadav, R. T., Kulkarni, G. U. and Rao, C. N. R. (1997). *J. Phys. Chem.* **B101**, 9876.
26. Sarathy, K. V., Kulkarni, G. U. and Rao, C. N. R. (1997). *J. Chem. Soc., Chem. Commun.*, 537.
27. Teranishi,T. and Miyake, M. (1998). *Chem. Mater.* **10**, 54; Teranishi, T., Hori, H. and Miyake, M. (1997). *J. Phys. Chem.* **B101**, 5774.

G. U. Kulkarni et al.

28. Fink, J., Kiely, C. J., Bethell, D. and Schiffrin, D. J. (1998). *Chem. Mater.* **10**, 922.

29. Schmidt, G., Bäumle, M. and Beyer, N. (2000). *Angew. Chem. Intl. Ed. Engl.* **1**, 39.

30. Thomas, P. J., Kulkarni, G. U. and Rao, C. N. R. (2001). *J. Phys. Chem.* **B104**, 8138.

31. Ohara, P. C., Leff, D. V., Heath, J. R. and Gelwart, W. A. (1995). *Phys. Rev. Lett.* **75**, 3466.

32. Bernal, J. D. (1964). *Proc. R. Soc. Lond. A* **280**, 299.

33. Heath, J. R., Knobler, M. C. and Leff, D. V. (1997). *J. Phys. Chem.* **B101**, 189.

34. Whetten, R. L., Shafigullin, M. M., Khoury, J. T., Schaaf, T. G., Vezmar, I., Alvarez, M. M. and Wilkinson, A. (1999). *Acc. Chem. Res.* **32**, 397.

35. Ohara, P. C., Heath, J. R. and Gelbart, W. (1997). *Angew. Chem. Intl. Ed. Engl.* **36**, 1078.

36. Bargeman, D. and Vader, F. V. V. (1972). *J. Electroanal. Chem.* **37**, 45.

37. Israelachvili, J. N. (1992). *Intermolecular and Surface Forces*. Academic Press, London.

38. Markovich, G., Collier, C.P., Hendricks, S. E., Ramacle, F., Levine, R. D. and Heath, J. R. (1999). *Acc. Chem. Res.* **32**, 415.

39. Medeiros-Ribeiro, G., Ohlberg, D. A. A., Williams, R. S. and Heath, J. R. (1999). *Phys. Rev.* **B59**, 1633.

40. Hornayak, G. L., Kröll, M., Pugin, R., Sanitowski, T., Schmid, G., Bovin, J. O., Karrson, G., Hofmeister, H. and Hopfe, S. (1997). *Eur. J. Chem.* **3**, 195.

41. Alivisatos, A. P., Johnsson, K. P., Peng, X., Wilson, T. E., Loweth, C. J., Burchez, M. P. Jr. and Schultz, P. G. (1996). *Nature* **382**, 609.

42. Chung, S. W., Markovich, G. and Heath, J. R. (1998). *J. Phys. Chem.* **B102**, 6685.

43. Sarathy, K. V., Thomas, P. J., Kulkarni, G.U. and Rao, C. N. R. (1999). *J. Phys. Chem.* **B103**, 399.

44. Brust, M., Bethell, D., Kiely, C.J. and Schiffrin, D.J. (1998). *Langmuir* **14**, 5425.

45. Blonder, R., Sheeney, L. and Willner, I. (1998). *J. Chem. Soc., Chem. Commun.*, 1393.

46. Liu, Y., Wany, Y. and Claus, R. O. (1998). *Chem. Phys. Lett.* **298**, 315.

47. Fritsche, H. G., Muller, H. and Fehrensen, B. (1997). *Z. Phy. Chem.* **199**, 87.

48. Feld, H., Leute, A., Rading, D., Benninghoven, A. and Schmid, G. (1990). *J. Am. Chem. Soc.* **112**, 8166.

49. Thomas, P. J., Kullarni, G. U. and Rao, C. N. R. (2001) *J. Phys. Chem.* **B105**, 2515.

9

Layered double hydroxides as templates for the formation of supramolecular structures

STEVEN P. NEWMAN AND WILLIAM JONES

9.1 Introduction

A key feature of the chemistry of layered materials is the location of molecules (guests) of interest within the relatively constrained region of the interlayer. As such, the incorporation of guest molecules into layered hosts affords a versatile approach towards obtaining novel nanocomposite materials. Once intercalated, the guests can demonstrate interesting physical and chemical properties. Numerous types of layered solids have therefore been studied, including clay minerals, graphite, transition metal dichalcogenides, and metal phosphates and phosphonates [1–3]. By appropriate choice of the host and guest, novel hybrid materials can be created, exhibiting enhanced properties compared with the two individual components [4]. An ideal property of the host matrix would be the ability to control the number, orientation and location of guests that can be incorporated.

Layered double hydroxides (LDHs), also known as anionic clays and hydrotalcite-like materials, are layered solids that are of increasing interest [5–8]. They consist of stacks of positively charged hydroxide layers with interlayer, charge-balancing, anions and are available as naturally occurring minerals [9] and as synthetic materials. They were prepared in the laboratory in 1942 when Feitknecht reacted dilute aqueous metal salt solutions with base [10, 11], although the first detailed structural analyses of LDHs were not performed until the late 1960s by Allmann and Taylor and their co-workers [12–15].

9.1.1 Structure of LDHs

One important sub-group of the LDH family may be represented by the general formula:

$$[M^{2+}_{1-x} M^{3+}_{x} (OH)_2]^{x+} A^{n-}_{x/n} \cdot m\,H_2O,$$

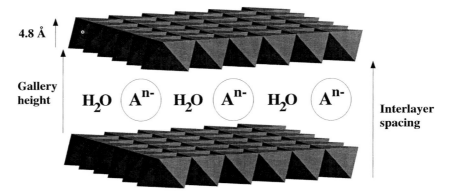

Figure 9.1. Schematic representation of the interlayer structure of a M^{2+}/M^{3+} LDH, based on the brucite structure. Interlayer anions (A^{n-}) and water are present.

where M^{2+} and M^{3+} are divalent and trivalent cations, respectively; x is equal to the ratio $M^{3+}/(M^{2+} + M^{3+})$ and A is an anion of valence n. The structure of this group may be described by considering the structure of brucite, $Mg(OH)_2$, which consists of charge-neutral layers of edge-sharing hydroxide octahedra, with Mg^{2+} cations occupying the octahedral vacancies. In an LDH, isomorphous substitution of a fraction of the Mg^{2+} ions with a trivalent cation, such as Al^{3+}, occurs and generates a positive charge on the layers which necessitates the presence of interlayer, charge balancing, anions. Water of crystallization also occupies the interlayer space of LDHs (Fig. 9.1).

A second sub-group of LDHs may be described on the basis of the gibbsite, $Al(OH)_3$, structure where the positive layer charge arises from the insertion of lithium cations into octahedral vacancies within the sheets of $Al(OH)_6$ octahedra. One third of the octahedral sites in gibbsite are vacant, which when filled with Li cations gives an LDH with the composition $[LiAl_2(OH)_6]^+ A^{n-}_{x/n} \cdot mH_2O$ [16]. This second group has been less studied.

The interlayer spacing of an LDH may be measured routinely by using X-ray diffraction techniques and is dependent on the size and disposition of the charge-balancing anion, as well as the interlayer water content. The gallery height of a LDH is determined by subtracting the hydroxide layer thickness of approximately 4.8 Å from the interlayer spacing. Whereas some naturally occurring LDHs have been studied by single crystal X-ray analysis, only polycrystalline samples, unsuitable for such analysis, have been obtained synthetically. The crystal chemistry of certain synthetic LDHs has been recently investigated, however, by powder X-ray diffraction (PXRD) methods, such as Rietveld structure refinement [17, 18]. LDHs generally crystallize with hexagonal symmetry, although different stacking arrangements of the hydroxide layers in the unit cell are observed, giving rise to polytypism [19, 20]. The naturally occurring LDH with the approximate composition $Mg_6Fe_2(OH)_{16}$ $(CO_3).4H_2O$, for example, may have either a two layer repeat (sjögrenite, the 2H polytype) or a three layer repeat (pyroaurite, the 3R polytype) [15]. The two forms can be distinguished only by X-ray analysis [19, 20].

9.1.2 Synthesis of LDHs

There are three general approaches to the preparation of LDHs: direct synthesis by coprecipitation; anion-exchange of a precursor LDH and rehydration of a calcined LDH precursor [5,6].

A common problem with all the methods is that in preparations of LDHs with anions other than carbonate it is important to avoid contact with carbonate (e.g. from atmospheric carbon dioxide), since this anion is readily incorporated and tenaciously held in the interlayer. Consequently, decarbonated and deionized water is often used for such preparations and exposure of the reacting material to the atmosphere is kept to a minimum.

Coprecipitation

In the coprecipitation method the M^{2+} (or mixtures of M^{2+} species) and M^{3+} (or mixtures) hydroxide layers are nucleated and grown from an aqueous solution containing the anion that is to be incorporated into the LDH. It has been used extensively for the one-pot direct synthesis of LDHs containing a variety of matrix cations and interlayer anions [21]. The anion that is to be introduced must have a high affinity for the hydroxide layers and be present in excess; otherwise the counter-anions of the metal salts may be incorporated. For this reason, metal nitrate and chloride salts are commonly utilized because of the low selectivity of LDHs toward these anions. In general, the coprecipitation is performed by adding an aqueous solution of the metal nitrate or chloride salts to an aqueous solution containing the desired interlayer anion. Usually, the coprecipitation is performed at constant pH through the simultaneous addition of base. A particular advantage of the coprecipitation method is that it allows for careful control of the charge density (M^{2+}/M^{3+} ratio) on the hydroxide layers of the product LDH. In order to ensure simultaneous precipitation of all the possible matrix cations present in the solution it is necessary to perform the synthesis at a pH at which both the M^{2+} and M^{3+} hydroxides precipitate and to control the pH during synthesis. To ensure that the M^{2+}/M^{3+} ratio in the starting solution is retained in the product LDH it is important to control the pH. Without pH control, the formation of impurity $M(OH)_2$ and/or $M(OH)_3$ phases, and consequently a product LDH with an undesired

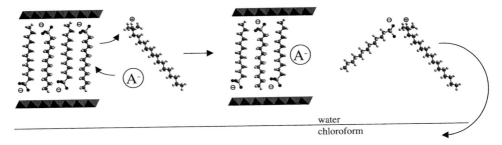

Figure 9.2. Schematic representation of the anion-exchange of dodecylsulfate via a phase separation mechanism. After Crepaldi *et al.* [26].

M^{2+}/M^{3+} ratio, is probable. Hydrothermal treatment is frequently used to increase yields and/or crystallinity of the product.

Anion-exchange

The anion-exchange properties of LDHs have been reported [22, 23]. Monovalent anions are easier to displace than divalent anions. LDHs containing nitrate anions are common precursors for anion-exchange syntheses owing to the relative ease with which the nitrate anion can be displaced from the interlayer. Organic-anion exchanged materials have also been used as precursors for the preparation of polyoxometalate derivatives [24]. In general, the anion-exchange reaction is performed by dispersing the precursor LDH in an aqueous solution containing an excess of the anion that is to be incorporated. The anion of interest must be stable at the pH of exchange and, if the M^{2+}/M^{3+} ratio within the LDH precursor is to be maintained during anion-exchange (i.e. no leaching of the cations occurs), the hydroxide layers must also be stable. This is especially the case when the source of the anion is the corresponding acid, e.g. HCl or HNO_3 [25]. An interesting anion-exchange method of inserting organic molecules in LDHs from a precursor ZnCr(dodecylsulfate) LDH has been reported recently [26]. The mechanism involved is based on the formation of a salt between dodecylsulfate (or an alternative anionic surfactant) and a cationic surfactant such as *N*-cetyl-*N,N,N*-trimethylammonium bromide (CTAB). The salt is then separated from the aqueous medium into an organic phase (dichloromethane or chloroform, for example), with concomitant incorporation of the anion of interest into the LDH (Figure 9.2). It was found that salt-formation and migration to the organic phase is rapid, which thus enhances the rate of anion-exchange. A variety of anions were successfully incorporated into the LDH using this procedure, including terephthalate and a sulfonated copper phthalocyanine.

Rehydration of a calcined LDH precursor

Miyata reported that a hydrotalcite-like, $MgAl(CO_3)$, LDH transforms into a mixed magnesium and aluminium oxide after heat-treatment at 500–800 °C that rehydrates and combines with anions to re-form an LDH in the presence of water and anion [27]. This reactivity of the mixed-oxide therefore provides an effective synthetic route for inserting inorganic and organic anions into LDHs. An advantage of this method is that the incorporation of competing counter anions into the LDH is prevented, although the inclusion of carbonate from atmospheric CO_2 remains a problem. The ease and extent of reconstruction of the heat-treated LDH are determined by the properties of the matrix cations and the temperature of the heat-treatment (i.e., the reactivity of the mixed-oxide). Sato *et al.* studied the rehydration in sodium carbonate solution for a variety of calcined LDHs [28]. The method was subsequently used by Chibwe and Jones for the incorporation of inorganic and organic anions [29, 30]. Solid-state ^{27}Al NMR measurements of the calcination/rehydration processes have been reported [31, 32].

Dimotakis and Pinnavaia showed that the use of glycerol as a swelling agent assisted the incorporation of organic guest via the rehydration route. In this procedure, a 1:2 (v/v) water/glycerol solution was used for the reconstruction of a calcined $MgAl(CO_3)$ precursor in a nitrogen atmosphere. The reaction products obtained were single phase and crystallographically well ordered. In the absence of glycerol, mixed phases were obtained [33]. Rehydration in pure water remains an attractive method for making hydroxide-containing LDHs (e.g. meixnerite) with enhanced properties [34].

9.2 Composition of the host hydroxide layers

Important variables related to the composition of the host hydroxide layers include the identity of the matrix cations, the charge on the layers (and hence the anion-exchange capacity) and any possible superstructure formed by ordering of the M(II) and M(III) cations. In several reviews the variety of cations that form LDHs has been discussed [6, 7, 35]. In general the cations should be of appropriate size (i.e. for brucite-based LDHs not too different from Mg^{2+}, which has an ionic radius of 0.72 Å) and charge.

The combination of Mg and Al has been the most frequently studied, with a variety of Mg/Al ratios and different charge-balancing anions. Figure 9.3 shows the variation of the hexagonal unit-cell parameters measured for $MgAl(CO_3)$ LDHs prepared with systematically varied Mg/Al ratio from 1.0 to 3.5, via the constant pH (= 10) coprecipitation method [36]. The *a*-parameter

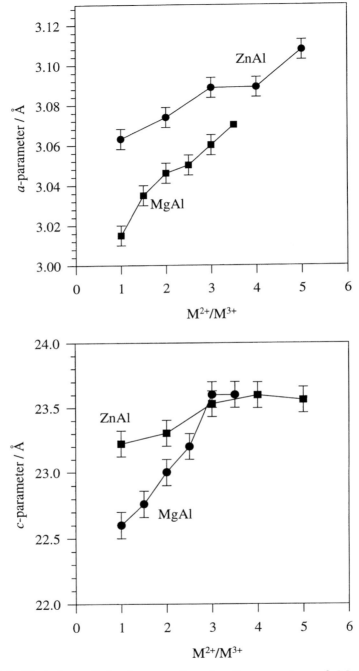

Figure 9.3. Variation of the hexagonal unit-cell parameters of $MgAl(CO_3)$ and $ZnAl(CO_3)$ LDHs with the M^{2+}/M^{3+} ratio used in their preparation.

of the hexagonal cell is equivalent to the mean metal cation nearest-neighbour separation in the plane of the hydroxide layers. Its value thus increases as the Mg/Al ratio is increased, owing to the larger size of Mg^{2+} compared to Al^{3+} (ionic radius $= 0.54$ Å). Furthermore, the coulombic charge on the hydroxide layers is reduced with increasing Mg/Al ratio. This leads to an increased separation of the layers (the overall coulombic attraction between adjacent hydroxide layers and the anionic layer sandwiched between them is reduced) and an increase in the c-parameter of the unit cell. The charge of an LDH is approximately $4e^-/nm^2$ and to be compared with a typical smectite clay with approximately $1e^-/nm^2$.

The reported Mg^{2+}/Al^{3+} ratio is usually in the range from approximately 1.0 to 5.0. Attempts to prepare LDHs with compositions outside of this range generally lead to the formation of impurity $M(OH)$ or $M(OH)_3$ phases. Shen *et al.*, however, have reported recently the synthesis of $MgAl(CO_3)$ LDHs with Mg/Al ratios from 3 to 12 without the formation of detectable separate phases [37]. The unit-cell data for these materials were not reported.

In general for cations other than Mg^{2+} and Al^{3+}, however, variable LDH stoichiometry has not been studied. M^{2+}/M^{3+} ratios close to 2 or 3 are generally reported although Kooli *et al.* have prepared a series of phase-pure $ZnAl(CO_3)$ LDHs with Zn/Al ratio from 2 to 5 by coprecipitation at pH 7 (Fig. 9.3) [38]. LDHs with $M^{3+} = Al$ and $M^{2+} = Mg$, Ni or Zn have recently been prepared by Costantino *et al.* using thermally induced hydrolysis of urea for pH control [39]. The products possessed a high degree of crystallinity and a narrow particle-size distribution. ZnCr, NiCr and MgCr combinations have also been studied [40–44], as well as LDHs containing Co in combination with Al^{3+}, Fe^{3+}, Mg^{2+} or Cr^{2+} [45–47]. For CoFe-LDHs, depending on the synthesis conditions, the Co^{2+} becomes partially oxidized to Co^{3+} [48]. With the combination of Mg and Co, approximately 23% of the original Co^{2+} is oxidized to Co^{3+}, giving an LDH of approximate composition $[(Mg^{2+}_{0.3}Co^{2+}_{0.6}Co^{3+}_{0.2}(OH)_2)]$ $(NO_3)_{0.2} \cdot H_2O$ [46]. Similarly, for a carbonate-containing MgMn-LDH [49] it was found that 84% of the manganese is oxidized to Mn^{3+} during synthesis, leading to the composition $(Mg^{2+}_{0.75}Mn^{2+}_{0.04}Mn^{3+}_{0.21}(OH)_2)(CO_3)_{0.11} \cdot nH_2O$. The synthesis of a CuCr-LDH has recently been described (Cu/Cr ratio close to 3) with carbonate as the charge balancing anion. The thermal characterization of this material in air and nitrogen has been studied [50].

V^{3+} cations have been incorporated along with Mg or Ni as the divalent cations [51, 52]. Although CuM^{3+}-LDHs have been reported, including CuAl-, CuMgAl- and CuCr-LDHs, in general they are poorly crystallized and contain impurity phases (such as malachite, $Cu_2OH_2CO_3$) [53–55]. The instability of Cu^{2+} (d^9) towards LDH formation may be attributed to a Jahn–Teller distortion of the hydroxide octahedra (well-crystallized Cu-containing LDHs may be formed when an additional divalent cation is present). Examples are not found in the literature for LDHs comprising Sc or Ti in the hydroxide layers.

LDHs containing matrix cations other than Mg, Al and the first-row transition metals may also be prepared. For example, well-crystallized $MgGa(CO_3)$ LDHs, with a variety of Mg/Ga ratios, have been described [18, 56]. Additionally, single phase $MgIn(CO_3)$ and $CdAl(CO_3)$ LDHs have also been prepared, with M^{2+}/M^{3+} ratios of approximately 2.6 and 2.0, respectively [57, 58].

A wide variety of LDHs containing more than one divalent or trivalent cation have been reported [59]. Basile et al., for example, prepared a series of $MgAl(CO_3)$ LDHs containing up to 5% of rhodium (Rh^{3+}) or ruthenium (Ru^{3+}) [60]. The presence of the noble metals specifically within the hydroxide layers of the LDH was inferred from measurements of the hexagonal unit-cell parameters and the absence of side-phases. Similarly, Fernandez et al. prepared $MgAl(CO_3)$ LDHs containing yttrium (Y^{3+}) using a coprecipitation method [61]. It was intended to prepare the LDHs with a constant M^{2+}/M^{3+} ratio but with varying Al^{3+}/Y^{3+} ratios (1, 2 and 4). In each case the LDH obtained, however, was enriched in aluminium compared with the composition of the starting solutions. Furthermore, the crystallinity of the product decreased as the Al^{3+}/Y^{3+} ratio increased, attributed to distortions introduced by the large difference in the ionic radii of the matrix cations (ionic radius of $Y^{3+} = 0.90$ Å).

Ternary LDH systems containing Mg, Al with either Sn^{4+} or Zr^{4+} in the hydroxide layers are also known [62–64]. For the tin-containing system, a single-phase LDH could be obtained in the composition range of Mg/Al/Sn from 3/1/0 to 3/0.7/0.3.

In general, the relative positions of the metal cations within the hydroxide layers are unknown. A single crystal X-ray diffraction study of the naturally occurring mineral hydrotalcite, $[(Mg_4Al_2(OH)_{12})]CO_3 \cdot 3H_2O$, suggested that the Mg and Al cations are randomly distributed [13]. The presence of long-range cation ordering has, however, frequently been suggested [18, 19, 65, 66]. If trivalent cations cannot be neighbors, then a M^{2+}/M^{3+} ratio of 2 should result in a completely ordered superlattice with a hexagonal unit-cell parameter in the plane of the hydroxide layers of $\sqrt{3}a$, rotated by 30° from the disordered lattice. The similar X-ray scattering power of Mg and Al, however, may be the reason for the absence of detectable cation-ordering for hydrotalcite. The existence of an ordered superlattice may lead to an ordered distribution of the interlayer anions [66]. Recently, a combined powder X-ray diffraction and X-ray absorption study of synthetic MgFe-LDHs found no correlation between

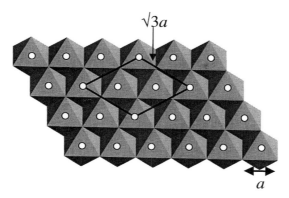

Figure 9.4. A view of the ordered unit-cell in the (001) plane of a Li_2Al-LDH. A similar ordering can occur in $M^{2+}M^{3+}$-LDHs with an M^{2+}/M^{3+} ratio of 2 if the trivalent cations are not neighbors.

Fe^{3+} cation positions over distances of tens of angstroms, although a high level of local ordering was detected [65]. In LiAl-LDHs, in which the Li cations fill octahedral vacancies in a gibbsite-like structure and thus are naturally ordered, the existence of a superlattice is regularly detected (Fig. 9.4) [16, 67, 68].

9.3 Identity of guest anions

In naturally occurring LDHs, the most commonly found interlayer anion is carbonate. In practice, however, there is no significant restriction to the identity of the charge balancing anion that can occupy the interlayer region. Examples include inorganic (such as halides, oxo-anions, silicates and polyoxometalates) as well as organic anions (such as alkyl or aryl carboxylates and sulfonates).

9.3.1 Inorganic guests

A variety of small inorganic anions can be incorporated into LDHs, either directly during synthesis or by subsequent anion-exchange. The ease of exchange has been studied by several groups and the generally accepted sequence is $OH^- > F^- > Cl^- > Br^- > NO_3^-$; divalent anions such as SO_4^{2-} and CO_3^{2-} have a higher affinity for LDHs than monovalent anions. The particular case of OH^- as the balancing anion is noteworthy because it is readily exchanged and also because on calcination no impurities are left or gases, except water, evolved (i.e. by-products such as SO_x or NO_x) – this is particularly beneficial for catalytic applications and environmental aspects.

A major use of LDHs is as catalyst precursors, with calcination of the LDH generating mixed metal oxides that are effective catalysts for a variety of reactions [7, 69]. In such cases the catalytic activity of the resulting material is primarily a function of the nature of the metal cations in the LDH. Pristine LDHs, however, have also been shown to possess catalytic properties, in which case the nature of the interlayer anion is important. Constantino and Pinnavaia, for example, have examined the basic catalytic properties of $MgAl(CO_3)$ LDHs at temperatures below the structural decomposition point ($<250°C$) [70]. The catalyst was found to exhibit specific activities approximately an order of magnitude larger than mixed oxides obtained from thermal activation at 450 and 890 °C. In addition, halide-exchange reactions producing alkyl iodides and chlorides have been catalysed by LDHs containing Cl^- and I^- [71]. Kaneda and co-workers have reported a series of oxidation reactions catalysed by $MgAl(CO_3)$ LDHs, such as Baeyer–Villiger oxidation of ketones, epoxidation of olefins and conversion of various pyridines to yield pyridine N-oxides using hydrogen peroxide and molecular oxygen as oxidants [72–74]. A highly active catalyst for a variety of base catalysed reactions can be produced by thermal decomposition of a $MgAl(CO_3)$ LDH and subsequent rehydration in N_2 to generate the hydroxide-containing form [75–77]. A highly basic and effective catalyst can be made by incorporation of *tert*-butoxide in the interlayer (incorporated via exchange of nitrate) [78, 79].

There has also been considerable interest in the incorporation of polyoxometalate anions such as $V_{10}O_{28}^{6-}$, $Mo_7O_{24}^{6-}$ and α-$(SiW_{11}O_{39})^{8-}$. The interest was generated in part by comparison with studies of the incorporation of large polyoxometalate cations of the type $(Al_{13}O_4(OH)_{24}(H_2O)_{12})^{7+}$ and $[(Zr_4(OH)_8(H_2O)_{16})]^{8+}$ into cationic clays whereby, with use of such large and robust guests, permanently expanded porous solids with useful adsorption and catalytic properties could be prepared. Upon calcination the polyoxometalate cations form oxide pillars, thus generating nanocomposite materials with open and rigid structures. This area has been recently reviewed by Rives and Ulibarri [80].

An early report of polyoxometalate incorporation was made by Kwon *et al.* [81]. The materials were prepared by the exchange of interlayer Cl^- anions in an aqueous solution of $(NH_4)_6(V_{10}O_{28})\cdot6H_2O$ at pH 4.5 and 25 °C. The interlayer spacing of the product (11.9 Å) indicated incorporation of $V_{10}O_{28}^{6-}$ with its C_2 axis parallel to the hydroxide layers (Fig. 9.5). It was demonstrated that the product catalyses the photo-oxidation of isopropyl alcohol to acetone. Similarly, Drezdzon demonstrated that the incorporation of $V_{10}O_{28}^{6-}$ and $Mo_7O_{24}^{6-}$ may be achieved via an intermediate MgAl-LDH containing terephthalate interlayer anions (in this case, the hydroxide layers are widely separated by the organic anion, thus facilitating exchange) [24]. Acidifying the

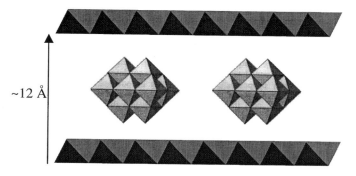

~12 Å

Figure 9.5. Schematic representation of a LDH containing the $V_{10}O_{28}{}^{6-}$ interlayer anion.

organo-LDH in the presence of $NaVO_3$, for example, leads to the simultaneous polymerization of the monovanadate and the displacement of terephthalate from the interlayer, resulting in the incorporation of the polyoxovanadate species. Evans *et al.* extended this approach by studying LDH precursors containing terephthalate, *p*-hydroxybenzoate or *p*-methylbenzoate and found that the latter is the most suitable anion for facile exchange with a variety of polyoxometalate anions [82].

Polyoxometalate anions have also been incorporated by the direct coprecipitation of a ZnAl-LDH with α-$(SiW_{11}O_{39})^{8-}$ under mildly acidic conditions [83], and a ZnCr-LDH with $V_{10}O_{28}^{6-}$ at pH 4.5 [84]. The reconstruction method has also been used to incorporate such anions into MgAl- and NiAl-LDHs [5, 29, 85, 86].

In general, LDHs containing polyoxometalate anions represent a class of pillared materials for selective adsorption and oxidation catalysis. Tatsumi *et al.*, for example, have reported that a MgAl-LDH containing $(Mo_7O_{24})^{6-}$ or $(W_{12}O_{41})^{10-}$ exhibits significant shape-selectivity for the epoxidation of alkenes with H_2O_2 [87].

LDHs containing a variety of anionic complexes may also be prepared [80, 88]. An early report was by Miyata and Kumura, who described a MgAl-LDH containing the hexacyanoferrate anion, $Fe(CN)_6^{3-}$ with an interlayer spacing of 10.9 Å [89]. More recently, Crespo *et al.* have studied the incorporation of hexacyano complexes ($Fe(CN)_6^{3-}$ and $Fe(CN)_6^{4-}$) into ZnAl-LDHs by the methods of anion exchange (from $ZnAl(NO_3)$ and ZnAl(terephthalate) LDHs), coprecipitation at pH 10 and rehydration of a calcined $ZnAl(CO_3)$ precursor [90]. The anion-exchange method was found to be the only method that produced a single crystallographic phase. In all cases, both species of hexacyano complex were identified in the interlayer of the LDH from FTIR measurements, regardless of which complex was used for the synthesis. Halo complexes have also been incorporated. Lopez-Salinas *et al.*, for example, have prepared a MgAl-LDH containing the $(NiCl_4)^{2-}$ ion via anion-exchange of a nitrate-containing precursor [91]. It was found that the chloride ligands can be exchanged with bromide from butylbromide in the liquid phase and that the reaction is strongly influenced by choice of solvent.

9.3.2 *Organic guests*

A wide variety of organic anions, such as alkyl or aryl carboxylates and sulfonates, have been incorporated between the hydroxide layers of LDHs [92, 93]. In a series of ZnAl-LDHs containing terminal linear aliphatic dicarboxylates, $C_nH_{2n}(CO_2^-)_2$ [89], it was found that the interlayer spacing of the organo-LDH increased linearly from 9.4 Å for oxalate ($n = 0$) to approximately 18–19 Å for sebacate ($n = 8$). The gallery height of the ZnAl(sebacate) LDH, obtained by subtracting the hydroxide layer thickness of 4.8 Å from the interlayer spacing, is consistent with an interlayer comprising a monolayer of the organic anions in an approximately vertical orientation with respect to the hydroxide layers. By contrast, whilst linear divalent anions generally form monolayers which bridge the hydroxide layers, linear monovalent anions commonly form bilayers. For the benzoate anion, for example, a gallery height of between 10.4 and 10.7 Å is generally observed [94] and a bilayer-like structure, with the carboxylate group attached to the hydroxide layers and the phenyl ring of the anions interleaved within the mid-plane of the gallery, has been proposed. A bilayer arrangement is to be expected in such cases because the interlayer has two ionic surfaces, top and bottom, to which the anionic functional group is equally probable to attach.

A large number of organic anions containing sulfate or sulfonate groups have been incorporated into LDHs. A series of alkyl sulfate anions, $C_nH_{2n+1}SO_4^-$, for example, was incorporated into a ZnCr-LDH via anion exchange by Boehm *et al.* [95]. The products of the exchange reaction were found to have interlayer spacings in the range from 21.1 ($n = 8$) to 32.6 Å ($n = 18$) and underwent substantial additional swelling in the presence of *n*-alkyl alcohols or *n*-alkylamines. Incorporation of dodecylsulfate anions into a series of NiAl-, MgAl- and ZnCr-LDHs with varying layer charge has also been studied (Fig. 9.6) [96].

Franklin *et al.* reported the presence of three distinct and interconvertible phases when the 5-benzoyl-4-hydroxy-2-methoxybenzenesulfonate anion was incorporated into a ZnAl-LDH [97]. The anion is dibasic, having both a strong sulfonate group and a substantially weaker phenolate group. It was found that anion exchange of the divalent anion resulted in an interlayer spacing of 13.8 Å.

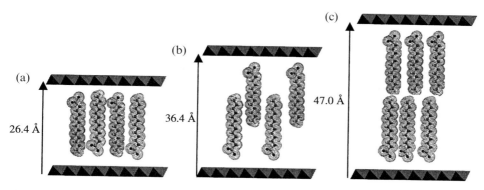

Figure 9.6. Schematic representation of a LDH containing dodecylsulfate interlayer anions in three different arrangements: monolayer (a); interdigitated (vertical) (b); and bilayer (c). After Clearfield *et al.* [96].

The monovalent form (in which only the sulfonic group is ionized) gave an interlayer spacing of 20 Å and readily converted to the divalent form within the LDH. The observations reveal that the guest molecule occupies two distinct conformations within the interlayer, depending upon the charge of the anion. Furthermore, it was found that the divalent phase may be converted into a 15.4 Å phase, which is intermediate between the monovalent and divalent forms (either by exposure of the LDH to water vapor, or dispersion in slightly acidic solution). The conformation of three isomers of naphthalenedisulfonate in the interlayer of ZnAl-LDHs has also been studied [98]. The shape of the guest anion was varied by changing the position of substitution (1, 5-, 2, 6- and 2, 7-) of the two sulfonate groups on the naphthalene unit.

It is known that the orientation of an organic anion within LDHs is strongly dependent upon the layer charge and the interlayer water content. The influence of water content and layer charge on the interlayer arrangement of terephthalate anions inside a MgAl-LDH has recently been studied [66, 94, 99]. For relatively high water content and layer charge an interlayer separation of approximately 14.0 Å is favored, corresponding to a vertical orientation of the terephthalate anion with respect to the hydroxide layers. For relatively low water content and layer charge, however, an interlayer spacing of approximately 8.2 Å is favored, corresponding to a horizontal terephthalate orientation. Interestingly, during cycles of dehydration–rehydration, the 14.0 and 8.2 Å interlayers coexist in varying proportions, depending both on the layer charge and water content of the LDH. In certain cases, the existence of a 22–23 Å phase, consisting of an ordered interstratification of the relatively hydrated (14.0 Å) and dehydrated (8.2 Å) component interlayers has been established (Fig. 9.7). Molecular dynamics computer simulations of this

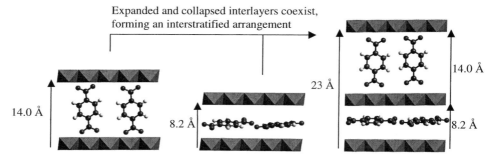

Figure 9.7. Schematic representation of three different interlayer arrangements of an LDH containing terephthalate.

system have also been used to study the influence of water content and layer charge on the interlayer arrangement of the LDH [99]. General agreement between the computer simulations and experimental measurements has been obtained. Composition variation within the galleries has been suggested by Kaneyoshi and Jones [100], with terephthalate and carbonate anions being co-exchanged and forming a super-structure. Similarly Ijdo and Pinnavaia have reported solid solution formation in amphiphilic organic-inorganic (cationic) clay heterostructures [101].

A different type of ordered interlayer occupancy has been observed for a LiAl-LDH, using time-resolved *in situ* energy-dispersive X-ray diffraction studies on the incorporation, via anion-exchange of Cl^-, of various dicarboxy-late anions (terephthalate or succinate, for example). The existence of a second-stage intermediate phase has been demonstrated. This second-stage intercalate transforms subsequently into the completely exchanged (first-stage) organo-LDH [102]. The observed second-stage LiAl-LDH phase consists of a regular alternation of interlayers containing either the dicarboxylate guest or unex-changed chloride (Fig. 9.8). The phenomenon of staging is well known for graphite and metal dichalcogenide intercalates. The second stage material could also be isolated by reaction of the LDH with the necessary stoichiomet-ric amount of anion.

Interestingly, the preferential incorporation of organic anions into LDHs may occur. A MgAl(Cl) LDH, for example, undergoes anion-exchange reac-tions in aqueous solutions with L- and D-histidine, showing apparent preference for the L-histidine optical isomer [103]. Selectivity towards anions with highest charge density is observed in the rehydration of calcined MgAl-LDHs in the presence of benzenecarboxylate anions [104]. Similarly, preferential incorpora-tion of isomers of naphthalenecarboxylate anions into calcined MgAl- and ZnAl-LDHs occurs [105]. Preferential incorporation of isomers of anthraqui-

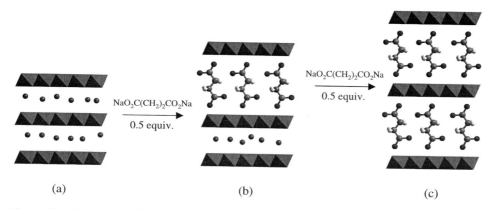

Figure 9.8. Schematic diagram showing the incorporation of succinate anions into a LiAl(Cl) LDH. After Fogg *et al.* [102].

none sulfonate anions in the direct synthesis by coprecipitation of ZnAl-LDHs has also been described [106]. More recently, it has been shown that a LiAl-LDH containing interlayer chloride will undergo selective anion-exchange reactions with a number of organic anions [107, 108]. It was found, for example, that the fumarate dianion (*trans*-$O_2CCHCHCO_2$)$^{2-}$ can be preferentially extracted by this LDH from an aqueous solution containing equal amounts of both fumarate and maleate (*cis*-$O_2CCHCHCO_2$)$^{2-}$. This is particularly interesting given that intercalation reactions are, in general, not selective processes – the interlamellar space simply expands to incorporate the guest. In the case of fumarate/maleate, however, the preference for fumarate was found to be in excess of 95% in water at room temperature. Similarly, terephthalate can be selectively removed from an aqueous solution containing all three benzenedicarboxylate isomers (terephthalate, phthalate and isophthalate). Similar selectivity in competitive exchange reactions with isomers of aromatic sulfonate anions has also been reported. Furthermore, it was found that the selectivity could be controlled by varying the temperature of the reaction. These processes have obvious importance to separation and chemical-sensing technologies. An important variable appears to be the solvent in which the exchange is taking place with the variation in solvation energies of the two competing ions being important, i.e. a competition between the solvent architecture around the anions in the bulk solvent and within the interlayer. Also important will be the strength with which the anions are held within the layers, and this will depend on the charge on the anions and the exact charge distribution around the anion.

The interlayer region of LDHs provides a novel environment for photochemical reactions of photoactive molecules. The controlled photodimerization of a variety of unsaturated carboxylates between the hydroxide layers of a

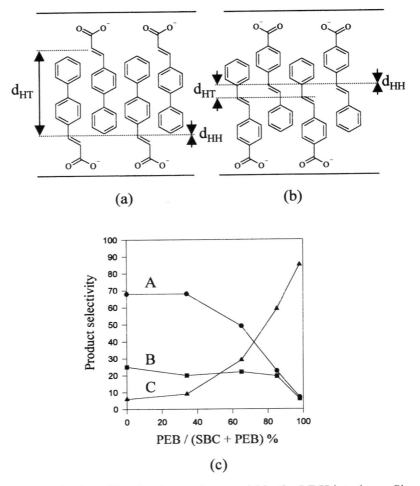

Figure 9.9. [2 + 2] photodimerization pathways within the LDH interlayer. Simplified drawing of anti-parallel packing of (a) *p*-phenylcinnamate and (b) stilbenecarboxylate anions (SBC). (c) Effect of coadsorbate, *p*-phenylbenzoate (PEB), on product selectivities of intercalated SBC in water. A: formation of head-to-head dimers. B: formation of head-to-tail dimers. C: isomerization to *cis*-SBC. After Sasai *et al.* [110].

MgAl-LDH has been reported [109–111] (see Fig. 9.9). *Syn* head-to-head cyclodimers were selectively formed in the irradiation of sodium cinnamates between the hydroxide layers. Using the known topochemical limit of between 4 and 5 Å for photochemical [2 + 2] cyclodimerization [112], the closest distance of approach for head-to-tail was too great for formation of the head-to-tail (HT) dimer (Fig. 9.9a). Clearly, useful comparisons can be made with the photoactivity of cationic photoactive molecules inside layered (cationic) aluminosilicates, e.g. stilbazolium ions. In clays the lower exchange value leads to lower guest density.

For the case of stilbenecarboxylate (SBC, Fig. 9.9b), by comparison, both *syn* head-to-head and *syn* head-to-tail cyclodimers were formed. The product selectivity may also be controlled by the presence of co-absorbed photo-inactive molecules e.g. phenylbenzoate (Fig. 9.9c).The nature of the photodimerisation products can to some extent be controlled by changing the M^{2+}/M^{3+} ratio, and hence the packing density of the incorporated monomer [113]. Norrish type II reactions of aromatic ketocarboxylates within the interlayer of MgAl-LDHs have also been studied [114].

The incorporation of photochromic molecules into LDHs may yield useful photoresponsive materials. Tagaya and co-workers have studied the photoisomerization of sulfonated indolinespirobenzopyran (SP-SO^{3-}) to merocyanine (MC) in the interlayer region of a MgAl-LDH [115, 116]. The photoisomerization was found to be irreversible for a MgAl-LDH containing only SP-SO^{3-} owing to the stability of MC in the polar environment of the hydroxide layers. In the presence of *p*-toluene sulfonate, however, the reversible photoisomerization between SP-SO^{3-} and MC in the interlayer region of the MgAl-LDH was observed. Subsequently, it was found that SP-SO^{3-} incorporated into a LiAl-LDH shows reversible photochromism even in the absence of an organic co-intercalate [117]. The incorporation of anionic dyes, such as indigo carmine, new coccine and methyl orange into LDHs may yield materials with potential applications as pigments [22, 118, 119].

An examination of a LiAl-LDH containing myrisate anions as a support for photochemical assemblies has recently been performed [120]. The goal of the study was to assemble an LDH containing TiO$_x$ in which the TiO$_x$ particles are able to interact with co-exchanged porphyrin anions. The incorporation of TiO$_x$ in the interlayer was achieved by taking advantage of the hydrophobic nature of a myristic acid exchanged LiAl-LDH, i.e. titanium butoxide was partitioned into the interlayer, followed by controlled hydrolysis under ambient conditions. As a result an assembled three-component system was created (Fig. 9.10) with the aim of sensitizing the semi-conductor particles. A suggested application for such a system is in the photodegradation of pollutants by acting as sacrificial electron donors to the metal porphyrin. The initial interlayer spacing of the myristate-LDH (approximately 21 Å) being sufficient to allow the porphyrin anions to adopt a vertical arrangement.

The incorporation of metallophthalocyanines into LDHs is of interest because these macrocyclic complexes can function as biomimetic catalysts for the autoxidation of organic molecules in aqueous solution [121–125]. Cobalt(II) phthalocyaninetetrasulfonate ([CoPcTs]$^{4-}$), for example, is active

visible light

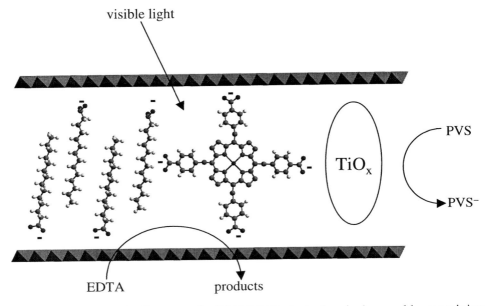

Figure 9.10. Schematic diagram of a LiAl-LDH photochemical assembly containing myristate anions, TiO$_x$ clusters and Zn-porphyrin. Sensitization of the TiO$_x$ by the porphyrin results from the close intralayer proximity of the species. (PVS = propyl violegen sulfonate After Robins and Dutta [120].

for the autoxidation of 1-decanethiol to the disulfide [124]. Incorporation of the phthalocyanine complex into a MgAl-LDH was found to improve significantly the catalyst activity and lifetime for this reaction, compared to the homogeneous catalyst (Fig. 9.11). In addition, the use of an LDH support affords easy removal of the catalyst from the reaction medium. X-ray data indicated a basal spacing of approximately 23 Å, suggesting a gallery height of 18.5 Å. This value is close to the van der Waals width of the sulfonated phenyl ring and strongly suggests a perpendicular arrangement of the anion with respect to the hydroxide layers. The same material is also active for the autoxidation of 2, 6-di-tert-butylphenol to the corresponding diphenoquinone [125], and the reductive dehalogenation of certain halogenated organic compounds [123]. The oxidation of cyclohexene over a ZnAl-LDH containing [CoPcTs]$^{4-}$ has also been reported [126]. Metallophthalocyanine complexes supported on LDHs, therefore, have a potential application to the treatment of waste water via the degradation of organic pollutants. Similar materials based on the incorporation of metalloporphyrins into LDHs have also been investigated [122, 127–129].

The hydrophobic nature and accessibility of the interlayer region of organo-LDHs makes these materials candidates for the adsorption of other organic

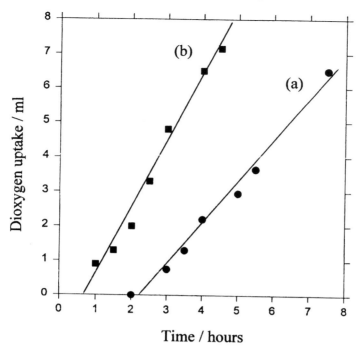

Figure 9.11. Dioxygen uptake plots for the autoxidation of 1-decanethiol in the presence of [CoPcTs]$^{4-}$ as catalyst: (a) unsupported and (b) supported on a MgAl-LDH. A shorter induction time and a larger turnover frequency were obtained for the LDH assembly compared with the homogeneous case. After Perez Bernal *et al.* [124].

molecules. As mentioned above, a ZnCr-LDH containing alkyl sulfate anions, for example, undergoes swelling perpendicular to the layers under *n*-alkyl alcohols or *n*-alkylamines [95]. In addition, it has been found that ZnCr-LDHs containing surfactant anions have the ability to take up a large variety of small organic molecules such as ethylene glycol, propanediol and glycerol into the interlayer [130]. Recently the exfoliation/delamination of a ZnAl-LDH has been reported by dispersion of a dodecyl sulfate exchanged material in butanol [131].

The formation of novel membrane-like materials based on LDHs has also been established. These materials are of relevance to separation and membrane technologies. A LiAl-LDH containing myrisate or hexanoate anions, for example, has the ability to partition pyrene from a methanol/water solution containing the polycyclic aromatic [132]. No sorption of pyrene was observed for a LiAl(succinate) LDH, however, and this was attributed to a sieving effect of this compound towards the pyrene molecule. The sorption of polyaromatic hydrocarbons, such as pyrene, is important from an environmental perspective.

In addition, the potential application of a LiAl(myrisate) LDH as a stationary phase in gas chromatography has been investigated, owing to the partitioning effect exhibited by this material [133].

Recently, the adsorption behavior of CaAl(dodecylsulfate) and CaAl (dodecylbenzenesulfate) LDHs towards liquid mixtures of organic molecules such as benzene, *n*-heptane and toluene has been studied [134]. The organo-LDHs adsorb these molecules with a considerable increase of the interlayer spacing and, in certain cases, act as selective adsorbents. C_{60} molecules may also be adsorbed into the hydrophobic interlayer of a MgAl(dodecylsulfate) LDH [135]. Heating the resulting compound under vacuum to decompose the dodecylsulfate left C_{60} molecules sandwiched between the hydroxide layers. The inclusion properties of guest organic molecules, such as benzene and *p*-, *o*- and *m*-xylene, by a host MgAl-LDH containing a modified β-cyclodextrin has also been investigated [136, 137]. The host–guest interaction was attributed to a partition process of the organic molecules into the cyclodextrin cavities as well as intermolecular pores.

An area of current interest for LDHs is as 'clay-modified' electrodes. Materials derived from LDHs containing organic electroactive molecules, such as anthraquinonesulfonate, have been prepared [138]. The electroactive anion was exchanged into a ZnCr-LDH and a thin film of the product deposited on a glassy carbon electrode. The electrochemical behavior of the substrate was then studied using cyclic voltammetry. The potential of ZnCr- and ZnAl-LDHs containing electroactive anions, such as the ferrocene sulfonate anion, as modified electrodes has also been established by Therias *et al.* [139, 140] and the electrochemical behavior of manganese porphyrins incorporated into a MgAl-LDH has also briefly been studied [141].

A LiAl-LDH intercalated with 4-nitrohippuric acid has been shown to exhibit second harmonic generation − 532 nm radiation from incident 1064 nm [142]. It was reported that the ability of the intercalated material to exhibit non-linear optical properties is due to a perpendicular monolayer packing of the acid molecules in the interlayer. Such an orientation leads to an ordered arrangement of dipoles, giving rise to a bulk dipole moment in the solid. Crystals of the pure acid exhibit no frequency-doubling characteristics due to a centrosymmetric packing in the crystal.

The surface modification of LDHs with organic molecules has recently been investigated [42, 143]. It has been reported, for example, that the product of the reaction of a ZnAl(OH) LDH with an organic oxychloride is an esterified LDH, with the organic molecules attached directly to the matrix cations (i.e. not through the surface OH groups) [144–146]. Similarly, the preparation of modified MgAl- and ZnAl-LDHs, in which alkyl and aryl phosphonate anions are connected directly to the matrix cations has recently been reported [147–151]. Such surface modification, also known as grafting, therefore generates organic–inorganic hybrid materials in which the organic molecule is covalently bonded to the inorganic lattice. These materials are expected to display higher chemical and thermal stability than conventional organo-LDHs.

Finally, it is noteworthy that the ability of LDHs to concentrate selectively and organize organic molecules has generated interest in their possible role in chemical evolution and the origins of life [152–156]. Glycolaldehyde phosphate incorporated into the interlayer of LDHs from highly dilute aqueous solution condenses to racemic aldotetrose-2,4-diphosphates and aldolhexose-2,4,6-triphosphates, for example [156]. No such reaction occurs in the absence of the LDH in otherwise identical conditions. Incorporation of biologically important molecules into LDHs, such as amino acids, has also been reported [157, 158].

9.4 Calcined LDHs

The thermal decomposition of LDHs with a range of compositions has been extensively studied [159]. The first stage (up to approximately 250 °C) is the removal of interlayer and external (interparticle) water. The second stage, from approximately 150 °C to 500 °C, depending upon the matrix cations present, is dehydroxylation of the layers. The temperature at which the interlayer anion is evolved varies considerably depending upon the nature of the anion as well as the composition of the layers, although for $M^{2+}M^{3+}(CO_3)$ LDHs carbonate expulsion is generally concomitant with dehydroxylation. Calcination of LDHs at or above the dehydroxylation temperature destroys the layered character of the materials and generates mixed oxides. Furthermore, because the matrix cations are intimately mixed in the precursor LDH the resulting oxide generally contains a homogeneous dispersion of metal cations, although its precise nature depends upon composition and calcination conditions.

In general, the sequence of oxide-formation from a $MgAl(CO_3)$ LDH proceeds as follows. Firstly, the LDH is converted to a mixed MgAl oxide with the MgO rock-salt type structure at approximately 400 °C. The lattice parameters of the mixed oxide are generally lower, however, than those measured for pure MgO, indicating that the Al^{3+} ions are inserted into the structure, which also introduces lattice defects. At higher temperatures the mixed oxide decomposes into MgO and spinel, $MgAl_2O_4$ [160–164]. A similar sequence has been observed for the thermal decomposition of a NiAl-LDH by Sato *et al.* [28].

The decomposition is more complex with LDHs containing alternative

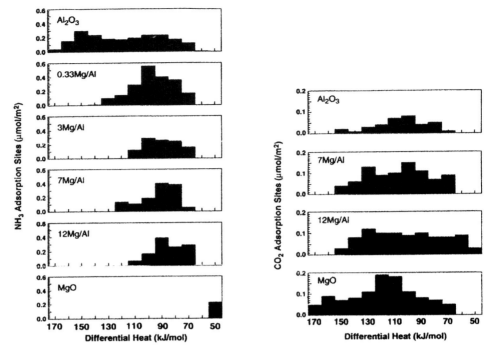

Figure 9.12. Histograms of the distribution of interaction strengths for ammonia (left) and carbon dioxide (right) adsorption of γ-Al$_2$O$_3$, MgO and mixed MgAl-oxides obtained from LDHs. After Shen *et al.* [183].

cations such as Co and Cr, due to the formation of phases with different oxidation states of these cations [43]. In certain cases (e.g. with CrO_4^{2-} and $Cr_2O_7^{2-}$ oxometalates) grafting of the interlayer anion to the hydroxide layers has been suggested to occur at moderate temperature [165].

Calcined MgAl-LDHs promote a wide variety of industrially important base-catalysed reactions [166] such as Aldol [167–169], Claisen–Schmidt [170], and Knoevenagel condensations [171, 172], Meerwein–Ponndorf–Verley reduction of carbonyl compounds [173], Henry reaction of aldehydes [174], isomerizations [175–178] and the polymerization of propylene oxide [179]. Details have been recently reviewed [180]. For catalysis applications, the high surface area of the calcined LDH and homogeneous dispersion of the metal cations are attractive properties [181]. Furthermore, the opportunity to control the composition of the hydroxide layers of the precursor LDH allows tuning of the catalytic properties of the resulting catalyst [182]. For MgAl mixed oxides obtained from LDHs with Mg/Al ratios of 0.5–9.0, generated by calcination in N$_2$ at 673 K, the nature, density and strength of surface basic sites is dependent on the Al content (Fig. 9.12) [37, 183]; with surface sites of low (OH$^-$ groups),

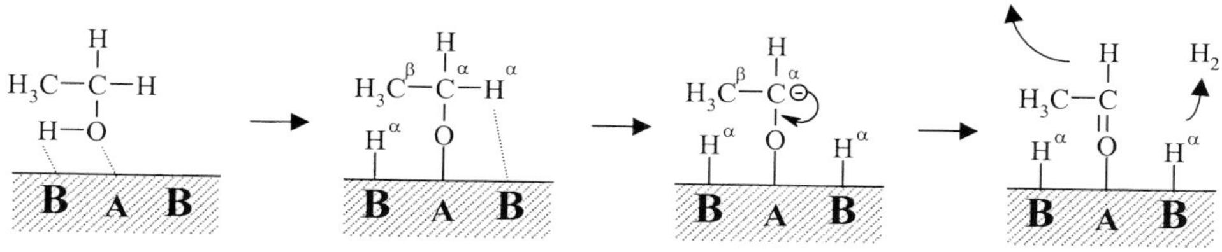

Figure 9.13. The acetaldehyde formation mechanism, where A and B are Lewis acid sites and Brønsted basic sites, respectively. Dehydration requires the combination of an acid and strong base site with an adjacent strong basic site. After Di Cosimo *et al.* [184].

medium (Mg–O pairs), and strong (O^{2-} anions) basicity [184]. In addition, surface Lewis acid sites are provided by both Mg^{2+} and Al^{3+}. Activity and selectivity of the catalysts for ethanol conversion reactions were found to be directly related to sample composition. Figure 9.13 schematically illustrates the conversion of ethanol to acetaldehyde. The dehydrogenation reaction involves the initial adsorption of ethanol on an acid–base pair site, resulting in the breaking of the OH-bond. The resulting ethoxy intermediate is converted by hydrogen abstraction at a strong base site. Pure MgO would possess only basic sites and therefore catalyses the dehydrogenation of ethanol to acetaldehyde at very low rates.

In addition to the combination of Mg and Al, catalysts containing redox-active metals have been obtained from LDHs. Oxides resulting from NiMgAl-LDHs have been studied recently for reactions such as the hydrogenation of acetonitrile and the partial oxidation of methane [185–187]. For the former reaction, selectivity for conversion to monoethylamine was found to depend on the $Mg/(Mg + Ni)$ ratio. Mixed CuMgAl oxides obtained from LDHs have been studied as potential catalysts for the coupling of phenylethyne [188]. It was found that Cu^{2+} embedded into the MgAl oxide matrix is a crucial redox site for this reaction. MgFe and MgCr mixed oxides obtained from LDHs have been also studied as potential catalysts [189–191]. MgAl- and CoAl-LDHs containing Ru^{3+} in the hydroxide layers have been used as precursors to mixed oxides for highly efficient oxidation of allylic and benzylic alcohols [192, 193].

Calcined LDHs also have application in the reduction of SO_x and NO_x emissions from the fluid catalytic cracking (FCC) units in oil refineries [194–196]. Corma *et al.* attempted to optimize the performance of mixed oxides produced from MgAl-LDHs as SO_x-removal additives to FCC catalysts [194]. Among the oxides studied, that obtained from a MgCuAl-LDH was found to be the most effective at catalysing both the oxidation of SO_2 to SO_3^{2-} in the FCC regenerator and the reduction of the sulfates to H_2S, which may be recovered, in the reductive atmosphere of the cracking zone. A similar material was also found to be active in the simultaneous removal of NO and SO_2 [195]. NiAl-, MgCo-, CoAl- and CuAl-LDHs, in certain cases also containing Rh, La or Pd, have also been studied as precursors to such de-NO$_x$ and de-SO$_x$ additives [196–202]. In general, the use of LDHs allows the concentration of each metal within the mixed oxide to be controlled and leads to well-defined, high surface area catalysts.

Calcined LDHs may also serve as effective catalyst supports. Davis and Derouane prepared, for example, a supported-platinum catalyst for aromatization of *n*-hexane, via the impregnation of a calcined MgAl-LDH with an aqueous solution of $Pt(NH_3)_4Cl_2$ [203]. More recently, Narayanan and Krishna have studied a system of palladium supported on calcined MgAl-LDHs for phenol hydrogenation [204–207].

It has been demonstrated that mixed oxides obtained from calcined LDHs have the ability to act as sorbents for a variety of anionic compounds from aqueous solution. This ability is because of the propensity for the mixed oxide to hydrate and re-form an LDH in such conditions and is of particular interest for the decontamination of waste-water. Hermosín *et al.* have found, for example, that MgAl-LDHs calcined at 500°C are potential sorbents for the pollutants trinitrophenol and trichlorophenol from water [208, 209]. The adsorption mechanism was shown, using PXRD, to involve reconstruction of the LDH, with the uptake of the phenolate anions into the interlayers. Similarly, the ability of calcined MgAl-LDHs to remove nitriloacetate anions from solution has been demonstrated [210]. Calcined LDHs have been utilized also for the sorption of radioactive anions, such as $^{131}I^-$, from aqueous solution [211]. A particularly attractive feature of the use of calcined LDHs for the remediation of waste-water is that the sorption capacity of the material may be regenerated via calcination of the rehydrated LDH.

9.5 Concluding remarks

LDHs are particularly useful as layered hosts because their synthesis is generally simple and the matrix is available with a variety of cations. In addition, the anion exchange capacity may be systematically varied over a wide range. Although not all cation combinations are equally appropriate, combinations such as Mg and Al are, and this combination has been successfully used to create novel organic–inorganic hybrids demonstrating attractive properties. Such hosts also provide the basis for well-dispersed catalysts for a variety of catalytic processes. Recent developments have also focussed on 'one-pot' preparations with attractive environmentally friendly chemistry, eliminating the product of unwanted gases and waste–water. Applications in the areas of de-SO$_x$ and de-NO$_x$ chemistry point to important new developments.

We have also illustrated the versatility of the system using examples of selective photoreactions, efficient separation processes, interlayer re-orientation effects and staging.

The interest in LDHs is much more recent than their cationic equivalents (cationic clays such as montmorillonite and saponite). This is, in part, because cationic clays are much more abundant in nature and have an important role in many soil processes, for example. It is clear, however, that the potential for creating novel supramolecular structures, either directly as intercalates or for orienting organic molecules at reactive surfaces, is likely to increase.

9.6 References

1. Whittingham, M. S. and Jacobson, A. J. (Eds.). (1982). *Intercalation Chemistry*, Academic Press, New York.
2. Alberti, G. and Bein, T. (1996). 'Solid-state supramolecular chemistry: two and three-dimensional inorganic networks' in *Comprehensive Supramolecular Chemistry* (J. L. Atwood, J. E. D. Davies, D. D. Macnicol, and F. Vogtle, Eds.). Pergamon Press, Oxford.
3. Newman, S. P. and Jones, W. (1999). Comparative study of some layered hydroxide salts containing exchangeable interlayer anions. *J. Solid State Chem.* **148**, 26.
4. Laine, R. M., Sanchez, C., Brinker, C. J. and Giannelis, E. (Eds.). (1998). *Hybrid Organic/Inorganic Materials*, volume 628. Materials Research Society, Warrendale, Pennsylvania.
5. Chibwe, M. and Jones, W. (1990). 'The synthesis, chemistry and catalytic applications of layered double hydroxides' in *Pillared Layered Solids* (I. V. Mitchell, Ed.). Elsevier Applied Sciences, London.
6. de Roy, A., Forano, C., El Malki, K. and Besse, J.-P. (1992). 'Anionic clays: trends in pillaring chemistry' in *Expanded Clays and other Microporous Solids* (M. L. Occeli and H. E. Robson, Eds.). Van Nostrand Reinhold, New York.
7. Cavani, F., Trifirò, F. and Vaccari, A. (1991). Hydrotalcite-type anionic clays: Preparation, properties and applications. *Catal. Today*, **11**, 173.
8. Vaccari, A. (1995). Synthesis and applications of anionic clays. *Appl. Clay Sci.* **10**, 1.
9. Frondel, C. (1941). Constitution and polymorphism of the pyroaurite and sjogrenite groups. *Am. Miner.* **26**, 295.
10. Feitknecht, W. (1942). Zür kenntnis der doppelhydroxyde und basischen doppelsalze. Über magnesium-aluminium doppelhydroxyde. *Helv. Chim. Acta* **25**, 131.
11. Feitknecht, W. (1942). Über die bilding von doppelhydroxyden zwischen zwei- und dreiwertigen metallen. *Helv. Chim. Acta* **25**, 555.
12. Allmann, R. (1967). The crystal structure of pyroaurite. *Acta Crystallogr.* **B24**, 972.
13. Allmann, R. and Jepsen, H. P. (1969). Die struktur des hydrotalkits. *Neues Jahrb. Miner. Monatsh.* **12**, 544.

14. Taylor, H. F. W. Segregation and cation-ordering in sjögrenite and pyroaurite. *Mineral. Mag*, **37**, 338.
15. Ingram, L. and Taylor, H. F. W. (1967). The crystal structures of sjögrenite and pyroaurite. *Mineral. Mag*. **36**, 465.
16. Serna, C. J., Rendon, J. L. and Iglesias, J. E. (1982). Crystal-chemical study of layered $[Al_2Li(OH)_6]^+X^-.nH_2O$. *Clays Clay Miner.* **30**, 180.
17. Besserguenev, A. V., Fogg, A. M., Francis, R. J., Price, S. J., O'Hare, D., Isupov, V. P. and Tolochko, B. P. (1997). Synthesis and structure of the gibbsite intercalation compounds $[LiAl_2(OH)_6]X$ {X = Cl, Br, NO_3} and $[LiAl_2(OH)_6]Cl.H_2O$ using synchrotron X-ray and neutron powder diffraction. *Chem. Mater.* **9**, 241.
18. Bellotto, M., Rebours, B., Clause, O., Lynch, L., Bazin, D. and Elkaim, E. (1996). A reexamination of hydrotalcite crystal chemistry. *J. Phys. Chem.* **100**, 8527.
19. Bookin, A. S., Cherkashin, V. I. and Drits, V. A. (1993). Polytype diversity of the hydrotalcite-like minerals. 2. Determination of the polytypes of experimentally studied varieties. *Clays Clay Miner.* **41**, 558.
20. Bookin, A. S. and Drits, V. A. (1993). Polytype diversity of the hydrotalcite-like minerals. 1. Possible polytypes and their diffraction features. *Clays Clay Miner.* **41**, 551.
21. Crepaldi, E. L., Pavan, P. C. and Valim, J. B. (2000). Comparative study of the co-precipitation methods for the preparation of layered double hydroxides. *J. Braz. Chem. Soc.* **11**, 64.
22. Miyata, S. (1983). Anion-exchange properties of hydrotalcite-like compounds. *Clays Clay Miner.* **31**, 305.
23. Brindley, G. W. and Kikkawa, S. (1980). Thermal behavior of hydrotalcite and of anion-exchanged forms of hydrotalcite. *Clays Clay Miner.* **28**, 87.
24. Drezdzon, M. A. (1988). Synthesis of isopolymetalate-pillared hydrotalcite via organic-anion-pillared precursors. *Inorg. Chem.* **27**, 4628.
25. Bish, D. L. (1980). Anion-exchange in takovite: application to other hydroxide minerals. *Bull. Mineral.* **103**, 170.
26. Crepaldi, E. L., Pavan, P. C. and Valim, J. B. (1999). A new method of intercalation by anion exchange in layered double hydroxides. *J. Chem. Soc., Chem. Commun.*, 155.
27. Miyata, S. (1980). Physico-chemical properties of synthetic hydrotalcites in relation to composition. *Clays Clay Miner.* **28**, 50.
28. Sato, T., Fujita, H., Endo, T., Shimada, M. and Tsunashima, A. (1988). Synthesis of hydrotalcite-like compounds and their physicochemical properties. *React. Solids* **5**, 219.
29. Chibwe, K. and Jones, W. (1989). Synthesis of polyoxometalate-pillared layered double hydroxides via calcined precursors. *Chem. Mater.* **1**, 489.
30. Chibwe, K. and Jones, W. (1989). Intercalation of organic and inorganic anions into layered double hydroxides. *J. Chem. Soc., Chem. Commun.*, 926.
31. Rocha, J., del Arco, M., Rives, V. and Ulibarri, M. A. (1999). Reconstruction of layered double hydroxides from calcined precursors: a powder XRD and Al-27 MAS NMR study. *J. Mater. Chem.* **9**, 2499.
32. Vucelic, M. and Jones, W. (1992). 'Solid state ^{27}Al NMR studies of LDH-intercalates' in *Multifunctional Mesoporous Inorganic Solids* (C. A. C. Sequeira and M. J. Hudson, Eds.). Kluwer Academic Publishers, Dordrecht.

33. Dimotakis, E. D. and Pinnavaia, T. J. (1990). New route to layered double hydroxides intercalated by organic-anions – precursors to polyoxometalate-pillared derivatives. *Inorg. Chem.* **29**, 2393.

34. Atkins, M. P., Chibwe, M. and Jones, W. 'Process for the preparation of catalysts'. *European Patent Application No. EP 0 421 678 A1.*, 1991.

35. Trifiro, F. and Vaccari, A. (1996). 'Hydrotalcite-like anionic clays (layered double hydroxides)' in *Comprehensive Supramolecular Chemistry* (G. Alberti and T. Bein, Eds.). Pergamon, Oxford.

36. Newman, S. P. (1999). 'Structural studies of layered anion exchangers: simulation and measurement'. *PhD Thesis, University Of Cambridge.*

37. Shen, J., Kobe, J. M., Chen, Y. and Dumesic, J. A. (1994). Synthesis and surface acid/base properties of magnesium-aluminium mixed oxides obtained from hydrotalcites. *Langmuir* **10**, 3902.

38. Kooli, F., Depege, C., Ennaqadi, A., de Roy, A. and Besse, J. P. (1997). Rehydration of Zn-Al layered double hydroxides. *Clays Clay Miner.* **45**, 92.

39. Costantino, U., Marmottini, F., Nocchetti, M. and Vivani, R. (1998). New synthetic routes to hydrotalcite-like compounds – characterisation and properties of the obtained materials. *Eur. J. Inorg. Chem.*, 1439.

40. del Arco, M., Rives, V., Trujillano, R. and Malet, P. (1996). Thermal behaviour of Zn-Cr layered double hydroxides with hydrotalcite-like structures containing carbonate or decavanadate. *J. Mater. Chem.* **6**, 1419.

41. Boclair, J. W., Braterman, P. S., Jiang, J. P., Lou, S. W. and Yarberry, F. (1999). Layered double hydroxide stability. 2. Formation of Cr(III)-containing layered double hydroxides directly from solution. *Chem. Mater.* **11**, 303.

42. Prevot, V., Forano, C. and Besse, J. P. (1998). Syntheses and thermal and chemical behaviors of tartrate and succinate intercalated Zn_3Al and Zn_2Cr layered double hydroxides. *Inorg. Chem.* **37**, 4293.

43. Labajos, F. M. and Rives, V. (1996). Thermal evolution of chromium(III) ions in hydrotalcite-like compunds. *Inorg. Chem.* **35**, 5313.

44. Boclair, J. W. and Braterman, P. S. (1998). One-step formation and characterization of Zn(II)-Cr(III) layered double hydroxides, $Zn_2Cr(OH)_6X$ (X = Cl, $1/2SO_4$). *Chem. Mater.* **10**, 2050.

45. Thompson, H. A., Parks, G. A. and Brown, G. E. (1999). Ambient-temperature synthesis, evolution, and characterization of cobalt-aluminum hydrotalcite-like solids. *Clays Clay Miner.* **47**, 425.

46. Zeng, H. C., Xu, Z. P. and Qian, M. (1998). Synthesis of non-Al-containing hydrotalcite-like compound $Mg_{0.3}Co(II)_{0.6}Co(III)_{0.2}(OH)_2(NO_3)_{0.2} \cdot H_2O$. *Chem. Mater.* **10**, 2277.

47. Ulibarri, M. A., Fernandez, J. M., Labajos, F. M. and Rives, V. (1991). Anionic clays with variable valence cations: synthesis and characterization of $[Co_{1-x}Al_x(OH)_2](CO_3)_{x/2} \cdot nH_2O$. *Chem. Mater.* **3**, 626.

48. del Arco, M., Trujillano, R. and Rives, V. (1998). Cobalt-iron hydroxycarbonates and their evolution to mixed oxides with spinel structure. *J. Mater. Chem.* **8**, 761.

49. Fernandez, J. M., Barriga, C., Ulibarri, M. A., Labajos, F. M. and Rives, V. (1994). Preparation and thermal-stability of manganese-containing hydrotalcite, $[Mg_{0.75}Mn(II)_{0.04}Mn(III)_{0.21}(OH)_2](CO_3)_{0.11} \cdot nH_2O$. *J. Mater. Chem.* **4**, 1117.

50. Crepaldi, E. L., Pava, P. C., Valim, J. B. and Jones, W. (2000). Synthesis and thermal behaviour of copper-chromium-carbonate layered double hydroxides, in preparation.

322		*S. P. Newman and W. Jones*

51. Rives, V., Labajos, F. M., Ulibarri, M. A. and Malet, P. (1993). A new hydrotalcite-like compound containing V^{3+} ions in the layers. *Inorg. Chem.* **32**, 5000.
52. Labajos, F. M., Sastre, M. D., Trujillano, R. and Rives, V. (1999). New layered double hydroxides with the hydrotalcite structure containing Ni(II) and V(III). *J. Mater. Chem.* **9**, 1033.
53. Challier, T. and Slade, R. C. T. (1994). Nanocomposite materials – polyaniline-intercalated layered double hydroxides. *J. Mater. Chem.* **4**, 367.
54. Alejandre, A., Medina, F., Salagre, P., Correig, X. and Sueiras, J. E. (1999). Preparation and study of Cu-Al mixed oxides via hydrotalcite-like precursors. *Chem. Mater.* **11**, 939.
55. Kooli, F., Rives, V. and Jones, W. (1997). Reduction of Ni^{2+}-Al^{3+} and Cu^{2+}-Al^{3+} layered double hydroxides to metallic Ni^o and Cu^o via polyol treatment. *Chem. Mater.* **9**, 2231.
56. Aramendia, M. A., Aviles, Y., Borau, V., Luque, J. M., Marinas, J. M., Ruiz, J. R. and Urbano, F. J. (1999). Thermal decomposition of Mg Al and Mg Ga layered-double hydroxides: a spectroscopic study. *J. Mater. Chem.* **9**, 1603.
57. Aramendia, M. A., Borau, V., Jimenez, C., Marinas, J. M., Romero, F. J. and Urbano, F. J. (1999). Synthesis and characterization of a novel Mg/In layered double hydroxide. *J. Mater. Chem.* **9**, 2291.
58. Vichi, M. F. and Alves, O. L. (1997). Preparation of Cd/Al layered double hydroxides and their intercalation reactions with phosphonic acids. *J. Mater. Chem.* **7**, 1631.
59. Kooli, F., Kosuge, K., Hibino, T. and Tsunashima, A. (1993). Synthesis and properties of Mg-Zn-Al-SO_4 hydrotalcite-like compounds. *J. Mater. Sci.* **28**, 2769.
60. Basile, F., Basini, L., Fornasari, M., Gazzano, M., Trifiro, F. and Vaccari, A. (1996). New hydrotalcite-type anionic clays containing noble metals. *J. Chem. Soc., Chem. Commun.*, 2435.
61. Fernandez, J. M., Barriga, C., Ulibarri, M. A., Labajos, F. M. and Rives, V. (1997). New hydrotalcite-like compounds containing yttrium. *Chem. Mater.* **9**, 312.
62. Velu, S., Suzuki, K., Okazaki, M., Osaki, T., Tomura, S. and Ohashi, F. (1999). Synthesis of new Sn-incorporated layered double hydroxides and their thermal evolution to mixed oxides. *Chem. Mater.* **11**, 2163.
63. Velu, S., Sabde, D. P., Shah, N. and Sivasanker, S. (1998). New hydrotalcite-like anionic clays containing Zr^{4+} in the layers: synthesis and physicochemical properties. *Chem. Mater.* **10**, 3451.
64. Velu, S., Ramaswamy, V., Ramani, A., Chanda, B. M. and Sivasanker, S. (1997). New hydrotalcite-like anionic clays containing Zr^{4+} in the layers. *J. Chem. Soc., Chem. Commun.*, 2107.
65. Vucelic, M., Jones, W. and Moggridge, G. D. (1997). Cation ordering in synthetic layered double hydroxides. *Clays Clay Miner.* **45**, 803.
66. Vucelic, M., Moggridge, G. D. and Jones, W. (1995). Thermal-properties of terephthalate-intercalated and benzoate-intercalated LDH. *J. Phys. Chem.* **99**, 8328.
67. Chisem, I. C. and Jones, W. (1994). Ion-exchange properties of lithium aluminum layered double hydroxides. *J. Mater. Chem.* **4**, 1737.
68. Dutta, P. K. and Puri, M. (1989). Anion exchange in lithium aluminate hydroxides. *J. Phys. Chem.* **93**, 376.

69. Vaccari, A. (1998). Preparation and catalytic properties of cationic and anionic clays. *Catal. Today* **41**, 53.
70. Constantino, V. R. L. and Pinnavaia, T. J. (1994). Structure–reactivity relationships for basic catalysts derived from a $Mg^{2+}/Al^{3+}/CO_3^{2-}$ layered double hydroxide. *Catal. Lett.* **23**, 361.
71. Martin, K. J. and Pinnavaia, T. J. (1986). Layered double hydroxides as supported anionic reagents. Halide ion reactivity in $[Zn_2Cr(OH)_6]X.nH_2O$. *J. Am. Chem. Soc.* **108**, 541.
72. Yamaguchi, K., Mizugaki,T., Ebitani, K. and Kaneda, K. (1999). Heterogeneous N-oxidation of pyridines using a combined oxidant of hydrogen peroxide and nitriles catalysed by basic hydrotalcites. *New J. Chem.* **23**, 799.
73. Ueno, S., Ebitani, K., Ookubo, A. and Kaneda, K. (1997). The active sites in the heterogeneous Baeyer–Villiger oxidation of cyclopentanone by hydrotalcite catalysts. *Appl. Surf. Sci.* **121**, 366.
74. Ueno, S., Yamaguchi, K., Yoshida, K., Ebitani, K. and Kaneda, K. (1998). Hydrotalcite catalysis: heterogeneous epoxidation of olefins using hydrogen peroxide in the presence of nitriles. *J. Chem. Soc., Chem. Commun.*, 295.
75. Kantam, M. L., Choudary, B. M., Reddy, C. V., Rao, K. K. and Figueras, F. (1998). Aldol and Knoevenagel condensations catalysed by modified Mg-Al hydrotalcite: a solid base as catalyst useful in synthetic organic chemistry. *J. Chem. Soc., Chem. Commun.*, 1033.
76. Rao, K. K., Gravelle, M., Valente, J. S. and Figueras, F. (1998). Activation of Mg-Al hydrotalcite catalysts for aldol condensation reactions. *J. Catal.* **173**, 115.
77. Kumbhar, P. S., Sanchez-Valente, J. and Figueras, F. (1998). Modified Mg-Al hydrotalcite: a highly active heterogeneous base catalyst for cyanoethylation of alcohols. *J. Chem. Soc., Chem. Commun.*, 1091.
78. Choudary, B. M., Kantam, M. L., Bharathi, B. and Reddy, C. V. (1998). Superactive Mg-Al-O-t-Bu hydrotalcite for epoxidation of olefins. *Synlett.*, 1203.
79. Choudary, B. M., Kantam, M. L., Kavita, B., Reddy, C. V., Rao, K. K. and Figueras, F. (1998). Aldol condensations catalysed by novel Mg-Al-O-t-Bu hydrotalcite. *Tetrahedron Letts.* **39**, 3555.
80. Rives, V. and Ulibarri, M. A. (1999). Layered double hydroxides (LDH) intercalated with metal coordination compounds and oxometalates. *Coord. Chem. Rev.* **181**, 61.
81. Kwon, T., Tsigdinos, G. A. and Pinnavaia, T. J. (1988). Pillaring of layered double hydroxides (LDHs) by polyoxometalate anions. *J. Am. Chem. Soc.* **110**, 3653.
82. Evans, J., Pillinger, M. and Zhang, J. (1996). Structural studies of polyoxometalate-anion-pillared layered double hydroxides. *J. Chem. Soc., Dalton Trans.*, 2963.
83. Narita, E., Kaviratna, P. D. and Pinnavaia, T. J. (1993). Direct synthesis of a polyoxometalate-pillared layered double hydroxide by coprecipitation. *J. Chem. Soc., Chem. Commun.*, 60.
84. Kooli, F. and Jones, W. (1995). Direct synthesis of polyoxovanadate-pillared layered double hydroxides. *Inorg. Chem.* **34**, 6237.
85. Kooli, F., Rives, V. and Ulibarri, M. A. (1995). Preparation and study of decavanadate-pillared hydrotalcite-like anionic clays containing transition metal cations in the layers. 1. Samples containing nickel-aluminium prepared by anionic exchange and reconstruction. *Inorg. Chem.* **34**, 5114.

86. Ulibarri, M. A., Labajos, F. M., Rives, V., Trujillano, R., Kagunya, W. and Jones, W. (1994). Comparative-study of the synthesis and properties of vanadate-exchanged layered double hydroxides. *Inorg. Chem.* **33**, 2592.

87. Tatsumi, T., Yamamoto, K., Tajima, H. and Tominaga, H. (1992). Shape selective epoxidation of alkenes catalysed by polyoxometalate-intercalated hydrotalcite. *Chem. Lett.*, 815.

88. Boclair, J. W., Braterman, P. S., Brister, B. D. and Yarberry, F. (1999). Layer-anion interactions in magnesium aluminum layered double hydroxides intercalated with cobalticyanide and nitroprusside. *Chem. Mater.* **11**, 2199.

89. Miyata, S. and Kumura, T. (1973). Synthesis of new hydrotalcite-like compounds and their physico-chemical properties. *Chem. Lett.*, 843.

90. Crespo, I., Barriga, C., Rives, V. and Ulibarri, M. A. (1997). Intercalation of iron hexacyano complexes in Zn,Al-hydrotalcite. *Solid State Ionics* **101**, 729.

91. Lopez-Salinas, E., Tomita, N., Matsui, T., Suzuki, E. and Ono, Y. (1993). Reactivity of interstitial $[NiCl_4]^{2-}$ complex anion in a Mg-Al hydrotalcite-like clay towards alkyl bromides in the liquid phase. *J. Mol. Catal.* **81**, 397.

92. Carlino, S. (1997). The intercalation of carboxylic acids into layered double hydroxides: a critical evaluation and review of the different methods. *Solid State Ionics* **98**, 73.

93. Newman, S. P. and Jones, W. (1998). Synthesis, characterization and applications of layered double hydroxides containing organic guests. *New J. Chem.* **22**, 105.

94. Kooli, F., Chisem, I. C., Vucelic, M. and Jones, W. (1996). Synthesis and properties of terephthalate and benzoate intercalates of Mg-Al layered double hydroxides possessing varying layer charge. *Chem. Mater.* **8**, 1969.

95. Boehm, H.-P., Steinle, J. and Vieweger, C. (1977). $[Zn_2Cr(OH)_6]X \cdot 2H_2O$, new layer compounds capable of anion exchange and intracrystalline swelling. *Angew. Chem. Int. Ed. Engl.* **16**, 265.

96. Clearfield, A., Kieke, M., Kwan, J., Colon, J. L. and Wang, R. C. (1991). Intercalation of dodecyl-sulfate into layered double hydroxides. *J. Inclusion Phenom. Mol. Recognit. Chem.* **11**, 361.

97. Franklin, K. R., Lee, E. and Nunn, C. C. (1995). Preparation and characterization of layered double hydroxides containing monovalent and divalent ions derived from 5-benzoyl-4-hydroxy-2-methoxybenzenesulfonic acid. *J. Mater. Chem.* **5**, 565.

98. Kanezaki, E., Kinugawa, K. and Ishikawa, Y. (1994). Conformation of intercalated aromatic molecular anions between layers of Mg/Al-hydrotalcites and Zn/Al-hydrotalcites. *Chem. Phys. Lett.* **226**, 325.

99. Newman, S. P., Williams, S. J., Coveney, P. V. and Jones, W. (1998). Interlayer arrangement of hydrated MgAl layered double hydroxides containing guest terephthalate anions: Comparison of simulation and measurement. *J. Phys. Chem. B* **102**, 6710.

100. Kaneyoshi, M. and Jones, W. (1998). Exchange of interlayer terephthalate anions from a Mg-Al layered double hydroxide: formation of intermediate interstratified phases. *Chem. Phys. Lett.* **296**, 183.

101. Ijdo, W. L. and Pinnavaia, T. J. (1999). Solid solution formation in amphiphilic organic-inorganic clay heterostructures. *Chem. Mater.* **11**, 3227.

102. Fogg, A. M., Dunn, J. S. and O'Hare, D. (1998). Formation of second-stage intermediates in anion-exchange intercalation reactions of the layered double hydroxide $[LiAl_2(OH)_6]Cl \cdot H_2O$ as observed by time-resolved, *in situ* X-ray diffraction. *Chem. Mater.* **10**, 356.

103. Ikeda, T., Amoh, H. and Yasunaga, T. (1984). Stereoselective exchange kinetics of L-histidine and D-histidine for Cl⁻ in the interlayer of a hydrotalcite-like compound by the chemical relaxation method. *J. Am. Chem. Soc.* **106**, 5772.

104. Sato, T. and Okuwaki, A. (1991). Intercalation of benzenecarboxylate ions into the interlayer of hydrotalcite. *Solid State Ionics* **45**, 43.

105. Tagaya, H., Sato, S., Morioka, H., Kadokawa, J., Karasu, M. and Chiba, K. (1993). Preferential intercalation of isomers of naphthalenecarboxylate ions into the interlayer of layered double hydroxides. *Chem. Mater.* **5**, 1431.

106. Kuk, W. K. and Huh, Y. D. (1997). Preferential intercalation of isomers of anthraquinone sulfonate ions into layered double hydroxides. *J. Mater. Chem.* **7**, 1933.

107. Fogg, A. M., Green, V. M., Harvey, H. G. and O'Hare, D. (1999). New separation science using shape-selective ion exchange intercalation chemistry. *Adv. Mater.* **11**, 1466.

108. Fogg, A. M., Dunn, J. S., Shyu, S. G., Cary, D. R. and O'Hare, D. (1998). Selective ion-exchange intercalation of isomeric dicarboxylate anions into the layered double hydroxide [LiAl$_2$(OH)$_6$]Cl .H$_2$O. *Chem. Mater.* **10**, 351.

109. Takagi, K., Shichi, T., Usami, H. and Sawaki, Y. (1993). Controlled photocycloaddition of unsaturated carboxylates intercalated in hydrotalcite clay interlayers. *J. Am. Chem. Soc.* **115**, 4339.

110. Sasai, R., Shinya, N., Shichi, T., Takagi, T. and Gekko, K. (1999). Molecular alignment and photodimerization of 4'-chloro-4-stilbenecarboxylic acid in hydrotalcite clays: bilayer formation in the interlayers. *Langmuir* **15**, 413.

111. Valim, J., Kariuki, B. M., King, J. and Jones, W. (1992). Photoactivity of cinnamate-intercalates of layered double hydroxides. *Mol. Cryst. Liq. Cryst.* **211**, 271.

112. Jones, W. (1997). 'Reactivity and crystal design in organic solid state chemistry' in *Organic Molecular Solids. Properties and Applications* (W. Jones, Ed.). CRC Press, Boca Raton, US.

113. Shichi, T., Takagi, K. and Sawaki, Y. (1996). Stereoselectivity control of [2 + 2] photocycloaddition by changing the site distances of hydrotalcite interlayers. *J. Chem. Soc., Chem. Commun.,* 2027.

114. Takagi, K., Harata, E., Shichi, T., Kanoh, T. and Sawaki, Y. (1997). Intercalation and control of the Norrish type II reactions of aromatic ketocarboxylates in hydrotalcite clay interlayers. *J. Photochem. Photobiol. A-Chem.* **105**, 47.

115. Tagaya, H., Kuwahara, T., Sato, S., Kadokawa, J., Karasu, M. and Chiba, K. (1993). Photoisomerization of indolinespirobenzopyran in layered double hydroxides. *J. Mater. Chem.* **3**, 317.

116. Tagaya, H., Sato, S., Kuwahara, T., Kadokawa, J., Masa, K. and Chiba, K. (1994). Photoisomerization of indolinespirobenzopyran in anionic clay matrices of layered double hydroxides. *J. Mater. Chem.* **4**, 1907.

117. Kuwahara, T., Tagaya, H. and Chiba, K. (1995). Photochromism of spiropyran dye in Li-Al layered double hydroxide. *Microporous Mater.* **4**, 247.

118. Park, I. Y., Kuroda, K. and Kato, C. (1990). Direct synthesis of intercalation compounds between a layered double hydroxide and an anionic dye. *J. Chem. Soc., Dalton Trans.,* 3071.

119. Costantino, U., Coletti, N., Nocchetti, M., Aloisi, G. G. and Elisei, F. (1999). Anion exchange of methyl orange into Zn-Al synthetic hydrotalcite and photophysical characterization of the intercalates obtained. *Langmuir* **15**, 4454.

120. Robins, D. S. and Dutta, P. K. (1996). Examination of fatty-acid exchanged layered double hydroxides as supports for photochemical assemblies. *Langmuir* **12**, 402.

121. Carrado, K. A., Forman, J. E., Botto, R. E. and Winans, R. E. (1993). Incorporation of phthalocyanines by cationic and anionic clays via ion-exchange and direct synthesis. *Chem. Mater.* **5**, 472.

122. Ukrainczyk, L., Chibwe, M., Pinnavaia, T. J. and Boyd, S. A. (1994). ESR study of cobalt(II) tetrakis(N-methyl-4-pyridiniumyl)porphyrin and cobalt(II) tetrasulfophthalocyanine intercalated In layered aluminosilicates and a layered double hydroxide. *J. Phys. Chem.* **98**, 2668.

123. Ukrainczyk, L., Chibwe, M., Pinnavaia, T. J. and Boyd, S. A. (1995). Reductive dechlorination of carbon-tetrachloride in water catalyzed by mineral supported biomimetic cobalt macrocycles. *Environ. Sci. Technol.* **29**, 439.

124. Perez Bernal, M. E., Ruano Casero, R. and Pinnavaia, T. J. (1991). Catalytic autoxidation of 1-decanethiol by cobalt(II) phthalocyaninetetrasulfonate intercalated in a layered double hydroxide. *Catal. Lett.* **11**, 55.

125. Chibwe, M. and Pinnavaia, T. J. (1993). Stabilization of a cobalt(II) phthalocyanine oxidation catalyst by intercalation in a layered double hydroxide host. *J. Chem. Soc., Chem. Commun.*, 278.

126. Shannon, I. J., Maschmeyer, T., Sankar, G., Thomas, J. M., Oldroyd, R. D., Sheehy, M., Madill, D., Waller, A. M. and Townsend, R. P. (1997). A new cell for the collection of combined EXAFS/XRD data *in situ* during solid/liquid catalytic reactions. *Catal. Lett.* **44**, 23.

127. Bonnet, S., Forano, C., de Roy, A. and Besse, J. P. (1996). Synthesis of hybrid organo-mineral materials: anionic tetraphenylporphyrins in layered double hydroxides. *Chem. Mater.* **8**, 1962.

128. Park, I. Y., Kuroda, K. and Kato, C. (1989). Preparation of a layered double hydroxide-porphyrin intercalation compound. *Chem. Lett.*, 2057.

129. Barloy, L., Lallier, J. P., Battioni, P., Mansuy, D., Piffard, Y., Tournous, M., Valim, J. B. and Jones, W. (1992). Manganese porphyrins adsorbed or intercalated in different mineral matrices – preparation and compared properties as catalyst for alkene and alkane oxidation. *New J. Chem.* **16**, 71.

130. Kopka, H., Beneke, K. and Lagaly, G. (1988). Anionic surfactants between double metal hydroxide layers. *J. Colloid Interface Sci.* **123**, 427.

131. Adachi-Pagano, M., Forano, C. and Besse, J.-P. (2000). Delamination of layered double hydroxides by use of surfactants. *J. Chem. Soc., Chem. Commun.*, 91.

132. Dutta, P. K. and Robins, D. S. (1994). Pyrene sorption in organic layered double-metal hydroxides. *Langmuir* **10**, 1851.

133. Jakupca, M. and Dutta, P. K. (1995). Thermal and spectroscopic analysis of a fatty acid-layered double-metal hydroxide and its application as a chromatographic stationary-phase. *Chem. Mater.* **7**, 989.

134. Dekany, I., Berger, F., Imrik, K. and Lagaly, G. (1997). Hydrophobic layered double hydroxides (LDHs): Selective adsorbents for liquid mixtures. *Colloid Polym. Sci.* **275**, 681.

135. Tseng, W. Y., Lin, J. T., Mou, C. Y., Cheng, S. F., Liu, S. B., Chu, P. P. and Liu, H. W. (1996). Incorporation of C_{60} in layered double hydroxide. *J. Am. Chem. Soc.* **118**, 4411.

136. Zhao, H. and Vance, G. F. (1998). Molecular inclusion properties of hydrophobic organic compunds by a modified β-cyclodextrin intercalated within a LDH. *J. Inclus. Phenom. Mol. Recog. Chem.* **31**, 305.

137. Zhao, H. and Vance, G. F. (1997). Intercalation of carboxymethyl-β-cyclodextrin into magnesium-aluminium layered double hydroxides. *J. Chem. Soc., Dalton Trans.*, 1961.

138. Mousty, C., Therias, S., Forano, C. and Besse, J. P. (1994). Anion-exchanging clay-modified electrodes – synthetic layered double hydroxides intercalated with electroactive organic-anions. *J. Electroanal. Chem.* **374**, 63.

139. Therias, S. and Mousty, C. (1995). Electrodes modified with synthetic anionic clays. *Appl. Clay Sci.* **10**, 147.

140. Therias, S., Lacroix, B., Schollhorn, B., Mousty, C. and Palvadeau, P. (1998). Electrochemical study of ferrocene and nitroxide derivatives intercalated in Zn-Cr and Zn-Al layered double hydroxides. *J. Electroanal. Chem.* **454**, 91.

141. Gaillon, L., Bedioui, F., Devynck, J. and Battioni, P. (1993). Electrochemical characterization of manganese porphyrins fixed onto silica and layered dihydroxide matrices. *J. Electroanal. Chem.* **347**, 435.

142. Cooper, S. and Dutta, P. K. (1990). 4-Nitrohippuric acid in layered lithium aluminates – onset of nonlinear optical-properties. *J. Phys. Chem.* **94**, 114.

143. Prevot, V., Forano, C. and Besse, J. P. (1999). Reactivity of oxalate with ZnAl layered double hydroxides through new materials. *J. Mater. Chem.* **9**, 155.

144. Tagaya, H., Morioka, H., Ogata, S., Karasu, M., Kadokawa, J. and Chiba, K. (1997). Surface modification of inorganic layer compound with organic compound and preparation of thin films. *Appl. Surf. Sci.* **121**, 476.

145. Tagaya, H., Ogata, S., Nakano, S., Kadokawa, J.-I., Karasu, M. and Chiba, K. (1998). Intercalation of azo compounds into layered aluminium dihydrogentriphosphate and a layered double hydroxide. *J. Inclu. Phenom. Mol. Recog. Chem.* **31**, 231.

146. Morioka, H., Tagaya, H., Karasu, M., Kadokawa, J. and Chiba, K. (1995). Preparation of new useful materials by surface modification of inorganic layered compound. *J. Solid State Chem.* **117**, 337.

147. Wang, J. D., Serrette, Y., Tian, Y. and Clearfield, A. (1995). Synthetic and catalytic studies of inorganically pillared and organically pillared layered double hydroxides. *Appl. Clay Sci.* **10**, 103.

148. Carlino, S., Hudson, M. J., Husain, S. W. and Knowles, J. A. (1996). The reaction of molten phenylphosphonic acid with a layered double hydroxide and its calcined oxide. *Solid State Ionics* **84**, 117.

149. Nijs, H., Clearfield, A. and Vansant, E. F. (1998). The intercalation of phenylphosphonic acid in layered double hydroxides. *Microp. Mesop. Mater.* **23**, 97.

150. Constantino, U., Casciola, M., Massinelli, L., Nocchetti, M. and Vivani, R. (1997). Intercalation and grafting of hydrogen phosphates and phosphonates into synthetic hydrotalcites and a.c.-conductivity of the compounds thereby obtained. *Solid State Ionics* **97**, 203.

151. Constantino, U., Clementi, S., Nocchetti, S. and Vivani, R. (1998). Intercalation and grafting of n-alkyl phosphonates into synthetic hydrotalcites. *Mol. Cryst. Liq. Cryst.* **311**, 207.

152. Krishnamurthy, R., Pitsch, S. and Arrhenius, G. (1999). Mineral induced formation of pentose-2,4-biphosphates. *Origins Life Evol. Biosphere* **29**, 139.

153. Kolb, V., Zhang, S., Xu, Y. and Arrhenius, G. (1997). Mineral induced phosphorylation of glycolate ion – a metaphor in chemical evolution. *Origins Life Evol. Biosphere* **27**, 485.

154. Kuma, K., Paplawsky, W., Gedulin, B. and Arrhenius, G. (1989). Mixed-valence hydroxides as bioorganic host minerals. *Origins Life Evol. Biosphere* **19**, 573.

155. DeGraaf, R. M., Visscher, J., Xu, Y., Arrhenius, G. and Schwartz, A. W. (1998). Mineral catalysis of a potentially prebiotic aldol condensation. *J. Mol. Evol.* **47**, 501.

156. Pitsch, S., Eschenmoser, A., Gedulin, B., Hui, S. and Arrhenius, G. (1995). Mineral induced formation of sugar phosphates. 15. Chemistry of alpha-aminonitrile from the Zurich group. *Origins Life Evol. Biosphere* **25**, 297.

157. Fudala, A., Palinko, I. and Kiricsi, I. (1999). Preparation and characterization of hybrid organic–inorganic composite materials using the amphoteric property of amino acids: amino acid intercalated layered double hydroxide and montmorillonite. *Inorg. Chem.* **38**, 4653.

158. Whilton, N. T., Vickers, P. J. and Mann, S. (1997). Bioinorganic clays: synthesis and characterization of amino- and polyamino acid intercalated layered double hydroxides. *J. Mater. Chem.* **7**, 1623.

159. Reichle, W. T., Kang, S. Y. and Everhardt, D. S. (1986). The nature of the thermal-decomposition of a catalytically active anionic clay mineral. *J. Catal.* **101**, 352.

160. Bellotto, M., Rebours, B., Clause, O., Lynch, L., Bazin, D. and Elkaim, E. (1996). Hydrotalcite decomposition mechanism: a clue to the structure and reactivity of spinel-like mixed oxides. *J. Phys. Chem.* **100**, 8535.

161. Hibino, T. and Tsunashima, A. (1997). Formation of spinel from a hydrotalcite-like compound at low temperature: reaction between edges of crystallites. *Clays Clay Miner.* **45**, 842.

162. Hibino, T. and Tsunashima, A. (1998). Characterization of repeatedly reconstructed Mg-Al hydrotalcite-like compounds: gradual segregation of aluminum from the structure. *Chem. Mater.* **10**, 4055.

163. Hibino, T. (1995). Decarbonation behavior of Mg-Al-CO_3 hydrotalcite-like compounds during heat treatment. *Clays Clay Miner.* **43**, 427.

164. Kagunya, W., Baddour Hadjean, R., Kooli, F. and Jones, W. (1998). Vibrational modes in layered double hydroxides and their calcined derivatives. *Chem. Phys.* **236**, 225.

165. Depage, C., Forano, C., de Roy, A. and Besse, J. P. (1994). [Cu-Cr] layered double hydroxides pillared by CrO_4^{2-} and $Cr_2O_7^{2-}$ oxometalates. *Mol. Cryst. Liq. Cryst.* **244**, 161.

166. Tanabe, K. and Holderich, W. F. (1999). Industrial application of solid-base catalysts. *Appl. Catal. A.* **181**, 399.

167. Reichle, W. T. (1985). Catalytic reactions by thermally activated, synthetic, anionic clay minerals. *J. Catal.* **94**, 547.

168. Dumitriu, E., Hulea, V., Chelaru, C., Catrinescu, C., Tichit, D. and Durand, R. (1999). Influence of the acid–base properties of solid catalysts derived from hydrotalcite-like compounds on the condensation of formaldehyde and acetaldehyde. *Appl. Catal. A* **178**, 145.

169. Reichle, W. T. (1980). Pulse microreactor examination of the vapor-phase aldol condensation of acetone. *J. Catal.* **63**, 295.

170. Climent, M. J., Corma, A., Iborra, S. and Primo, J. (1995). Base catalysis for fine chemicals production: Claisen Schmidt condensation on zeolites and hydrotalcites for the production of chalcones and flavanones of pharmaceutical interest. *J. Catal.* **151**, 60.

171. Climent, M. J., Corma, A., Guil-Lopez, R., Iborra, S. and Primo, J. (1999). Solid catalysts for the production of fine chemicals: the use of ALPON and hydrotalcite base catalysts for the synthesis of arylsulfones. *Catal. Lett.* **59**, 33.

172. Corma, A., Fornes, V., Martin-Aranda, R. M. and Rey, F. (1992). Determination of base properties of hydrotalcites: condensation of benzaldehyde with ethyl acetoacetate. *J. Catal.* **134**, 1992.

173. Kumbhar, P. S., Sanchez-Valente, J., Lopez, J. and Figueras, F. (1998). Meerwein–Ponndorf–Verley reduction of carbonyl compounds catalysed by Mg-Al hydrotalcite. *J. Chem. Soc., Chem. Commun.*, 535.

174. Bulbule, V. J., Deshpande, V. H., Velu, S., Sudalai, A., Sivasankar, S. and Sathe, V. T. (1999). Heterogeneous Henry reaction of aldehydes: diastereoselective synthesis of nitroalcohol derivatives over Mg-Al hydrotalcites. *Tetrahedron* **55**, 9325.

175. McKenzie, A. L., Fishel, C. T. and Davis, R. J. (1992). Investigation of the surface structure and properties of calcined hydrotalcites. *J. Catal.* **138**, 547.

176. Fishel, C. T. and Davis, R. J. (1994). Characterization of Mg-Al mixed oxides by temperature-programmed reaction of 2-propanol. *Langmuir* **10**, 159.

177. Fishel, C. T. and Davis, R. J. (1994). Use of catalytic reactions to probe Mg-Al mixed oxide surfaces. *Catal. Lett.* **25**, 87.

178. Davis, R. J., McKenzie, A. L. and Fishel, C. T. (1992). Catalytic reactions of 2-propanol on calcined hydrotalcites. *Abstr. Pap. Am. Chem. Soc.* **204**, 471.

179. Laycock, D. E., Collacott, R. J., Skelton, D. A. and Tchir, M. F. (1991). Stereospecific polymerisation of propylene oxide on thermally activated synthetic hydrotalcite. *J. Catal.* **130**, 354.

180. Vaccari, A. (1999). Clays and catalysis: a promising future. *Appl. Clay Sci.* **14**, 161.

181. Tichit, D., Lhouty, M. H., Guida, A., Chiche, B. H., Figueras, F., Auroux, A., Bartalini, D. and Garrone, E. (1995). Textural properties and catalytic activity of hydrotalcites. *J. Catal.* **151**, 50.

182. Corma, A., Fornes, V. and Rey, F. (1994). Hydrotalcites as base catalysts: influence of the chemical composition and synthesis conditions on the dehydrogenation of isopropanol. *J. Catal.* **148**, 205.

183. Shen, J. Y., Tu, M. and Hu, C. (1998). Structural and surface acid/base properties of hydrotalcite-derived MgAlO oxides calcined at varying temperatures. *J. Solid State Chem.* **137**, 295.

184. Di Cosimo, J. I., Diez, V. K., Xu, M., Iglesia, E. and Apesteguia, C. R. (1998). Structure and surface and catalytic properties of Mg-Al basic oxides. *J. Catal.* **178**, 499.

185. Basile, F., Basini, L., Damore, M., Fornasari, G., Guarinoni, A., Matteuzzi, D., DelPiero, G., Trifiro, F. and Vaccari, A. (1998). Ni/Mg/Al anionic clay derived catalysts for the catalytic partial oxidation of methane – residence time dependence of the reactivity features. *J. Catal.* **173**, 247.

186. Dung, N. T., Tichit, D., Chiche, B. H. and Coq, B. Influence of the thermal treatments of a (Ni + Mg)/Al layered double hydroxide in the hydrogenation of acetonitrile. *Appl. Catal. A* **169**, 179.

187. Cabello, F. M., Tichit, D., Coq, B., Vaccari, A. and Dung, N. T. (1997). Hydrogenation of acetonitrile on nickel-based catalysts prepared from hydrotalcite-like precursors. *J. Catal.* **167**, 142.

188. Auer, S. M., Wandeler, R., Gobel, U. and Baiker, A. (1997). Heterogeneous coupling of phenylethyne over Cu-Mg-Al mixed oxides – influence of catalyst composition and calcination temperature on structural and catalytic properties. *J. Catal.* **169**, 1.

189. Tu, M., Shen, J. Y. and Chen, Y. (1997). Microcalorimetric studies of surface acid/base properties of magnesium-iron catalysts prepared from hydrotalcite-type precursors. *J. Solid State Chem.* **128**, 73.

190. Kumbhar, P. S., Sanchez-Valente, J. and Figueras, F. (1998). Reduction of aromatic nitro compounds with hydrazine hydrate in the presence of iron(III) oxide-MgO catalyst prepared from a Mg-Fe hydrotalcite precursor. *Tetrahedron Letts.* **39**, 2573.

191. Velu, S. and Swamy, C. S. (1997). Effect of substitution of Fe^{3+}/Cr^{3+} on the alkylation of phenol with methanol over magnesium-aluminium calcined hydrotalcite. *Appl. Catal. A* **162**, 81.

192. Kaneda, K., Yamashita, T., Matsushita, T. and Ebitani, K. (1998). Heterogeneous oxidation of allylic and benzylic alcohols catalyzed by Ru-Al-Mg hydrotalcites in the presence of molecular oxygen. *J. Org. Chem.* **63**, 1750.

193. Matsushita, T., Ebitani, K. and Kaneda, K. (1999). Highly efficient oxidation of alcohols and aromatic compounds catalysed by the Ru-Co-Al hydrotalcite in the presence of molecular oxygen. *J. Chem. Soc., Chem. Commun.*, 265.

194. Corma, A., Palomares, A. E. and Rey, F. (1994). Optimization of SO_x additives of FCC catalysts based on $MgO-Al_2O_3$ mixed oxides produced from hydrotalcites. *Appl. Catal. B* **4**, 29.

195. Corma, A., Palomares, A. E., Rey, F. and Marquez, F. (1997). Simultaneous catalytic removal of SO_x and NO_x with hydrotalcite-derived mixed oxides containing copper, and their possibilities to be used in FCC units. *J. Catal.* **170**, 140.

196. Shannon, I. J., Rey, F., Sankar, G., Thomas, J. M., Maschmeyer, T., Waller, A. M., Palomares, A. E., Corma, A., Dent, A. J. and Greaves, G. N. (1996). Hydrotalcite-derived mixed oxides containing copper: catalysts for the removal of nitric oxide. *J. Chem. Soc., Faraday Trans.* **92**, 4331.

197. Kannan, S. (1998). Decomposition of nitrous oxide over the catalysts derived from hydrotalcite-like compounds. *Appl. Clay Sci.* **13**, 347.

198. Qian, M. and Zeng, H. C. (1997). Synthesis and characterization of Mg-Co catalytic oxide materials for low-temperature N_2O decomposition. *J. Mater. Chem.* **7**, 493.

199. Perez Ramirez, J., Kapteijn, F. and Moulijn, J. A. (1999). High activity and stability of the Rh-free Co-based ex-hydrotalcite containing Pd in the catalytic decomposition of N_2O. *Catal. Lett.* **60**, 133.

200. Perez Ramirez, J., Overeijnder, J., Kapteijn, F. and Moulijn, J. A. (1999). Structural promotion and stabilizing effect of Mg in the catalytic decomposition of nitrous oxide over calcined hydrotalcite-like compounds. *Appl. Catal. B* **23**, 59.

201. Oi, J., Obuchi, A., Ogata, A., Bamwenda, G. R., Tanaka, R., Hibino, T. and Kushiyama, S. (1997). Zn, Al, Rh-mixed oxides derived from hydrotalcite-like compound and their catalytic properties for N_2O decomposition. *Appl. Catal. B* **13**, 197.

202. Trombetta, M., Ramis, G., Busca, G., Montanari, B. and Vaccari, A. (1997). Ammonia adsorption and oxidation on Cu/Mg/Al mixed oxide catalysts prepared via hydrotalcite-type precursors. *Langmuir* **13**, 4628.

203. Davis, R. J. and Derouane, E. G. (1991). A non-porous supported-platinum catalyst for aromatization of n-hexane. *Nature* **349**, 313.
204. Narayanan, S. and Krishna, K. (1997). Structural influence of hydrotalcite on Pd dispersion and phenol hydrogenation. *J. Chem. Soc., Chem. Commun.*, 1991.
205. Narayanan, S. and Krishna, K. (1998). Hydrotalcite-supported palladium catalysts Part I: preparation, characterization of hydrotalcites and palladium on uncalcined hydrotalcites for CO chemisorption and phenol hydrogenation. *Appl. Catal. A* **174**, 221.
206. Narayanan, S. and Krishna, K. (1998). Palladium mixed hydroxide catalysts for phenol hydrogenation. *Stud. Surf. Sci. Catal.* **113**, 359.
207. Narayanan, S. and Krishna, K. (1999). Structure activity relationship in Pd/hydrotalcite: effect of calcination of hydrotalcite on palladium dispersion and phenol hydrogenation. *Catal. Today* **49**, 57.
208. Hermosin, M. C., Pavlovic, I., Ulibarri, M. A. and Cornejo, J. (1996). Hydrotalcite as sorbent for trinitrophenol: sorption capacity and mechanism. *Wat. Res.* **30**, 171.
209. Ulibarri, M. A., Pavlovic, I., Hermosin, M. C. and Cornejo, J. (1995). Hydrotalcite-like compounds as potential sorbents of phenols from water. *Appl. Clay Sci.* **10**, 131.
210. Kaneyoshi, M. and Jones, W. (1999). Formation of Mg-Al layered double hydroxides intercalated with nitrilotriacetate anions. *J. Mater. Chem.* **9**, 805.
211. Olguin, M. T., Bosch, P., Acosta, D. and Bulbulian, S. (1998). $^{131}I^-$ sorption by thermally treated hydrotalcites. *Clays Clay Miner.* **46**, 567.

10

Molecular machines

FRANÇISCO M. RAYMO AND J. FRASER STODDART

10.1 Introduction

Transistors [ref. 1] are solid state electronic devices capable of amplification and switching operations. By varying an input signal (current or voltage), they are switched reversibly between a *State 0* and a *State 1*. The two states differ in the magnitude of an output signal (current or voltage) that is amplified on going from *State 0* to *State 1*. By interconnecting appropriately several transistors, electronic circuits able to solve logic functions [ref. 2] can be fabricated. The complexity of a logic function and the speed at which it is carried out by an electronic circuit are related to the number of transistors integrated within the circuit. Thus, one of the major goals of the electronic industry is to build ultra-densely integrated electronic circuits containing the largest possible number of transistors. This objective can only be achieved by reducing as much as possible the sizes of these switching devices. Indeed, over the past decades the dimensions of transistors have shrunk at an exponential rate and the smallest commercially available transistors have reached micrometer sizes. Stimulated by this tremendous pace toward miniaturization, a number of researchers have envisaged [ref. 3] the possibility of designing and fabricating molecular-sized electronic devices. In particular, bistable molecular and supramolecular [ref. 4] systems that can be switched [ref. 5] reversibly between a *State 0* and a *State 1* by using external control are expected to lead to a new generation of sub-nanometer-sized transistors ideal for the fabrication of ultradensely integrated electronic circuits and the computers of the future.

Molecules incorporating interlocked [refs. 6, 7] components (Fig. 10.1) and their supramolecular analogs are suitable candidates for the generation of bistable chemical systems. A [2]pseudorotaxane is a *supramolecular complex* composed of a macrocyclic host encircling a linear guest. The two components are held together solely by noncovalent bonding interactions and they can

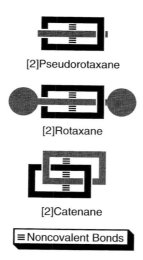

[2]Pseudorotaxane

[2]Rotaxane

[2]Catenane

≡ Noncovalent Bonds

Figure 10.1. Schematic representations of a [2]pseudorotaxane, a [2]rotaxane, and a [2]catenane.

become dissociated from one another in solution. By contrast, bulky stoppers attached covalently to the termini of the linear component of a [2]rotaxane prevent dethreading. Similarly, the two interlocked macrocyclic components of a [2]catenane cannot dissociate from one another unless at least one covalent bond is cleaved. Thus, even although they are composed of two 'unbound' components, [2]rotaxanes and [2]catenanes behave as *single molecules*. These intriguing superstructural and structural features possessed by [2]pseudorotaxanes and by [2]rotaxanes and [2]catenanes, respectively, could pave the way for the realization of bistable systems. In many instances, the relative movements [ref. 8] of the interlocked components can be triggered by chemical, electrochemical, and photochemical stimuli (input signal) forcing the molecule or the complex to switch between two nondegenerate states (*State 0* and *State 1*) that can be distinguished spectroscopically (output signal).

10.2 [2]Pseudorotaxanes

The complexation/decomplexation of a linear guest by a macrocyclic host can be exploited [ref. 7] (Fig. 10.2) to switch reversibly between two states. When the two components are dissociated, the system is in its *State 0*. When they are associated in the form of a [2]pseudorotaxane, the system is in its *State 1*. Switching between these two states can be achieved [ref. 7] by protonation/deprotonation, by oxidation/reduction, or by photoinduced isomerization of the guest. Alternatively, the addition/removal of a sequestering agent able to

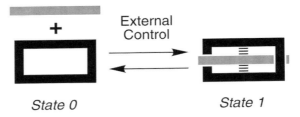

Figure 10.2. Externally controlled switching between a [2]pseudorotaxane (*State 1*) and its separate components (*State 0*).

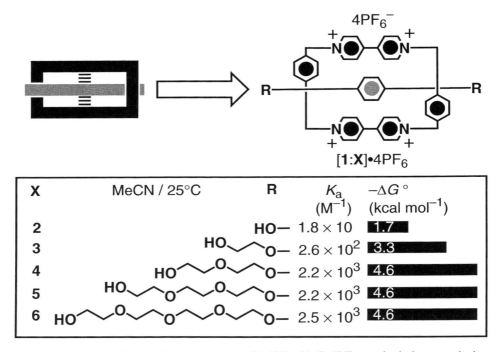

X	MeCN / 25°C	R	K_a (M^{-1})	$-\Delta G°$ (kcal mol^{-1})
2		HO—	1.8×10	1.7
3	HO$\diagdown$O—		2.6×10^2	3.3
4	HO$\diagdown$O$\diagdown$O—		2.2×10^3	4.6
5	HO$\diagdown$O$\diagdown$O$\diagdown$O—		2.2×10^3	4.6
6	HO$\diagdown$O$\diagdown$O$\diagdown$O$\diagdown$O—		2.5×10^3	4.6

Figure 10.3. The [2]pseudorotaxanes [**1:2**]•4PF$_6$–[**1:6**]•4PF$_6$ and their association constants (K_a) and free energies of complexation ($-\Delta G°$) in MeCN at 25 °C.

form a strong adduct with the guest can be exploited [ref. 7] to switch the system between *State 0* and *State 1*. An ideal macrocyclic host for the realization of such bistable systems is the tetracationic cyclophane **1**•4PF$_6$ illustrated in Fig. 10.3. This host incorporates two dicationic bipyridinium recognition sites and is able to bind [refs. 9, 10] π-electron rich aromatic polyethers inside its cavity. The association constants (K_a) and the free energies of complexation ($-\Delta G°$) for the binding of the guests **2**–**6** reveal [refs. 9a, b] that the number of oxygen atoms incorporated in the substituents attached to the aromatic ring of the guest influences dramatically the binding event. The K_a value increases by one

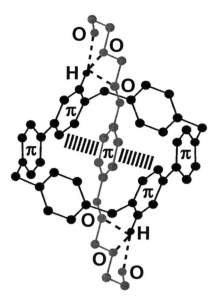

Figure 10.4. The geometry adopted by the [2]pseudorotaxane [**1:4**]$^{4+}$ in the solid state and the intercomponent [C–H···O] and [π···π] stacking interactions.

order of magnitude approximately on going from **2** to **3** and from **3** to **4**. Thereafter, however, it remains more or less constant on going from **4** to **5** and from **5** to **6**. These results indicate that the first three oxygen atoms are essential to the realization of strong binding between host and guest. Indeed, the solid state structure of the [2]pseudorotaxane [**1:4**]$^{4+}$ reveals [ref. 9b] (Fig. 10.4) intercomponent [C−H ··· O] interactions [ref. 11] between some of the bipyridinium hydrogen atoms in the α-position with respect to the nitrogen atoms and the three oxygen atoms of each polyether chain. These noncovalent bonding interactions are accompanied by [π ··· π] stacking [ref. 12] between the π-electron deficient bipyridinium units and the π-electron rich 1,4-dioxybenzene ring. When the π-electron density of the aromatic ring of the guest is reduced by attaching four (**7**) or two (**8**) electron withdrawing fluorine substituents, no or only weak complexation is detected [refs. 9c, e], respectively (Fig. 10.5). By contrast, when a 1,5-dioxynaphthalene recognition site (**9**) is incorporated in the guest a much stronger complex is formed [ref. 9d]. The remarkable stability of the complex [**1:9**]•4PF$_6$ was exploited [ref. 9d] to template [ref. 13] (Fig. 10.6) the formation of **1**•4PF$_6$ from its precursors. When **10**•2PF$_6$ and **11** are reacted under high pressure conditions in the presence of **9**, **1**•4PF$_6$ is obtained in a yield of 81%, after counterion exchange.

The host **1**•4PF$_6$ and the guest **12** form [ref. 14] (step 1 in Fig. 10.7) spontaneously the [2]pseudorotaxane [**1:12**]•4PF$_6$ when mixed in CD$_3$CN. Protonation

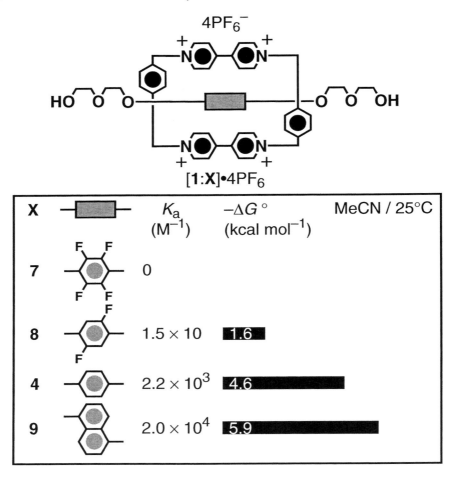

Figure 10.5. The [2]pseudorotaxanes [**1:4**]•4PF$_6$ and [**1:7**]•4PF$_6$–[**1:9**]•4PF$_6$ and their association constants (K_a) and free energies of complexation ($-\Delta G°$) in MeCN at 25 °C.

of the amino groups of the guest occurs after the addition (step 2 in Fig. 10.7) of CF$_3$CO$_2$H. The resulting dicationic guest is expelled (step 3 in Fig. 10.7) from the cavity of the tetracationic host as a result of electrostatic repulsion. Deprotonation of the ammonium centers of the guest occurs after the addition (step 4 in Fig. 10.7) of *i*-Pr$_2$NEt. The regenerated neutral form of the guest reinserts itself inside the cavity of the host, yielding (step 1 in Fig. 10.7) the original [2]pseudorotaxane [**1:12**]•4PF$_6$. As a result of [π ⋯ π] stacking interactions between the π-electron deficient bipyridinium units and the π-electron rich 1,4-dioxybenzene ring, the occurrence of complexation and the decomplexation steps are accompanied by the appearance and disappearance, respectively, of the charge transfer band. As a result, the acid/base-controlled

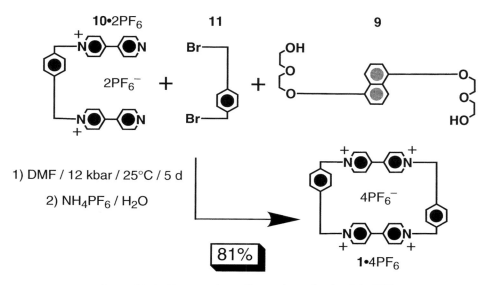

Figure 10.6. The template-directed synthesis of **1**•4PF$_6$.

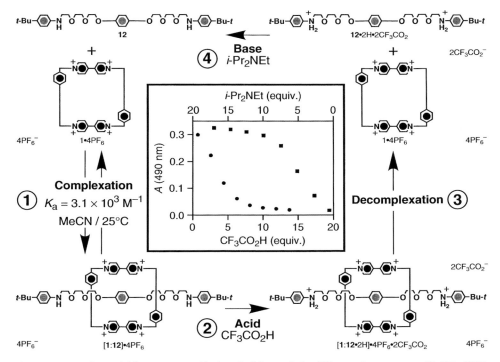

Figure 10.7. The acid/base-controlled switching of the [2]pseudorotaxane [**1**:**12**]•4PF$_6$ and the change in the absorbance (*A*) of the charge transfer band (λ_{max} = 490 nm) associated with [**1**:**12**]•4PF$_6$ upon addition of CF$_3$CO$_2$H and of *i*-Pr$_2$NEt is shown.

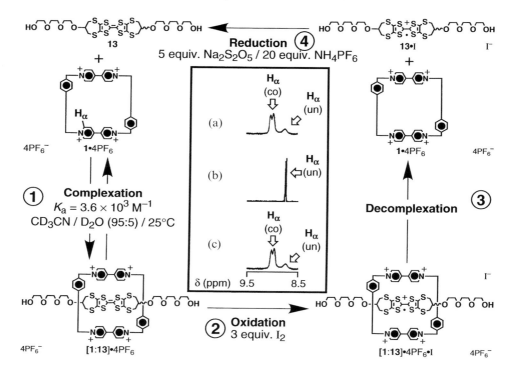

Figure 10.8. The redox-controlled switching of the [2]pseudorotaxane [**1**:**13**]•4PF$_6$ and the partial ^{1}H-NMR spectra [400 MHz, CD$_3$CN/D$_2$O (95:5), 25°C] of an equimolar mixture of **1**•4PF$_6$ and **13** (a) after complexation, (b) after oxidation and decomplexation, and (c) after reduction and complexation. The bipyridinium hydrogen atoms (H$_\alpha$) in the α-positions with respect to the nitrogen atoms were used as the probe protons and (co) and (un) stand for complexed and uncomplexed species, respectively.

switching [ref. 15] of this system can be followed (inset in Fig. 10.7) by absorption spectroscopy. In the absence of CF$_3$CO$_2$H, a charge transfer band (λ_{max} = 490 nm) is observed. The gradual addition of CF$_3$CO$_2$H is accompanied by the decrease of the absorbance (A) associated with the charge transfer band as the guest is protonated and subsequently expelled from the cavity of the host. After the addition of *c*. 20 equivalents of CF$_3$CO$_2$H, no charge transfer band can be detected. The gradual addition of *i*-Pr$_2$NEt is accompanied by the increase of the A value as the guest is deprotonated and subsequently complexed by the host.

The host **1**•4PF$_6$ and the guest **13** form [ref. 16] (step 1 in Fig. 10.8) spontaneously the [2]pseudorotaxane [**1**:**13**]•4PF$_6$ when mixed in CD$_3$CN/D$_2$O (95:5). Oxidation of the tetrathiafulvalene unit of the guest occurs after the addition (step 2 in Fig. 10.8) of three equivalents of I$_2$. The resulting monocationic guest is expelled (step 3 in Fig. 10.8) from the cavity of the tetracationic host as a result of electrostatic repulsion. Reduction of the tetrathiafulvalene unit of the guest occurs after the addition (step 4 in Fig. 10.8) of five equivalents of $Na_2S_2O_5$ and a large excess of NH_4PF_6. The regenerated neutral form of the guest reinserts itself inside the cavity of the host, yielding (step 1 in Fig. 10.8) the original [2]pseudorotaxane **[1:13]**•$4PF_6$. The redox-controlled switching [ref. 17] of this system can be followed (inset in Fig. 10.8) by ^{1}H-NMR spectroscopy using the bipyridinium hydrogen atoms (H_α) in the α-positions with respect to the nitrogen atoms as the probe protons. As the [2]pseudorotaxane **[1:13]**•$4PF_6$ and its separate components are in slow exchange on the ^{1}H-NMR timescale, the spectrum shows ((a) in the inset in Fig. 10.8) signals for the H_α protons associated with the complexed (co) and the uncomplexed (un) tetracationic cyclophane. After the oxidation of the guest, decomplexation occurs and only the signals for the complexed cyclophane can be observed ((b) in the inset in Fig. 10.8). Reduction of the guest is followed by complexation and the original signals for the complexed and the uncomplexed tetracationic cyclophane appear ((c) in the inset in Fig. 10.8) once again in the ^{1}H-NMR spectrum.

The *trans*-isomer (*trans*-**14**) of the guest **14** is bound [ref. 18] (step 1 in Fig. 10.9) by **1**•$4PF_6$ to form the [2]pseudorotaxane [**1**:*trans*-**14**]•$4PF_6$ in CD_3CN. By contrast, the *cis*-isomer (*cis*-**14**) of **14** is not complexed by **1**•$4PF_6$. Thus, when a solution of **1**•$4PF_6$ and *trans*-**14** is irradiated ($\lambda = 360$ nm) for 1 h, isomerization (step 2 in Fig. 10.9) of the free guest from *trans* to *cis* occurs. As a result, the equilibrium between the [2]pseudorotaxane [**1**:*trans*-**14**]•$4PF_6$ and its separate components is displaced in favor of the uncomplexed species. Further irradiation ($\lambda = 440$ nm) for 1 h induces the reisomerization (step 3 in Fig. 10.9) of the free guest from *cis* to *trans*, restoring the original equilibrium between the [2]pseudorotaxane [**1**:*trans*-**14**]•$4PF_6$ and its separate components. The light-controlled switching [ref. 19] of this system can be followed (inset in Fig. 10.9) by ^{1}H-NMR spectroscopy using the bipyridinium hydrogen atoms (H_α) in the α-positions with respect to the nitrogen atoms as the probe protons. The ^{1}H-NMR spectrum of a CD_3CN solution of **1**•$4PF_6$ and the *trans*-isomer of the guest **14** reveals ((a) in the inset in Fig. 10.9) signals for H_α in the complexed (co) and the uncomplexed (un) species which are undergoing slow exchange on the ^{1}H-NMR timescale. After the photoinduced isomerization of the free guest from *trans* to *cis*, the intensities of the signals associated with the H_α protons in the complexed species decrease ((b) in the inset in Fig. 10.9) relative to those of the complexed species. However, after the photoinduced re-isomerization of the free guest from the *cis*- to *trans*-isomer, the original ratio between the intensities of the signals for the H_α protons associated with the complexed and the uncomplexed species is restored ((c) in the inset in Fig. 10.9).

The host **1·4PF₆** binds [ref. 20] (step 1 in Fig. 10.10) the guest **15** in CD₃CN to form the [2]pseudorotaxane [**1:15**]·4PF₆. Upon addition of two equivalents of the sequestering agent, *o*-chloroanil (**16**), the adduct [**15:16**] is formed (step 2 in Fig. 10.10). As the free guest **15** is removed from the equilibrium, the [2]pseudorotaxane [**1:15**]·4PF₆ dissociates completely into its separate components. Reduction of the sequestering agent **16** destroys (step 3 in Fig. 10.10) the adduct [**15:16**], releasing the guest **15** and restoring the original equilibrium between the [2]pseudorotaxane [**1:15**]·4PF₆ and its separate components. The supramolecularly controlled switching [ref. 21] of this system can be followed (inset in Fig. 10.10) by ¹H-NMR spectroscopy using the phenoxy hydrogen atoms (H$_a$) in the *ortho*-positions with respect to the *t*-butyl substituents as the probe protons. The ¹H-NMR spectrum of a CD₃CN solution of **1·4PF₆** and **15** reveals ((a) in the inset in Fig. 10.10) signals for the H$_\alpha$ protons in the

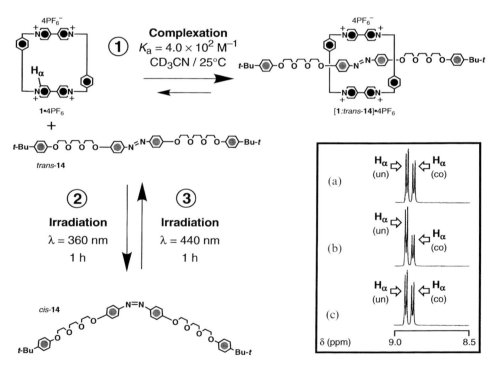

Figure 10.9. The light-controlled switching of the [2]pseudorotaxane [**1**:*trans*-**14**]·4PF₆ and the partial ¹H-NMR spectra [400 MHz, CD₃CN, 25 °C] of an equimolar mixture of **1·4PF₆** and **14** (a) after complexation, (b) after irradiation (λ = 360 nm, 1 h) and decomplexation, and (c) after irradiation (λ = 440 nm, 1 h) and complexation. The bipyridinium hydrogen atoms (H$_\alpha$) in the α-positions with respect to the nitrogen atoms were used as the probe protons and (co) and (un) stand for complexed and uncomplexed species, respectively.

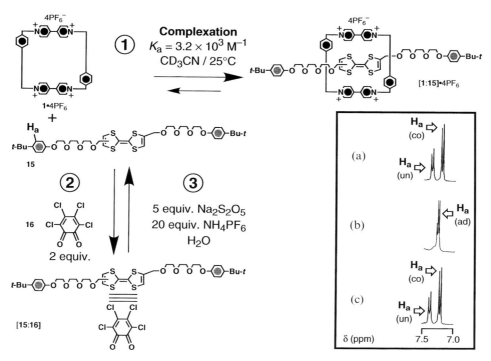

Figure 10.10. The supramolecularly controlled switching of the [2]pseudorotaxane [**1**:**15**]•4PF$_6$ and the partial ^{1}H-NMR spectra [400 MHz, CD$_3$CN, 25 °C] of an equimolar mixture of **1**•4PF$_6$ and **15** (a) after complexation, (b) after addition of **16**, and (c) after reduction of **16** are shown. The phenoxy hydrogen atoms (H$_a$) in the *ortho*-positions with respect to the *t*-butyl substituents as the probe protons and (co) and (un) stand for complexed and uncomplexed species, respectively.

complexed (co) and the uncomplexed (un) species which are in slow exchange on the ^{1}H-NMR timescale. After the addition of **16**, only the signals of the adduct [**15**:**16**] are observed ((b) in the inset in Fig. 10.10). However, after the reduction of **16** and release of the guest **15**, the ^{1}H-NMR spectrum shows ((c) in the inset in Fig. 10.10) the original signals for the H$_\alpha$ protons in the [2]pseudorotaxane [**1**:**15**]•4PF$_6$ and its separate components.

The redox-controlled switching between two [2]pseudorotaxanes was realized [ref. 22] (Fig. 10.11) using the host **1**•4PF$_6$ and the guests **9** and **17**. The 1,5-dioxynaphthalene-based guest **9** is fluorescent in its free form. Upon insertion of this guest inside the cavity of **1**•4PF$_6$ ($K_a = 10^5$ M^{-1} in H$_2$O at 25 °C), its fluorescence is quenched as a result of charge transfer interactions between the 1,5-dioxynaphthalene ring system and the sandwiching bipyridinium units. The binding of **17** by **1**•4PF$_6$ ($K_a \gg 10^5$ M^{-1} in H$_2$O at 25 °C) is accompanied by the appearance of a charge transfer band in the absorption spectrum as a result of

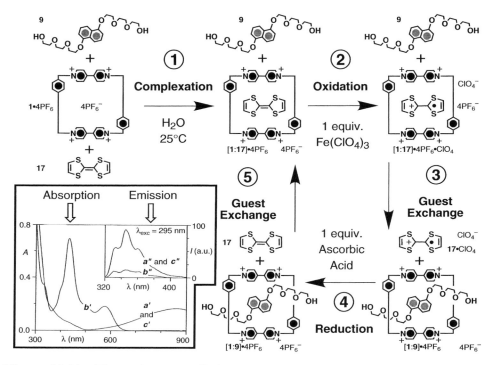

Figure 10.11. The redox-controlled switching between the [2]pseudorotaxanes [**1:9**]•4PF$_6$ and [**1:17**]•4PF$_6$ and the absorption and emission spectra of an equimolar mixture of **1**•4PF$_6$, **9**, and **17** (*a′/a″*) after the formation of [**1:17**]•4PF$_6$, (*b′/b″*) after oxidation and guest exchange, and (*c′/c″*) after reduction and guest exchange.

the [π ⋯ π] stacking interactions between the tetrathiafulvalene unit and the sandwiching bipyridinium units. Upon oxidation of the tetrathiafulvalene unit, the resulting monocationic guest is expelled from the cavity of the tetracationic host and the charge transfer band disappears. Thus, the complexation/decomplexation of the guests **9** and **17** by the host **1**•4PF$_6$ can be followed by emission and absorption spectroscopies, respectively. Upon mixing the host **1**•4PF$_6$ and the guests **9** and **17**, the [2]pseudorotaxane [**1:17**]•4PF$_6$ is formed (step 1 in Fig. 10.11) exclusively. The absorption and emission spectra show (*a′* and *a″* in the inset in Fig. 10.11) respectively the charge transfer band ($\lambda_{max} = 850$ nm) of the complex [**1:17**]•4PF$_6$ and the fluorescence of the free guest **9**. After the addition of one equivalent of Fe(ClO$_4$)$_3$, **17** is oxidized (step 2 in Figure 10.11) to its radical cation **17**$^{+•}$ that is then expelled from the cavity of the tetracationic host and replaced (step 3 in Fig. 10.11) by the neutral guest **9**. As a result, the charge transfer band of the [2]pseudorotaxane [**1:17**]•4PF$_6$ disappears (*b′* in the inset in Fig. 10.11) and is replaced by the characteristic absorption bands of the radical monocation **17⁺·**. Furthermore, the fluorescence of the 1,5-dioxynaphthalene ring system of **9** is quenched (*b″* in the inset in Fig. 10.11) upon formation of the [2]pseudorotaxane **[1:9]·4PF$_6$**. Reduction (step 4 in Fig. 10.11) of **17⁺·** back to the neutral form **17** is followed (step 5 in Figure 10.11) by the exchange of the guests to yield the original [2]pseudorotaxane **[1:17]·4PF$_6$**. As a result, the charge transfer band of the complex **[1:17]·4PF$_6$** and the fluorescence of the free guest **9** reappear (*c′* and *c′* in the inset in Fig. 10.11).

The redox-controlled switching between two [2]pseudorotaxanes was also realized [ref. 12] (Fig. 10.12) using the hosts **1·4PF$_6$** and **18** and the guest **17**. The redox potentials for the tetrathiafulvalene semicouples **17⁺·/17** and **17²⁺/17⁺·** are +0.321 and +0.714 V, respectively, in MeCN. In the presence of the π-electron deficient host **1·4PF$_6$**, the redox potential of the semicouple **17⁺·/17** shifts to +0.391 V while that of the semicouple **17²⁺/17⁺·** is unaffected. These observations indicate that only the neutral form of the guest **17** is bound by the host **1·4PF$_6$** ($K_a = 10^5$ M⁻¹ in MeCN at 25 °C). In the presence of the π-electron rich host **18**, the redox potential of the semicouple **17⁺·/17** is unaffected, while that of the semicouple **17²⁺/17⁺·** shifts to +0.700 V, suggesting that only the dicationic form **17²⁺** is bound by **18** ($K_a = 4.1 \times 10^3$ M⁻¹ in MeCN at 25 °C). When the two hosts **1·4PF$_6$** and **18** are mixed with the neutral

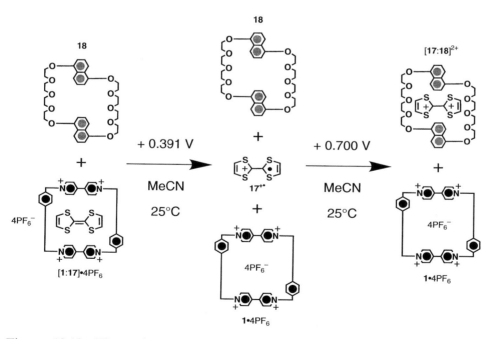

Figure 10.12. The redox-controlled switching between the [2]pseudorotaxanes **[1:17]·4PF$_6$** and **[17:18]⁴⁺**.

form **17** in MeCN, only the [2]pseudorotaxane [**1**:**17**]•4PF$_6$ is obtained. Oxidation (+0.391 V) of the guest **17** to the radical monocation **17**$^{+•}$ is accompanied by the dissociation of the complex. Further oxidation of **17**$^{+•}$ to the dication **17**$^{2+}$ is followed by the formation of the [2]pseudorotaxane [**17**:**18**]$^{2+}$.

10.3 [2]Rotaxanes

The relative movement of the macrocyclic with respect to the dumbbell-shaped component of a [2]rotaxane can be exploited [ref. 7] (Fig. 10.13) to perform switching operations. Indeed, two states are associated with a [2]rotaxane incorporating two different recognition sites within its dumbbell-shaped component. The macrocyclic component encircles the recognition site *0* in *State 0*, while it is located around the recognition site *1* in *State 1*. Switching between these two states can be achieved [ref. 7] by protonation/deprotonation and/or by oxidation/reduction of one of the two recognition sites incorporated in the dumbbell-shaped component. A [2]rotaxane incorporating a π-electron deficient macrocyclic component and two π-electron rich recognition sites within its dumbbell-shaped component has been synthesized [ref. 24] as illustrated in Fig. 10.14. Reaction of **10**•2PF$_6$ and **11** in the presence of the dumbbell-shaped compound **19** gives the [2]rotaxane **20**•4PF$_6$ in a yield of 19%, after counterion exchange. At room temperature in CD$_3$CN, the 'shuttling' of the macrocyclic component from one π-electron rich recognition site to the other is fast on the ^{1}H-NMR timescale and averaged signals for probe protons in the two translational isomers associated with **20**•4PF$_6$ are observed in the ^{1}H-NMR spectrum. At −40°C, this dynamic process becomes slow on the ^{1}H-NMR timescale and separate signals for the probe protons in the two translational isomers can be distinguished in a ratio of 84:16 in favor of the one having the more π-electron rich

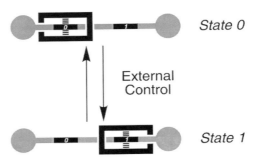

Figure 10.13. Externally controlled switching between the two states (*State 0* and *State 1*) associated with a [2]rotaxane incorporating two different recognition sites within the linear portion of its dumbbell-shaped component.

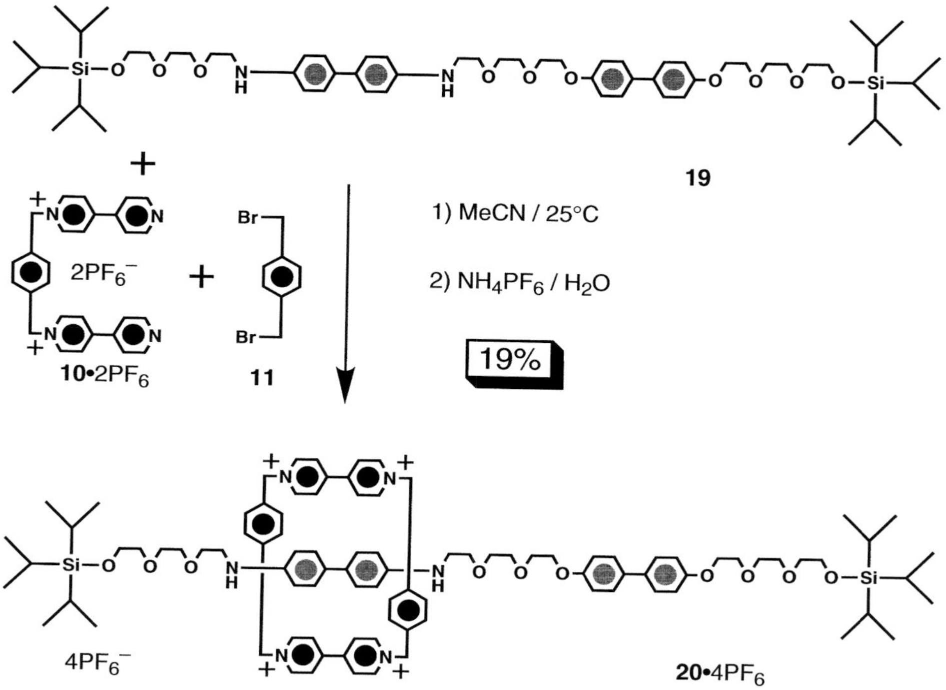

Figure 10.14. The template-directed synthesis of the [2]rotaxane **20•4PF**$_6$.

benzidine recognition site encircled by the tetracationic cyclophane component. Consistently, the absorption spectrum of **20•4PF**$_6$ reveals (**a** in the inset
in Fig. 10.15) an absorption band ($\lambda_{max} = 690$ nm) characteristic of charge
transfer interactions between a benzidine unit and two sandwiching bipyridinium units. Upon addition of CF_3CO_2H, the nitrogen atoms of the benzidine recognition site become protonated (step 1 in Fig. 10.15). Electrostatic
repulsion between the newly formed dicationic unit and the tetracationic
cyclophane force the macrocyclic component to 'shuttle' (step 2 in Fig. 10.15)
to the biphenol recognition site. As a result, the absorption band associated
with the charge transfer interactions between the benzidine ring system and
the bipyridinium units in the tetracationic cyclophane disappears (**b** in the
inset in Fig. 10.15). After the addition of C_5D_5N, deprotonation occurs (step
3 in Fig. 10.15) and the original equilibrium between the two translational
isomers of **20•4PF**$_6$ is restored (step 4 in Fig. 10.15). Consistently, the absorption spectrum reveals (**c** in the inset in Fig. 10.15) once again the charge transfer band associated with the interactions between the benzidine ring system
and the bipyridinium units. Switching [ref. 25] of the [2]rotaxane **20•4PF**$_6$ can
also be performed (Fig. 10.16) electrochemically by oxidizing/reducing the

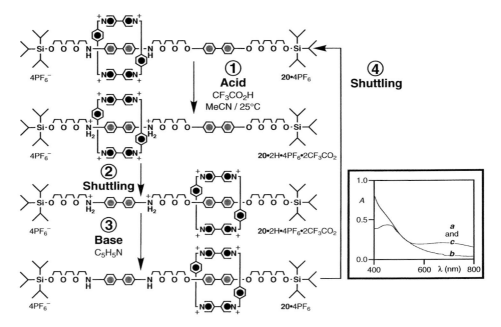

Figure 10.15. The acid/base-controlled switching of the [2]rotaxane **20•**4PF$_6$ and the absorption spectra (MeCN, 25°C) of **20•**4PF$_6$ (**a**) before the addition of CF$_3$CO$_2$H, (**b**) after the addition of CF$_3$CO$_2$H and shuttling, and (**c**) after the addition of C$_5$H$_5$N and shuttling.

benzidine recognition site. The oxidation (step 1 in Fig. 10.16) of the benzidine unit to a radical cation forces the tetracationic cyclophane to 'shuttle' (step 2 in Fig. 10.16) to the biphenol recognition site as a result of electrostatic repulsion between the two positively-charged entities. Reduction (step 3 in Fig. 10.16) of the radical cation back to the neutral species restores (step 4 in Fig. 10.16) the original equilibrium between the two translational isomers of the [2]rotaxane **20•**4PF$_6$.

An alternative synthetic approach [ref. 26] to a [2]rotaxane incorporating two different recognition sites within the linear portion of its dumbbell-shaped component is illustrated in Fig. 10.17. Reaction of **21•**2PF$_6$ with **22** in the presence of the crown ether host **23** gives the [2]rotaxane **24•**3PF$_6$ in a yield of 38%, after counterion exchange. In (CD$_3$)$_2$CO at 25°C, the ^{1}H-NMR spectrum of the [2]rotaxane **24•**3PF$_6$ shows the selective binding of the ammonium recognition site by the macrocyclic polyether. This co-conformation [ref. 27] is stabilized by a combination of [$^+$N−H $\cdots$ O] and [C−H $\cdots$ O] hydrogen bonds between the acidic hydrogen atoms of the secondary dialkylammonium recognition site and the polyether oxygen atoms. Irradiation of the protons in positions 2 and 6 on the 3,5-di-t-butylbenzyl ring of the stopper adjacent to the

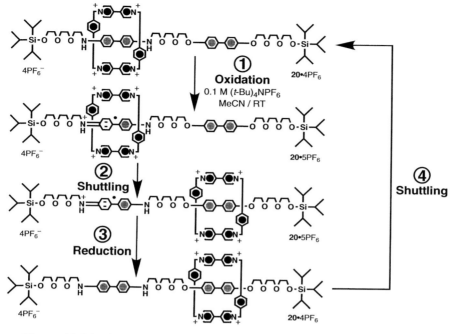

Figure 10.16. The redox-controlled switching of the [2]rotaxane **20**•4PF$_6$.

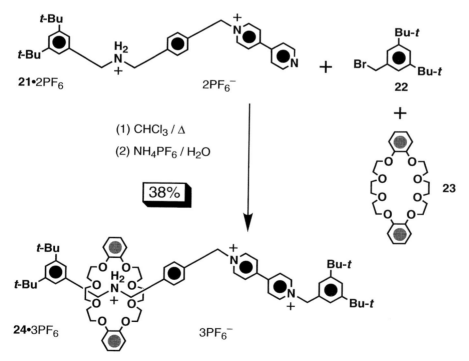

Figure 10.17. The template-directed synthesis of the [2]rotaxane **24**•3PF$_6$.

ammonium recognition site shows NOE enhancements of some of the reso-
nances associated with the *O*-methylene protons in the crown ether. Addition
of *i*-Pr$_2$NEt deprotonates (step 1 in Fig. 10.18) the ammonium recognition site.
As a result, the macrocyclic polyether component 'shuttles' (step 2 in Fig.
10.18) to embrace one of the pyridinium rings of the bipyridinium recognition
site. This step is accompanied by the appearance of a red color associated with
charge transfer interactions between the pyridinium ring and the sandwiching
catechol rings. Addition of CF$_3$CO$_2$H regenerates (step 3 in Fig. 10.18) the
ammonium recognition site. Thus, the macrocyclic component 'shuttles' (step 4
in Fig. 10.18) back to encircle this $^+$NH$_2$ center and the orange color associated
with the charge transfer interaction disappears. The acid/base-controlled
switching of the [2]rotaxane **24**•3PF$_6$ can be followed (inset in Fig.10.18) by ^{1}H-
NMR spectroscopy using the bipyridinium hydrogen atoms (H$_\beta^1$ and H$_\beta^2$) in
the β-positions with respect to the nitrogen atoms as the probe protons. These
protons give rise ((a) in the inset in Fig. 10.18) to one set of signals only in the

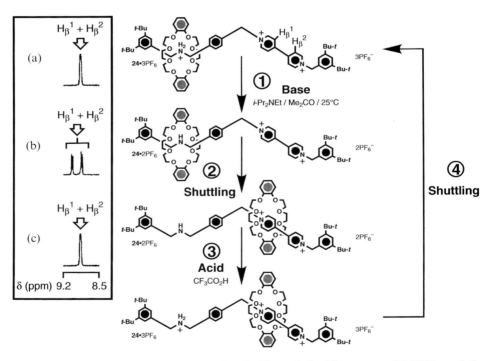

Figure 10.18. The acid/base-controlled switching of the [2]rotaxane **24**•3PF$_6$ and the
partial ^{1}H-NMR spectra [400 MHz, (CD$_3$)$_2$CO, 25 °C] of **24**•3PF$_6$ (a) before the addition of *i*-Pr$_2$NEt, (b) after the addition of *i*-Pr$_2$NEt and shuttling, and (c) after the addition of CF$_3$CO$_2$H. The bipyridinium hydrogen atoms (H$_\beta^1$ and H$_\beta^2$) in the β-positions
with respect to the nitrogen atoms were used as the probe protons.

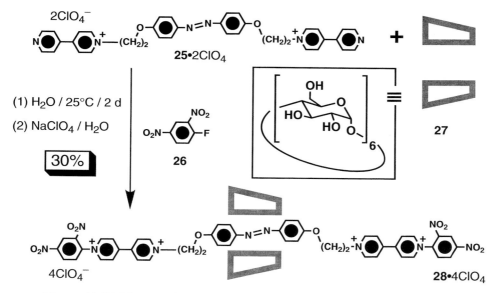

Figure 10.19. The template-directed synthesis of the [2]rotaxane **28**·4ClO$_4$.

^{1}H-NMR spectrum of the [2]rotaxane **24**·3PF$_6$ recorded in CD$_3$CN at 25°C. After the addition of *i*-Pr$_2$NEt and the 'shuttling' of the macrocyclic component to the bipyridinium recognition site, two distinct sets of signals are observed ((b) in the inset in Fig. 10.18) for H$_\beta^1$ and H$_\beta^2$. After the addition of CF$_3$CO$_2$H and the 'shuttling' of the macrocyclic component back to the ammonium recognition site, once again one set of signals is observed ((c) in the inset in Fig. 10.18) for H$_\beta^1$ and H$_\beta^2$.

The synthesis [ref. 28] of a [2]rotaxane incorporating a photoactive azobenzene unit in the linear portion of its dumbbell-shaped component is illustrated in Fig. 10.19. Reaction of **25**·2ClO$_4$ with **26** in the presence of α-cyclodextrin (**27**) gives the [2]rotaxane **28**·4ClO$_4$ in a yield of 30% after counterion exchange. In this [2]rotaxane, the cyclodextrin component encircles preferentially the *trans*-azobiphenoxy unit. However, upon irradiation (λ = 360 nm) of an aqueous solution of **28**·4ClO$_4$, isomerization (step 1 in Fig. 10.20) of the azobiphenoxy unit from *trans* to *cis* occurs. As a result, the macrocyclic component moves to encircle the bismethylene spacer bridging the *cis*-azobiphenoxy unit and one of the two bipyridinium units. Upon further irradiation (λ = 430 nm), the isomerization (step 2 in Fig. 10.20) of azobiphenoxy unit from *cis* back to *trans* occurs and is accompanied by the 'shuttling' of the cyclodextrin component back to the *trans*-azobiphenoxy unit. The light-controlled switching [ref. 29] of the [2]rotaxane **28**·4ClO$_4$ can be followed by circular dichroism measurements. Before irradiation, a positive circular dichroism induced band

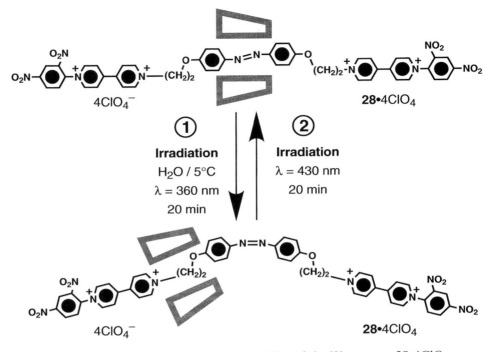

Figure 10.20. The light-controlled switching of the [2]rotaxane $28 \cdot 4\text{ClO}_4$.

($\lambda = 360$ nm), corresponding to the π–π^* transition associated with the azobi-phenoxy unit, is observed. After irradiation ($\lambda = 360$ nm), the intensity of this band decreases but is restored after irradiation at a longer wavelength ($\lambda = 430$ nm). Since the sign of a π–π^* transition associated with a guest encircled by an α-cyclodextrin is positive, the observed results indicate that the azobiphenoxy unit is inserted originally inside the cyclodextrin cavity and that it is expelled after irradiation at a wavelength of 360 nm.

10.4 [2]Catenanes

The circumrotation of one of the two macrocyclic components of a [2]catenane through the cavity of the other can be exploited [ref. 7] (Fig. 10.21) to perform switching operations. Indeed, two states are associated with a [2]catenane incorporating two different recognition sites within one of its two macrocyclic components. In *State 0*, the recognition site *0* comprised within one macrocyclic component is located inside the cavity of the other and the recognition site *1* is 'alongside'. In *State 1*, the recognition site *1* is 'inside' and the recognition site *0* is 'alongside'. Switching between these two states can be achieved [ref. 7]

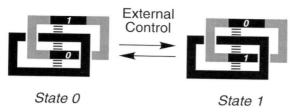

State 0   *State 1*

Figure 10.21. Externally controlled switching between the two states (*State 0* and *State 1*) associated with a [2]catenane incorporating two different recognition sites within one of its two macrocyclic components.

by oxidation/reduction of the recognition site *0* or *1* or by the addition/removal of a 'locking' component. [2]Catenanes incorporating a π-electron deficient macrocyclic component and two π-electron rich recognition sites within the second macrocycle can be synthesized [refs. 20, 30] as illustrated in Fig. 10.22. Reaction of **10·2PF$_6$** and **11** in the presence of either **29** or **30** gives the [2]catenanes **31·4PF$_6$** or **32·4PF$_6$**, respectively, in yields of 43 or 23%, respectively, after counterion exchange. The X-ray crystallographic analysis of the [2]catenane **32·4PF$_6$** reveals (Fig. 10.23) that the tetrathiafulvalene recognition site is located exclusively inside the cavity of the tetracationic cyclophane in the solid state, while the 1,5-dioxynaphthalene recognition site is positioned alongside. In both [2]catenanes, absorption and ^{1}H-NMR spectroscopic studies show (*vide infra*) that the tetrathiafulvalene unit is also located preferentially 'inside' in solution. This selectivity offers the possibility of switching these [2]catenanes by oxidation/reduction of their tetrathiafulvalene units. In the case of **31·4PF$_6$**, oxidation of the tetrathiafulvalene unit is achieved (step 1 in Fig. 10.24) upon addition of one equivalent of $Fe(ClO_4)_3$. The resulting monocationic unit is expelled from the cavity of the tetracationic cyclophane and it is replaced (step 2 in Fig. 10.24) by the neutral 1,4-dioxybenzene recognition site. Upon addition of one equivalent of $Na_2S_2O_5$ and H_2O, the tetrathiafulvalene unit is reduced (step 3 in Fig. 10.24) back to its neutral state. Circumrotation of the macrocyclic polyether through the cavity of the tetracationic cyclophane follows (step 4 in Fig. 10.24) immediately, resulting in the reinsertion of the tetrathiafulvalene recognition site inside the cavity of the π-electron deficient macrocycle. The redox-controlled switching [ref. 31] of the [2]catenane **31·4PF$_6$** can be monitored (inset in Fig. 10.24) by absorption spectroscopy. Initially, an absorption band ($\lambda_{max} = 850$ nm) associated with charge transfer interactions between the tetrathiafulvalene unit and the sandwiching bipyridinium units is observed (***a*** in the inset in Fig. 10.24). After oxidation of the tetrathiafulvalene unit and circumrotation of the macrocyclic polyether, the charge transfer band disappears (***b*** in the inset in Fig. 10.24) and is replaced by the characteristic

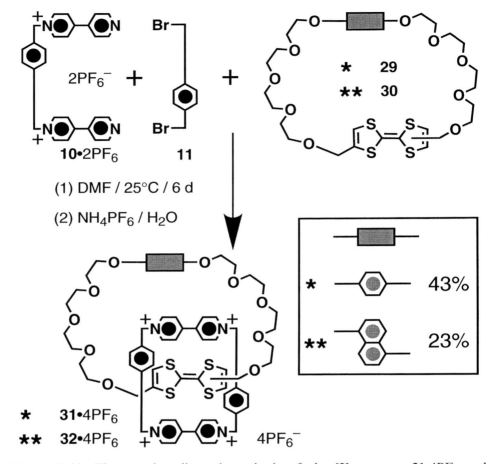

Figure 10.22. The template-directed synthesis of the [2]catenanes **31•4PF$_6$** and **32•4PF$_6$**.

absorption bands of the monocationic form of tetrathiafulvalene. After reduction of the tetrathiafulvalene unit back to its neutral form and circumrotation of the macrocyclic polyether, the original charge transfer band is restored (*c* in the inset in Fig. 10.24). The [2]catenane **31•4PF$_6$** can be also switched by supramolecular control. In CD$_3$CN at 25 °C, two translational isomers in equilibrium are associated (step 1 in Fig. 10.25) with the [2]catenane **31•4PF$_6$**. The major translational isomer incorporates the tetrathiafulvalene recognition site 'inside' and the 1,4-dioxybenzene ring 'alongside'. The minor translational isomer incorporates the 1,4-dioxybenzene recognition site 'inside' and the tetrathiafulvalene unit 'alongside'. Upon addition of two equivalents of *o*-chloroanil (**16**), the equilibrium between the two translational isomers of

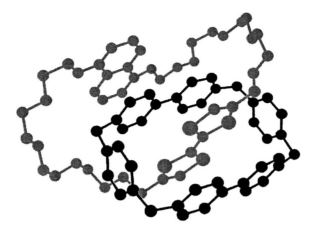

Figure 10.23. The geometry adopted by the [2]catenane **32**⁴⁺ in the solid state.

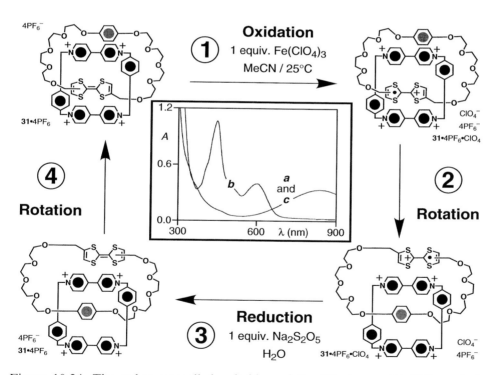

Figure 10.24. The redox-controlled switching of the [2]catenane **31**•4PF₆ and the absorption spectra (MeCN, 25 °C) of **31**•4PF₆ (*a*) before oxidation, (*b*) after oxidation and rotation, and (*c*) after reduction and rotation.

31•4PF$_6$ is displaced completely in favor of the one incorporating the 1,4-dioxybenzene recognition site 'inside'. Indeed, the tetrathiafulvalene unit is 'locked' (step 2 in Fig. 10.25) 'alongside' by [π ··· π] stacking interactions with the *o*-chloroanil unit to form the adduct [**16**:**31**]•4PF$_6$. Upon addition of five equivalents of Na$_2$S$_2$O$_5$ and a large excess of NH$_4$PF$_6$/H$_2$O, the *o*-chloroanil unit is reduced, the adduct is destroyed (step 3 in Fig. 10.25), and the original equilibrium between the two translational isomers of the [2]catenane **31**•4PF$_6$ is restored. The supramolecularly controlled switching of the [2]catenane **31**•4PF$_6$ can be monitored (inset in Fig. 10.25) by ^{1}H-NMR spectroscopy using the tetrathiafulvalene and the 1,4-dioxybenzene hydrogen atoms (H$_b$ and H$_c$) as the probe protons. As a result of the presence of the *cis* and the *trans* forms (Fig. 10.25) of the tetrathiafulvalene unit, the ^{1}H-NMR spectrum (CD$_3$CN, 25 °C) of the [2]catenane **31**•4PF$_6$ shows ((a) in the inset in Fig. 10.25) two singlets of different intensities for H$_b$ and two more for H$_c$. Upon addition of **16**, the relative intensities of the two singlets for H$_b$ change ((b) in the inset in Fig.

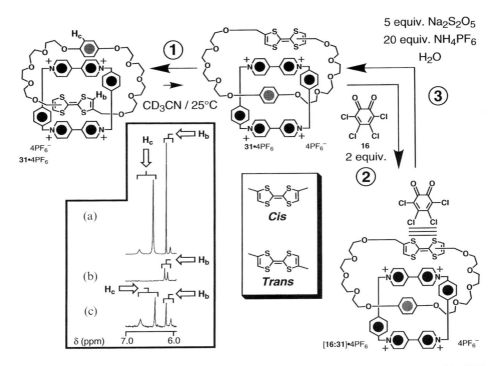

Figure 10.25. The supramolecularly controlled switching of the [2]catenane **31**•4PF$_6$ and the partial ^{1}H-NMR spectra (400 MHz, CD$_3$CN, 25 °C) of **31**•4PF$_6$ (a) before the addition of **16**, (b) after the addition of **16** and the formation of [**16**:**31**]•4PF$_6$, and (c) after the reduction of **16** and rotation. The tetrathiafulvalene and the 1,4-dioxybenzene hydrogen atoms (H$_b$ and H$_c$) were used as the probe protons.

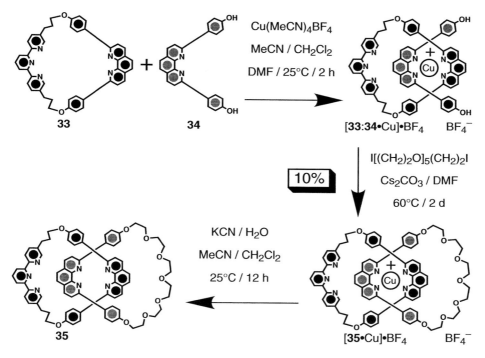

Figure 10.26. The template-directed synthesis of the [2]catenane **35**.

10.25) but their chemical shifts are not affected significantly, indicating that the tetrathiafulvalene unit is not oxidized by **16**. By contrast, the two singlets for H_c are shifted by *c.* −3.0 p.p.m. as the 1,4-dioxybenzene ring becomes encircled by the tetracationic cyclophane. After the reduction of the *o*-chloroanil ring incorporated within the adduct [**16**:**31**]•$4PF_6$ and the subsequent circumrotation of the macrocyclic polyether, the original singlets associated with H_c are observed ((c) in the inset in Fig. 10.25) once again.

The ability of transition metals to coordinate organic ligands can be exploited [ref. 32] (Fig. 10.26) to template the formation of [2]catenanes. Upon mixing equimolar amounts of the phenanthroline-based ligands **33** and **34** with $Cu(MeCN)_4BF_4$ in solution, the complex [**33**:**34**•Cu]•BF_4 self-assembles spontaneously. Reaction of [**33**:**34**•Cu]•BF_4 with $I[(CH_2)_2O]_5(CH_2)_2I$ gives the [2]catenate [**35**•Cu]•BF_4 in a yield of 10%. Demetalation [ref. 33] can be achieved by treating [**35**•Cu]•BF_4 with KCN in solution to afford the [2]catenane **35** quantitatively. The redox-controlled switching [ref. 34] of the [2]catenate [**35**•Cu]•BF_4 can be realized by oxidation/reduction of the transition metal, since Cu^+ and Cu^{2+} have different coordination requirements. Indeed, Cu^+ prefers tetracoordination while Cu^{2+} is stabilized by pentacoordination.

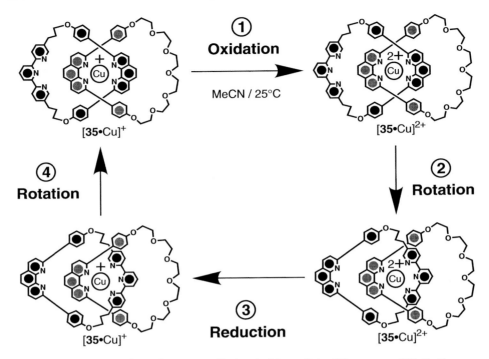

Figure 10.27. The redox-controlled switching of the [2]catenane [**35•Cu**]⁺.

The electrochemical oxidation of [**35•Cu**]⁺ generates (step 1 in Fig. 10.27) a
Cu^{2+} center in a tetracoordination environment. As a result, circumrotation of
the terpyridine-containing macrocycle through the cavity of the other macrocycle occurs (step 2 in Fig. 10.27) to afford a more stable pentacoordinated
Cu^{2+} center. The electrochemical reduction of [**35•Cu**]²⁺ affords (step 3 in Fig.
10.27) a pentacoordinated Cu^+ center. However, the circumrotation of the terpyridine-containing macrocycle yields back (step 4 in Fig. 10.27) the original
co-conformation in which the Cu^+ center can enjoy its much preferred tetra-
coordinating environment. The switching of this [2]catenate is accompanied by
a change in color from red to green on going from [**35•Cu**]⁺ to [**35•Cu**]²⁺ and
vice versa. Also, the redox potential for the Cu^{2+}/Cu^+ semicouple varies from
+0.63 V to –0.07 V on going from a tetra- to a pentacoordination environment
and *vice versa*.

10.5 Conclusions and future perspectives

As a result of their superstructural and structural properties, [2]pseudorotax-
anes and [2]rotaxanes and [2]catenanes, respectively, are ideal candidates for

the design of molecular-sized switches. Indeed, a number of bistable chemical systems, that can be switched reversibly between a *State 0* and a *State 1*, have been realized in the shape of such interpenetrating and interlocked supramolecular and molecular species. In all instances, the switching processes are associated with the relative movement of one of the two interpenetrating or interlocked components relative to the other. These molecular level motions can be induced by chemical, electrochemical, and/or photochemical stimuli (input signal) that alter the recognition properties of an appropriate unit incorporated in one of the two components. In response to the input signal, the chemical system alters its co-conformation and this change can be read spectroscopically (output signal). Although these bistable systems have been synthesized and their switching properties have been characterized fully in solution, considerable progress has to be made to incorporate them into ultra-densely integrated electronic circuits. Methods for their efficient and precise attachment to solid supports need to be developed. Furthermore, techniques to fabricate arrays of interconnected molecular-sized switches and to operate them individually have to be devised. Nonetheless, the remarkable progress achieved so far in the field is extremely encouraging. The recent realization of electronically-configurable logic gates, based on rotaxanes incorporating π-electron deficient dumbbell-shaped components and π-electron rich macrocycles, demonstrates [ref. 35] that electronic circuits made of organic molecules can, indeed, be fabricated.

10.6 References

1. Muller, R. S. and Kamins, T. I. (1986). *Device Electronics for Integrated Circuits*. Wiley, New York.
2. Mitchell, R. J. (1995). *Microprocessor Systems*. Macmillan, London.
3. Jortner, J. and Ratner, M. (Eds.). (1997). *Molecular Electronics*. Blackwell Science, Oxford.
4. (a) Vögtle, F. (1991). *Supramolecular Chemistry*. Wiley, New York. (b) Lehn, J.-M. (1995). *Supramolecular Chemistry*. VCH, Weinheim. (c) Lehn, J.-M., Atwood, J. L., Davies, J. E. D., MacNicol, D. D. and Vögtle, F. (Eds.). (1996). *Comprehensive Supramolecular Chemistry*. Pergamon, Oxford.
5. For reviews on switchable molecular and supramolecular systems, see: (a) Balzani, V. and Scandola, F. (1991). *Supramolecular Photochemistry*. Horwood, Chichester. (b) Balzani, V. (1992). *Tetrahedron* **48**, 10443–10514. (c) Bissell, R. A., de Silva, A. P., Gunaratne, H. Q. N., Lynch, P. L. M., Maguire, G. E. M., McCoy, C. P. and Sandanayake, K. R. A. S. (1993). *Top. Curr. Chem.* **168**, 223–264. (d) de Silva, A. P. and McCoy, C. P. (1994). *Chem. Ind.* 992–996. (e) Fabbrizzi, L. and Poggi, A. (1995). *Chem. Soc. Rev.* **24**, 197–202. (f) Shinkai, S. (1996). *Comprehensive Supramolecular Chemistry* (Atwood, J. L., Davies, J. E. D., Macnicol, D. D. and Vögtle, F., Eds.), Vol. 1, pp. 671–700. Pergamon, Oxford. (g) de Silva, A. P., Gunaratne, H. Q. N., Gunnlaugsson, T., Huxley,

A. J. M., McCoy, C. P., Rademacher, J. T. and Rice, T. E. (1997). *Chem. Rev.* **97**, 1515–1566. (h) Ward, M. D. (1997). *Chem. Ind.*, 640–645. (i) Beer, P. D. (1998). *Acc. Chem. Res.* **31**, 71–80. (j) Boulas, P. L., Gómez-Kaifer, M. and Echegoyen, L. (1998). *Angew. Chem., Int. Ed. Engl.* **37**, 216–247. (k) Bryce, M. R. (1999). *Adv. Mater.* **11**, 11–23. (l) Niemz, A. and Rotello, V. M. (1999). *Acc. Chem. Res.* **32**, 42–52.

6. For monographs and reviews on molecules incorporating interlocked components, see: (a) Schill, G. (1971). *Catenanes, Rotaxanes and Knots.* Academic Press, New York. (b) Walba, D. M. (1985). *Tetrahedron* **41**, 3161–3212. (c) Dietrich-Buchecker, C. O. and Sauvage, J.-P. (1987). *Chem. Rev.* **87**, 795–810. (d) Dietrich-Buchecker, C. O. and Sauvage, J.-P. (1991). *Bioorg. Chem. Front.* **2**, 195–248. (e) Chambron, J.-C., Dietrich-Buchecker, C. O. and Sauvage, J.-P. (1993). *Top. Curr. Chem.* **165**, 131–162. (f) Gibson, H. W. and Marand, H. (1993). *Adv. Mater.* **5**, 11–21. (g) Gibson, H. W., Bheda, M. C. and Engen, P. T. (1994). *Prog. Polym. Sci.* **19**, 843–945. (h) Amabilino, D. B., Parsons, I. W. and Stoddart, J. F. (1994). *Trends Polym. Sci.* **2**, 146–152. (i) Amabilino, D. B. and Stoddart, J. F. (1995). *Chem. Rev.* **95**, 2725–2828. (j) Gibson, H. W. (1996). *Large Ring Molecules* (Semlyen, J. A., Ed.), pp. 191–202. Wiley, New York. (k) Belohradsky, M., Raymo, F. M. and Stoddart, J. F. (1996). *Collect. Czech. Chem. Commun.* **61**, 1–43. (l) Raymo, F. M. and Stoddart, J. F. (1996). *Trends Polym. Sci.* **4**, 208–211. (m) Belohradsky, M., Raymo, F. M. and Stoddart, J. F. (1997). *Collect. Czech. Chem. Commun.* **62**, 527–557. (n) Jäger, R. and Vögtle, F. (1997). *Angew. Chem., Int. Ed. Engl.* **36**, 930–944. (o) Clarkson, G. J., Leigh, D. A. and Smith, R. A. (1998). *Curr. Op. Solid State Mater. Sci.* **3**, 579–584. (p) Leigh, D. A. and Murphy, A. (1999). *Chem. Ind.* 178–183. (q) Breault, G. A., Hunter, C. A. and Mayers, P. C. (1999). *Tetrahedron* **55**, 5265–5293. (r) Dietrich-Buckecher, C. O. and Sauvage, J.-P. (Eds.). (1999). *Catenanes, Rotaxanes and Knots.* VCH-Wiley, Weinheim. (p) Raymo, F. M. and Stoddart, J. F. (1999). *Chem. Rev.* **99**, 1643–1663.

7. For reviews on switchable pseudorotaxanes, rotaxanes and catenanes, see: (a) Benniston, A. C. (1996). *Chem. Soc. Rev.* **25**, 427–435. (b) Balzani, V., Gómez-López, M. and Stoddart, J. F. (1998). *Acc. Chem. Res.* **31**, 405–414. (c) Sauvage, J.-P. (1998). *Acc. Chem. Res.* **31**, 611–619. (d) Sauvage, J.-P. (1998). *Bull. Pol. Acad. Sci. Chem.* **46**, 289–307. (e) Chambron, J.-C. and Sauvage, J.-P. (1998). *Chem. Eur. J.* **4**, 1362–1366. (f) Kaifer, A. E. (1999). *Acc. Chem. Res.* **32**, 62–71.

8. For examples of molecular machines based of the relative movements of their components, see: (a) Iwamura, H. and Mislow, K. (1988). *Acc. Chem. Res.* **21**, 175–182. (b) Mislow, K. (1989). *Chemtracts* **2**, 151–174. (c) Kelly, T. R., Bowyer, M. C., Bhaskar, K. V., Bebbington, D., Gracia, A., Lang, F., Kim, M. H. and Jette, M. P. (1994). *J. Am. Chem. Soc.* **116**, 3657–3658. (d) Bedard, T. C. and Moore, J. S. (1997). *J. Am. Chem. Soc.* **117**, 10662–10671. (e) Canevet, C., Libman, J. and Shanzer, A. (1996). *Angew. Chem., Int. Ed. Engl.* **35**, 2657–2660. (f) de Santis, G., Fabbrizzi, L., Iacopino, D., Pallavicini, P., Perotti, A. and Poggi, A. (1997). *Inorg. Chem.* **36**, 827–832. (g) Willner, I. (1997). *Acc. Chem. Res.* **30**, 347–356. (h) Ikeda, A., Tsudera, T. and Shinkai, S. (1997). *J. Org. Chem.* **62**, 3568–3574. (i) Kelly, T. R., Tellita, I. and Sestelo, J. P. (1997). *Angew. Chem., Int. Ed. Engl.* **36**, 1866–1868. (j) Deans, R., Niemz, A., Breinlinger, E. C. and Rotello, V. M. (1997). *J. Am. Chem. Soc.* **119**, 10863–10864. (k) Stevens, A. N. and Richards, C. J. (1997). *Tetrahedron Lett.* **38**, 7805–7808. (l) Malpass, J. R., Sun, G., Fawcett, J. and Wartner, R. N.

(1998). *Tetrahedron Lett.* **38**, 3083–3086. (m) Bryce, M. R., de Miguel, P. and Devonport, W. (1998). *J. Chem. Soc., Chem. Commun.*, 2565–2566. (n) Takashita, M. and Irie, M. (1998). *J. Org. Chem.* **63**, 6643–6649. (o) Archut, A., Azzellini, G. C., Balzani, V., De Cola, L. and Vögtle, F. (1998). *J. Am. Chem. Soc.* **120**, 12187–12191. (p) de Silva, A. P., Dixon, I. M., Gunaratne, H. Q. N., Gunnlaugsson, T., Maxwell, P. R. S. and Rice, T. E. (1999). *J. Am. Chem. Soc.* **121**, 1393–1394. (q) Fabbrizzi, L., Gatti, F., Pallavicini, P. and Zambarbieri, E. (1999). *Chem. Eur. J.* **5**, 682–690.

9. For the binding properties of **1·4PF₆**, see: (a) Ashton, P. R., Goodnow, T. T., Kaifer, A. E., Reddington, M. V., Slawin, A. M. Z., Spencer, N., Stoddart, J. F., Vicent, C. and Williams, D. J. (1989). *Angew. Chem., Int. Ed. Engl.* **28**, 1396–1399. (b) Anelli, P.-L., Ashton, P. R., Ballardini, R., Balzani, V., Delgado, M., Gandolfi, M. T., Goodnow, T. T., Kaifer, A. E., Philp, D., Pietraszkiewicz, M., Prodi, L., Reddington, M. V., Slawin, A. M. Z., Spencer, N., Stoddart, J. F., Vicent, C. and Williams, D. J. (1992). *J. Am. Chem. Soc.* **114**, 193–218. (c) Gillard, R. E., Stoddart, J. F., White, A. J. P., Williams, B. J. and Williams, D. J. (1996). *J. Org. Chem.* **61**, 4504– 4505. (d) Asakawa, M., Dehaen, W., L'abbé, G., Menzer, S., Nouwen, J., Raymo, F. M., Stoddart, J. F. and Williams, D. J. (1996). *J. Org. Chem.* **61**, 9591–9595. (e) Ballardini, R., Balzani, V., Brown, C. L., Credi, A., Gillard, R. E., Montalti, M., Philp, D., Stoddart, J. F., Venturi, M., White, A. J. P., Williams, B. J. and Williams, D. J. (1997). *J. Am. Chem. Soc.* **119**, 12 503–12 513.

10. For accounts and reviews on the use of this recognition motif to template the synthesis of catenanes and rotaxanes, see: (a) Amabilino, D. B. and Stoddart, J. F. (1993). *Pure Appl. Chem.* **65**, 2351–2359. (b) Pasini, D., Raymo, F. M. and Stoddart, J. F. (1995). *Gazz. Chim. Ital.* **125**, 431–435. (c) Langford, S. J. and Stoddart, J. F. (1996). *Pure Appl. Chem.* **68**, 1255–1260. (d) Amabilino, D. B., Raymo, F. M. and Stoddart, J. F. (1996). *Comprehensive Supramolecular Chemistry* (Hosseini, M. W. and Sauvage, J.-P., Eds.), Vol. 9, pp. 85–130. Pergamon, Oxford. (e) Raymo, F. M. and Stoddart, J. F. (1997). *Pure Appl. Chem.* **69**, 1987–1997. (f) Gillard, R. E., Raymo, F. M. and Stoddart, J. F. (1997). *Chem. Eur. J.* **3**, 1933–1940. (g) Raymo, F. M. and Stoddart, J. F. (1998). *Chemtracts* **11**, 491–511.

11. For accounts and reviews on [C–H ⋯ O] hydrogen bonds, see: (a) Desiraju, G. R. (1991). *Acc. Chem. Res.* **24**, 290–296. (b) Desiraju, G.R. (1996) *Acc. Chem. Res.* **29**, 441–449. (c) Steiner, T. (1997). *J. Chem. Soc., Chem. Commun.*, 727–734. (d) Berger, I. and Egli, M. (1997). *Chem. Eur. J.* **3**, 1400–1404.

12. For accounts and reviews on [π ⋯ π] stacking interactions, see: (a) Schwartz, M. H. (1990). *J. Inclusion Phenom.* **9**, 1–35. (b) Williams, J. H. (1993). *Acc. Chem. Res.* **26**, 593–598. (c) Hunter, C. A. (1993). *Angew. Chem., Int. Ed. Engl.* **32**, 1584–1586. (d) Hunter, C. A. (1993). *J. Mol. Biol.* **230**, 1025–1054. (e) Dahl, T. (1994). *Acta Chem. Scand.* **48**, 95–116. (f) Cozzi, F. and Siegel, J. S. (1995). *Pure Appl. Chem.* **67**, 683–689. (g) Claessens, C. G. and Stoddart, J. F. (1997). *J. Phys. Org. Chem.* **10**, 254–272.

13. For accounts and reviews on template-directed syntheses, see: (a) Busch, D. H. and Stephenson, N. A. (1990). *Coord. Chem. Rev.* **100**, 119–154. (b) Lindsey, J. S. (1991). *New J. Chem.* **15**, 153–180. (c) Whitesides, G. M., Mathias, J. P. and Seto, C. T. (1991). *Science* **254**, 1312–1319. (d) Philp, D. and Stoddart, J. F. (1991). *Synlett* 445–458. (e) Busch, D. H. (1992). *J. Inclusion Phenom.* **12**, 389–395. (f) Anderson, S., Anderson, H. L. and Sanders, J. K. M. (1993). *Acc.*

Chem. Res. **26**, 469–475. (g) Cacciapaglia, R. and Mandolini, L. (1993). *Chem. Soc. Rev.* **22**, 221–231. (h) Hoss, R. and Vögtle, F. (1994). *Angew. Chem., Int. Ed. Engl.* **33**, 375–384. (i) Schneider, J. P. and Kelly, J. W. (1995). *Chem. Rev.* **95**, 2169–2187. (j) Philp, D. and Stoddart, J. F. (1996). *Angew. Chem., Int. Ed. Engl.* **35**, 1155–1196. (k) Raymo, F. M. and Stoddart, J. F. (1996). *Pure Appl. Chem.* **68**, 313–322. (l) Fyfe, M. C. T. and Stoddart, J. F. (1997). *Acc. Chem. Res.* **30**, 393–401. (m) Hubin, T. J., Kolchinski, A. G., Vance, A. L. and Busch, D. L. (1999). *Adv. Supramol. Chem.* **5**, 237–357.

14. Matthews, O. A., Raymo, F. M., Stoddart, J. F., White, A. J. P. and Williams, D. J. (1998). *New J. Chem.* 1131–1134.

15. For related examples of switchable [2]pseudorotaxanes, see: (a) Ashton, P. R., Gómez-López, M., Iqbal, S., Preece, J. A. and Stoddart, J. F. (1997). *Tetrahedron Lett.* **38**, 3635–3638. (b) Asakawa, M., Iqbal, S., Stoddart, J. F. and Tinker, N. D. (1996). *Angew. Chem., Int. Ed. Engl.* **35**, 976–978. (c) Ashton, P. R., Iqbal, S., Stoddart, J. F. and Tinker, N. D. (1996). *J. Chem. Soc., Chem. Commun.*, 479–481. (d) Montalti, M., Ballardini, R., Prodi, L. and Balzani, V. (1996). *J. Chem. Soc., Chem. Commun.*, 2011–2012. (e) Ashton, P. R., Ballardini, R., Balzani, V., Gómez-López, M., Lawrence, S. E., Martínez-Díaz, M.-V., Montalti, M., Piersanti, A., Prodi, L., Stoddart, J. F. and Williams, D. J. (1997). *J. Am. Chem. Soc.* **119**, 10641–10651. (f) Montalti, M. and Prodi, L. (1998). *J. Chem. Soc., Chem. Commun.*, 1461–1462. (g) Ashton, P. R., Ballardini, R., Balzani, V., Fyfe, M. C. T., Gandolfi, M. T., Martínez-Díaz, M. V., Morosini, M., Schiavo, C., Shibata, K., Stoddart, J. F., White, A. J. P. and Williams, D. J. (1998). *Chem. Eur. J.* **4**, 2332–2341.

16. Asakawa, M., Ashton, P. R., Balzani, V., Boyd, S. E., Credi, A., Mattersteig, G., Menzer, S., Montalti, M., Raymo, F. M., Ruffilli, C., Stoddart, J. F., Venturi, M. and Williams, D. J. (1999). *Eur. J. Org. Chem.*, 985–994.

17. For related examples of switchable [2]pseudorotaxanes, see: (a) Diaz, A., Quintela, P. A., Schuette, J. M. and Kaifer, A.E. (1988). *J. Phys. Chem.* **92**, 3537–3542. (b) Fonseca, R. J., Colina, J. T. and Smith, D. K. (1992). *J. Electroanal. Chem.* **340**, 341–348. (c) Smith, E. A., Lilienthal, R. R., Fonseca, R. J. and Smith, D. K. (1994). *Anal. Chem.* **66**, 3013–3020. (d) Córdova, E., Bissell, R. A. and Kaifer, A. E. (1995). *J. Org. Chem.* **60**, 1033–1038. (e) Devonport, W., Blower, M. A., Bryce, M. R. and Goldenberg, L. M. (1997). *J. Org. Chem.* **62**, 885–887. (f) Asakawa, M., Ashton, P. R., Balzani, V., Credi, A., Mattersteig, G., Matthews, O. A., Montalti, M., Spencer, N., Stoddart, J. F. and Venturi, M. (1997). *Chem. Eur. J.* **3**, 1992–1996. (g) Ashton, P. R., Ballardini, R., Balzani, V., Boyd, S. E., Credi, A., Gandolfi, M. T., Gómez-López, M., Iqbal, S., Philp, D., Preece, J. A., Prodi, L., Ricketts, H. G., Stoddart, J. F., Tolley, M. S., Venturi, M., White, A. J. P. and Williams, D. J. (1997). *Chem. Eur. J.* **3**, 152–168. (h) Collin, J.-P., Gaviña, P. and Sauvage, J.-P. (1996). *J. Chem. Soc., Chem. Commun.*, 2005–2006. (i) Mirzoian, A. and Kaifer, A.E. (1997). *Chem. Eur. J.* **3**, 1052–1058. (j) Castro, R., Godinez, L. A., Criss, C. M. and Kaifer, A. E. (1997). *J. Org. Chem.* **62**, 4928–4935.

18. Asakawa, M., Ashton, P. R., Balzani, V., Brown, C. L., Credi, A., Matthews, O. A., Newton, S. P., Raymo, F. M., Shipway, A. N., Spencer, N., Quick, A., Stoddart, J. F., White, A. J. P. and Williams, D. J. (1999). *Chem. Eur. J.* **5**, 860–875.

19. For related examples of switchable [2]pseudorotaxanes, see: (a) Ballardini, R., Balzani, V., Gandolfi, M. T., Prodi, L., Venturi, M., Philp, D., Ricketts, H. G.

and Stoddart, J. F. (1993). *Angew. Chem., Int. Ed. Engl.* **32**, 1301–1303. (b) Seiler, M., Dürr, H., Willner, I., Joselevich, E., Doron, A. and Stoddart, J. F. (1994). *J. Am. Chem. Soc.* **116**, 3399–3404. (c) Kropf, M., Joselevich, E., Dürr, H. and Willner, I. (1996). *J. Am. Chem. Soc.* **118**, 655–665. (d) David, E., Born, R., Kaganer, E., Joselevich, E., Dürr, H. and Willner, I. (1997). *J. Am. Chem. Soc.* **119**, 7778–7790. (e) Benniston, A. C., Harriman, A. and Yufit, D. S. (1997). *Angew. Chem., Int. Ed. Engl.* **36**, 2356–2358. (f) Ashton, P. R., Balzani, V., Kocian, O., Prodi, L., Spencer, N. and Stoddart, J. F. (1998). *J. Am. Chem. Soc.* **120**, 11 190–11 191.

20. Balzani, V., Credi, A., Mattersteig, G., Matthews, O. A., Raymo, F. M., Stoddart, J. F., Venturi, M., White, A. J. P. and Williams, D. J. (2000). *J. Org. Chem.* **65**, 1924–1936.

21. For related examples of switchable [2]pseudorotaxanes, see: (a) Ballardini, R., Balzani, V., Credi, A., Gandolfi, M. T., Langford, S. J., Menzer, S., Prodi, L., Stoddart, J. F., Venturi, M. and Williams, D. J. (1996). *Angew. Chem., Int. Ed. Engl.* **35**, 978–981. (b) Credi, A., Balzani, V., Langford, S. J. and Stoddart, J. F. (1997). *J. Am. Chem. Soc.* **119**, 2679–2681.

22. Credi, A., Balzani, V., Langford, S. J ., Montalti, M., Raymo, F. M. and Stoddart, J. F. (1998). *New J. Chem.*, 1061–1065.

23. Ashton, P. R., Balzani, V., Becher, J., Credi, A., Fyfe, M. C. T., Mattersteig, G., Menzer, S., Nielsen, M. B., Raymo, F. M., Stoddart, J. F., Venturi, M. and Williams, D. J. (1999). *J. Am. Chem. Soc.* **121**, 3951–3957.

24. Bissell, R. A., Córdova, E., Kaifer, A. E. and Stoddart, J. F. (1994). *Nature* **369**, 133–137.

25. For related examples of switchable [2]rotaxanes, see: (a) Collin, J.-P., Gaviña, P. and Sauvage, J.-P. (1997). *New J. Chem.* **21**, 525–528. (b) Gaviña, P. and Sauvage, J.-P. (1997). *Tetrahedron Lett.* **38**, 3521–3524 (b) Armaroli, N., Balzani, V., Collin, J.-P., Gaviã, P., Sauvage, J.–P. and Ventura, B. (1999). *J. Am. Chem. Soc.* **121**, 4397–4408.

26. (a) Martínez-Díaz, M.-V., Spencer, N. and Stoddart, J. F. (1997). *Angew. Chem., Int. Ed. Engl.* **36**, 1904–1907. (b) Ashton, P. R., Ballardini, R., Balzani, V., Baxter, I., Credi, A., Fyfe, M. C. T., Gandolfi, M. T., Gómez-López, M., Martínez-Díaz, M.-V., Piersanti, A., Spencer, N., Stoddart, J. F., Venturi, M., White, A. J. P. and Williams, D. J. (1998). *J. Am. Chem. Soc.* **120**, 11932–11942.

27. For a definition of the term 'co-conformation', see: Fyfe, M. C. T., Glink, P. T., Menzer, S., Stoddart, J. F., White, A. J. P. and Williams, D. J. (1997). *Angew. Chem., Int. Ed. Engl.* **36**, 2068–2070.

28. Murakami, H., Kawabuchi, A., Kotoo, K., Kunitake, M. and Nakashima, N. (1997). *J. Am. Chem. Soc.* **119**, 7605–7606.

29. For related examples of switchable [2]rotaxanes, see: (a) Benniston, A. C. and Harriman, A. (1993). *Angew. Chem.* **32**, 1459–1461. (b) Benniston, A. C., Harriman, A. and Lynch, V. M. (1994). *Tetrahedron Lett.* **35**, 1473–1476. (c) Benniston, A. C., Harriman, A. and Lynch, V. M. (1995). *J. Am. Chem. Soc.* **117**, 5275–5291.

30. Asakawa, M., Ashton, P. R., Balzani, V., Credi, A., Hamers, C., Mattersteig, G., Montalti, M., Shipway, A. N., Spencer, N., Stoddart, J. F., Tolley, M. S., Venturi, M., White, A. J. P. and Williams, D. J. (1998). *Angew. Chem., Int. Ed. Engl.* **37**, 333–337.

31. For related examples of switching [2]catenanes, see: (a) Gunter, M. J. and Johnston, M. R. (1994). *J. Chem. Soc., Chem. Commun.*, 829–830. (b) Ashton,

P. R., Ballardini, R., Balzani, V., Gandolfi, M. T., Marquis, D. J.-F., Pérez-García, L., Stoddart, J. F. and Venturi, M. (1994). *J. Chem. Soc., Chem. Commun.*, 177–180. (c) Ballardini, R., Balzani, V., Credi, A., Gandolfi, M. T., Prodi, L., Venturi, M., Pérez-García, L. and Stoddart, J. F. (1995). *Gazz. Chim. Ital.* **125**, 353–359. (d) Ashton, P. R., Ballardini, R., Balzani, V., Credi, A., Gandolfi, M. T., Menzer, S., Pérez-García, L., Prodi, L., Stoddart, J. F., Venturi, M. White, A. J. P. and Williams, D. J. (1995). *J. Am. Chem. Soc.* **117**, 11171–11197.

32. (a) Livoreil, A., Dietrich-Buchecker, C. O. and Sauvage, J.-P. (1994). *J. Am. Chem. Soc.* **116**, 9399–9400. (b) Baumann, F., Livoreil, A., Kaim, W. and Sauvage, J.-P. (1997). *J. Chem. Soc., Chem. Commun.*, 35–36. (c) Livoreil, A., Sauvage, J.-P., Armaroli, N., Balzani, V., Flamigni, L. and Ventura, B. (1997). *J. Am. Chem. Soc.* **119**, 12114–12124.

33. The terms '[2]catenate' and '[2]catenand' are often used to indicate the metalated and demetalated forms, respectively, of [2]catenanes such as **35**. In related [2]catenanes, the demetalation/metalation processes are accompanied by the circumrotation of one macrocyclic component through the cavity of the other. These systems can be regarded as chemically controllable molecular switches. For examples, see: (a) Dietrich-Buchecker, C. O., Sauvage, J.-P. and Kern, J.-M. (1984). *J. Am. Chem. Soc.* **106**, 3043–3045. (b) Albrecht-Gary, A.-M., Saad, Z., Dietrich-Buchecker, C. O. and Sauvage, J.-P. (1985). *J. Am. Chem. Soc.* **107**, 3205–3209. (c) Cesario, M., Dietrich-Buchecker, C. O., Guilhem, J., Pascard, C. and Sauvage, J.-P. (1985). *J. Chem. Soc., Chem. Commun.*, 244–247. (d) Cesario, M., Dietrich-Buchecker, C. O., Edel, A., Guilhem, J., Kintzinger, J.-P., Pascard, C. and Sauvage, J.-P. (1986). *J. Am. Chem. Soc.* **108**, 6250–6254. (e) Albrecht-Gary, A.-M., Dietrich-Buchecker, C. O., Saad, Z., Sauvage, J.-P. and Weiss, J. (1986). *J. Chem. Soc. Chem. Commun.*, 1325–1327. (f) Dietrich-Buchecker, C. O., Sauvage, J.-P. and Weiss, J. (1986). *Tetrahedron Lett.* **27I**, 2257–2260. (g) Albrecht-Gary, A.-M., Dietrich-Buchecker, C. O., Saad, Z. and Sauvage, J.-P. (1988). *J. Am. Chem. Soc.* **110**, 1467–1472. (h) Dietrich-Buchecker, C. O., Sauvage, J.-P. (1990). *Tetrahedron.* **46**, 503–512. (i) Albrecht-Gary, A.-M., Dietrich-Buchecker, C. O., Saad, Z. and Sauvage, J.-P. (1992). *J. Chem. Soc., Chem. Commun.*, 280–282. (j) Armaroli, N., De Cola, L., Balzani, V., Sauvage, J.-P., Dietrich-Buchecker, C. O., Kern, J.-M. and Bailal, A. (1992). *J. Chem. Soc., Dalton Trans.*, 3242–3246. (k) Sauvage, J.-P. and Weiss, J. (1985). *J. Am. Chem. Soc.* **107**, 6108–6110. (l) Dietrich-Buchecker, C. O., Khemiss, A. and Sauvage, J.-P. (1986). *J. Chem. Soc. Chem., Commun.* 1376–1378. (m) Armaroli, N., Balzani, V., De Cola, L., Hemmert, C. and Sauvage, J.-P. (1994). *New J. Chem.* **18**, 775–782. (n) Amabilino, D. B., Dietrich-Buchecker, C. O., Livoreil, A., Pérez-García, L., Sauvage, J.-P. and Stoddart, J. F. (1996). *J. Am. Chem. Soc.* **118**, 3905–3913.

34. For a related redox switchable [2]catenane, see: Cárdenas, D. J., Livoreil, A. and Sauvage, J.-P. (1996). *J. Am. Chem. Soc.* **118**, 11980–11981.

35. (a) Collier, C. P., Wong, E. W., Behloradsky, M., Raymo, F. M., Stoddart, J. F., Kuekes, P. J., Williams, R. S. and Heath, J. R. (1999). *Science*, **285**, 391–394. (b) Wong, E. W., Collier, C. P., Behloradsky, M., Raymo, F. M., Stoddart, J. F. and Heath, J. R. (2000). *J. Am. Chem. Soc.* **122**, 5831–5840. (c) see also, Collier, C. P., Mattersteig, G., Wong, E. W., Luo, Y., Beverly, K., Sampaio, J., Raymo, F. M., Stottart, J. F. and Heath, J. R. (2000). *Science* **289**, 1172–1175.

11

Some aspects of supramolecular design of organic materials

UDAY MAITRA AND R. BALASUBRAMANIAN

11.1 Introduction

Chambers 21st Century Dictionary defines 'material' as 'any substance out of which something is, or may be, made' [1]. For the purpose of this chapter a somewhat different statement will be used: any organic substance that may have potential applications will be termed an organic material. Superimposed on this definition we would like to have another constraint: such materials should be supramolecularly synthesizable (i.e. noncovalently assembled from molecular components).

'Chemistry beyond the molecule' [2], or supramolecular chemistry as we know it, has evolved over the past three decades, and has now become a hot area of research in chemistry. One of the characteristics of supramolecular chemistry is that the traditional barriers of organic, inorganic, physical and analytical chemistry do not exist here – it is a unique multidisciplinary area in which all areas of chemistry (and some other disciplines) are blended nicely.

Supramolecular chemistry originally developed with the design of receptors for metal ions and small organic molecules (the discovery of crown ethers and synthesis of cryptands, cyclophanes, molecular tweezers, etc.), but has now grown beyond molecular recognition and catalysis. From studies on simple molecular recognition we have learnt how complex molecular species can be designed and built – systems which will not only be useful for understanding molecular interactions, but will also function in a predictable manner. The concept of self-assembly [3, 4] is increasingly being used towards achieving this. However, it would be appropriate to mention here that at present the design principles and our understanding of such materials are still in infancy, primarily because of the complex nature of molecular interactions in an ensemble of molecules.

Owing to the rather broad nature of the subject no attempt will be made to

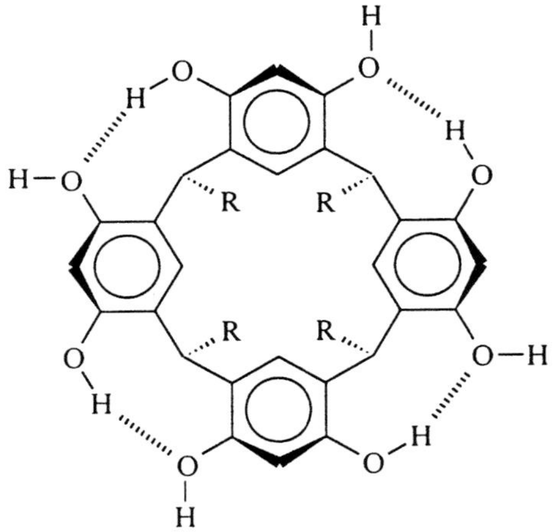

Figure 11.1. Calix[4]resorcinarene, which forms a snub cube with 60 hydrogen bonds.

provide a comprehensive review of the subject [5]. A few selected examples from the literature, as well as some current work carried out in the authors' laboratory, will be discussed.

11.2 Molecular forces

Among the different types of non-covalent forces that have been utilized for the design and study of intermolecular complexes, hydrogen bonding played a very major role. Hydrogen bonding from its directionality and relatively higher strength forms the very basis of structural integrity and functions in biology [6]. As this is a relatively well understood phenomenon, it has been employed in designing myriad number of supramolecular systems. Utilizing this interaction, materials have been designed that encompass a variety of dimensions; starting from the nanoscopic regime to mesoscopic recognition to the design of organic molecular solids/materials. For example, Atwood *et al.* have recently reported the structure of a hexameric calix[4]resorcinarene which is held together by 60 hydrogen bonds with an internal volume of 1375 Å^3, resembling a virus in its topology and appearing as a snub cube (one of the Archimedean solids) (Fig. 11.1) [7].

Pi-stacking and electron donor–acceptor interactions have also been extensively used for designing host–guest complexes in a predictable manner [8].

Metal–ligand coordination, although not exactly a non-covalent interaction, has been used extensively to design a variety of supramolecular complexes [9]. Many such systems have used the metal ion essentially as a control element for organizing ligands around it [10]. Constable and coworkers

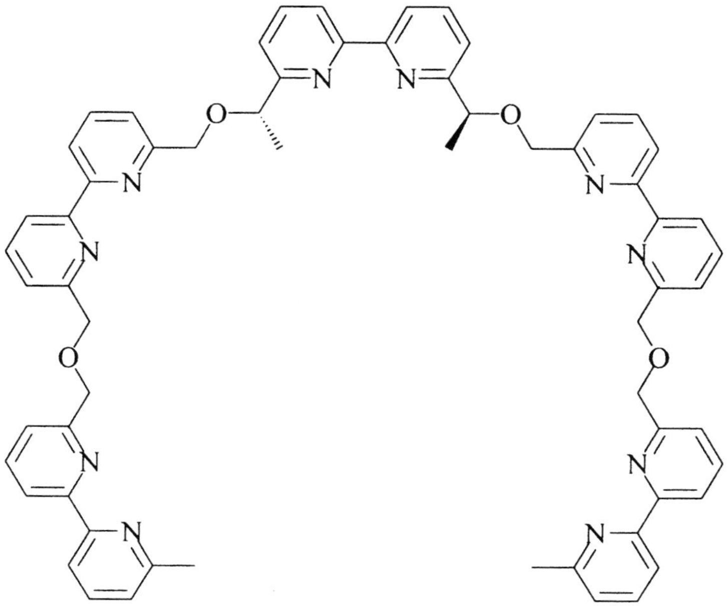

Figure 11.2. Chiral ligand that forms a chiral helicate with metal ions.

thoroughly exploited pyridine–metal ion binding to generate helices; for example, complexation of sexipyridine with Mn(II) and Cd(II) generated double helices [11]. Among the initial reports, Lehn's chiral helical induction by the complexation of chiral pentakis(bipyridine) (Fig. 11.2) with Cu(I) and Ag(I) is noteworthy [12]. Fujita [13] and Stang [9b] have elegantly used metal coordination to create large molecular structures via self-assembly of a large number of metal ions and ligands.

The existence of the hydrophobic effect [14] in the formation of micelles, liposomes and other aggregates in water has been recognized and elaborately studied for several decades. Detergents, lipids and many derivatives made from them have been extensively studied and there are numerous reviews which document these in greater detail [15]. Synthetic bilayer membranes (BLMs) have been studied for the past three decades by using a variety of physical techniques [16]. Hydrophobic interaction has also been utilized for the design of hosts that bind guest molecules in water [17].

11.3 Supramolecular design and construction

This section is broadly divided into four sections. We start with a discussion on dendrimers, followed by supramolecular polymers. Subsequently we discuss various types of organic network structure, and finally end with a discussion of

surface end groups

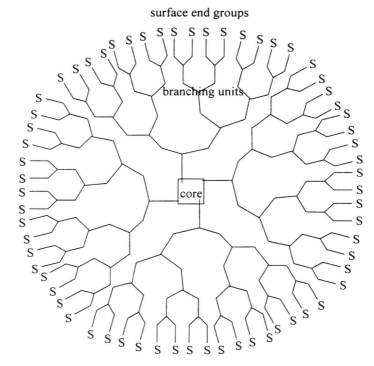

Figure 11.3. General architecture of a dendrimer.

organo and aqueous gels in which complex network structures are also present. Networks present in gels probably represent the most complex (and least understood) molecular organization.

11.3.1 Dendrimers

Dendrimers, also referred to as Cascade molecules, Arborals, Unimolecular micelles, Starburst species and Cauliflower [18], are unidisperse molecules of nanoscopic dimensions that were first reported in 1978 [19]. These molecules consist of three distinguishing architectural features: (a) a core, (b) interior regions consisting of branching units connecting the external surface and the core, and (c) the end groups (Fig. 11.3). The unabated interest in these macro-molecules is the result of a variety of reasons, such as: (a) these systems are rel-atively easy to construct, (b) they have good solubility, (c) they can be easily characterized, and (d) they allow one to position a large number of similar groups in the surface that could be easily utilized/modified to make them func-tional [20]. The incorporation of chirality in these species was reported almost immediately after their maiden appearance [21]. We have recently reported the design and construction of chiral dendrons based on bile acid; they are a hep-tamer and a nonamer containing seven and nine bile acid units respectively (Fig. 11.4) [22]. This work represents the utilization of the largest chiral repeat units in a dendritic species.

The initial focus in dendrimer research was largely on their synthesis, but recently more importance has been given towards their functional aspects [23]. The successful blending of dendrimer chemistry with several contemporary themes such as host–guest chemistry [24], metallo-organic chemistry [25], lumi-nescent materials [26], catalysis [20a], medicinal chemistry [18d] and poly-mers [27] has contributed enormously over the years to a rich chemistry with potential applications. As a detailed survey of this area is beyond the scope of this chapter, we will restrict ourselves to two topics involving these molecules: (a) dendritic self-assembly [28] and (b) metallodendrimers [25, 26].

Zimmerman has summarized how various facets of supramolecular chemis-try with crown ethers, rotaxanes, cyclophanes, bolaamphiphiles, gelators, hydrogen-bond receptors and self-assembly have been congenially fused with dendrimers [24]. His group has demonstrated the prowess of self-assembly by the construction of a hexameric species (Fig. 11.5) based on hydrogen bonding (molecular mass of 34 kD; diameter and height of 9 and 2 nm, respec-tively) [29]. The dimensions of this supramolecularly assembled dendrimer could be compared to a small protein. Self-assembly of dendrimers giving rise to supramolecular entities of cylindrical and spherical shapes, some of which are liquid crystalline, have also been documented in the literature [30]. Percec and co-workers have revealed the synthesis of a series of dendrons, whose shape varied with generation numbers. The fourth-generation dendron adopted a disc-like shape, whereas the fifth-generation dendron adopted a globular shape. Metallodendrimers, whose origin may be traced to well-studied coordination chemistry, are dendritic molecules with metal in one or more of the following domains; namely, core (porphyrin, ferrocene), branching units (or even as a branching point) and peripheral units (quite often ferrocene units), which promise myriad applications [26]. Polypyridine–metal complexa-tion is again widely used to construct a variety of species. A representative example of homo- and heterometallic branching centers by Balzani and coworkers is shown here (Fig. 11.6).

11.3.2 *Supramolecular polymers*

Molecules held together by tailor-made non-covalent interactions came into existence during the past decade and are called *supramolecular polymers* [31].

U. Maitra and R. Balasubramanian

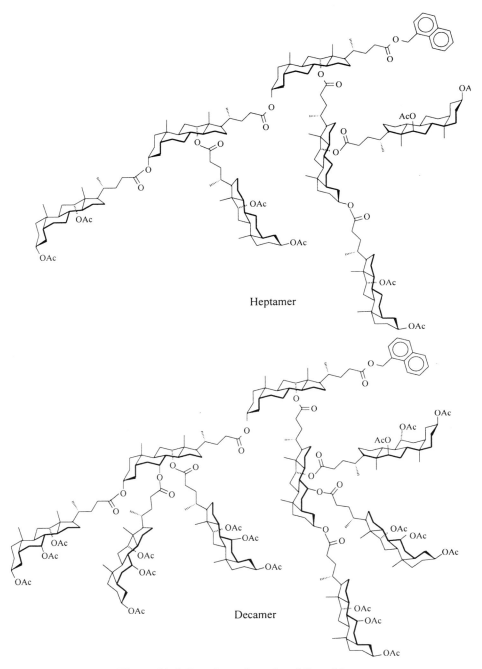

Heptamer

Decamer

Figure 11.4. Dendrons based on bile acid.

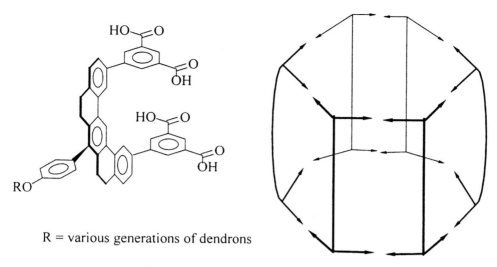

R = various generations of dendrons

Figure 11.5. A self-assembled dendron.

The first report of this kind is credited to Lehn and co-workers [32]. They reported the design and synthesis based on the hydrogen bonding capability of uracil to 2,6-diaminopyridine [33]. *Bis*(uracil) and *bis*(2,6-diamino pyridine) units were synthesized with a tartaric acid spacer (Fig. 11.7). The choice of tartaric acid did the magic in dictating properties of the supramolecular assembly. The 1:1 mixture was examined by SEM, to establish the concentration dependence on the length and orientation of the helical fibres. The D-mixtures and L-mixtures showed opposite helical patterns as expected. By mixing D-pyridine and L-uracil derivatives, and vice versa, they showed that it was the uracil component that dictated the macromolecular chirality. The above mentioned supramolecular polymer exhibited excellent liquid crystalline properties, whereas the individual components were solids. Multi pyridine/carboxylic acid systems were also studied for their liquid crystalline behavior [34]. In spite of the existence of rich literature on liquid crystals, the importance of hydrogen bonding in designing such aggregates has been appreciated only recently [35].

Meijer *et al.* reported the self-aggregation behavior of a molecule in chloroform (Fig. 11.8) containing two units of 2-ureido-4-pyrimidone linked through a spacer (which self-associates in the DDAA–AADD pattern strongly with a dimerization constant $> 10^6$ M^{-1}) [36]. This compound formed viscous solutions in chloroform, and the viscosity observed was dependent on the concentration and temperature.

Gibson *et al.* recently demonstrated that a pseudorotaxane-like linear assembly can be synthesized from a *bis* crown ether moiety and a *bis* ammonium salt

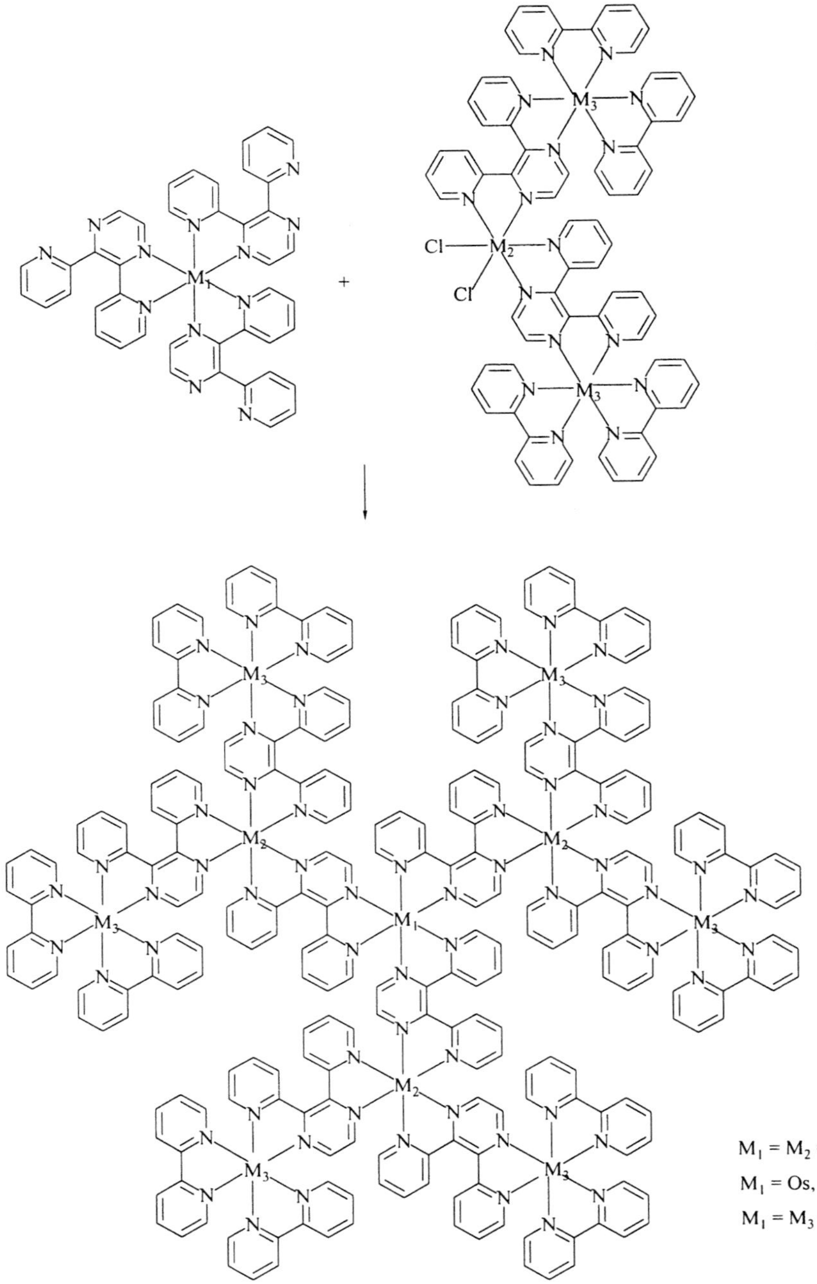

U. *Maitra and R. Balasubramanian*

$M_1 = M_2 = M_3 = Ru$

$M_1 = Os, \ M_2 = M_3 = Ru$

$M_1 = M_3 = Os, \ M_2 = Ru$

Figure 11.6. A metallodendrimer.

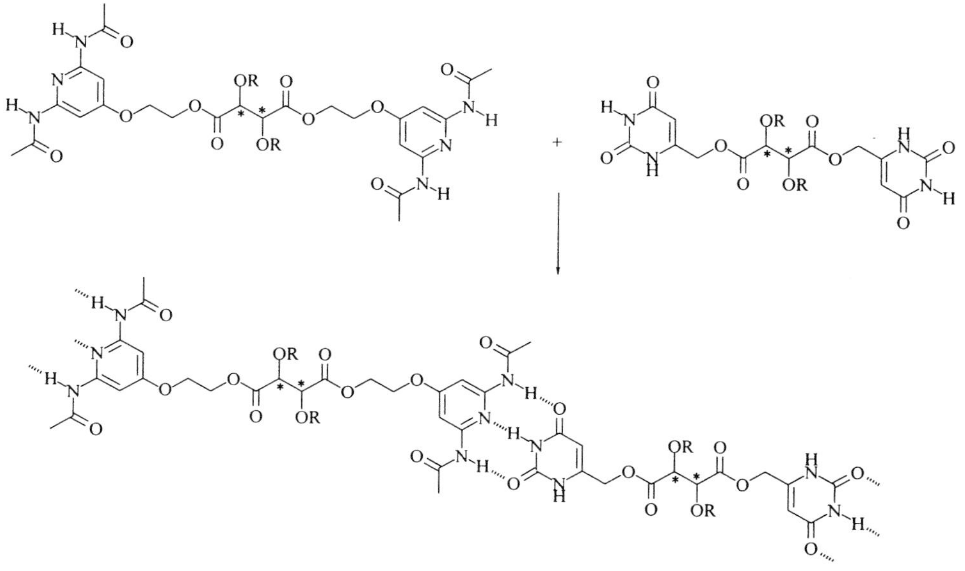

Figure 11.7. A liquid crystalline supramolecular polymer.

2-ureido-4-pyrimidone

dimer of 2-ureido-4-pyrimidone

R = H or Me

Monomer which forms Supramolecular Polymer

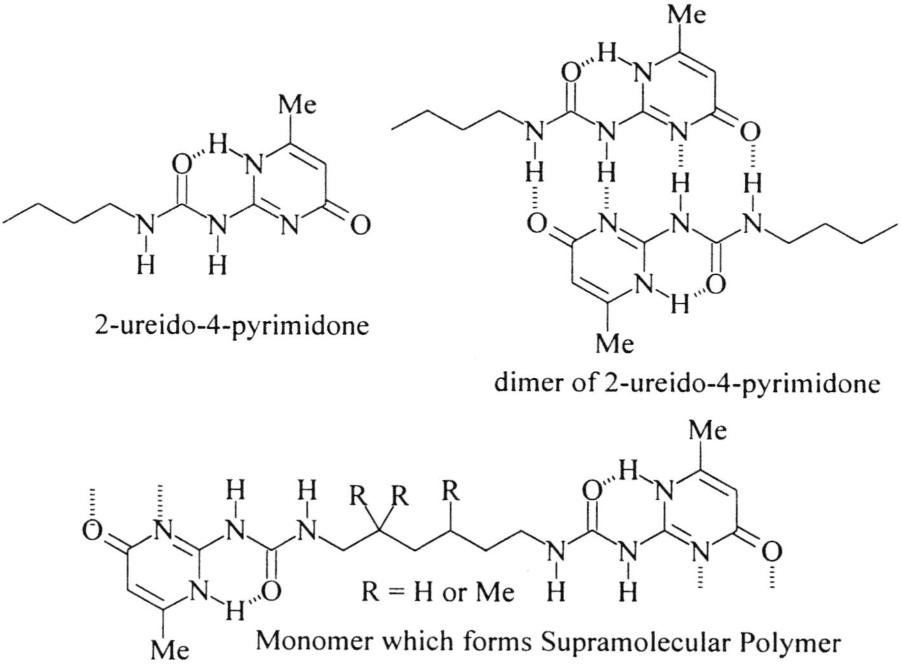

Figure 11.8. Example of a supramolecular polymer.

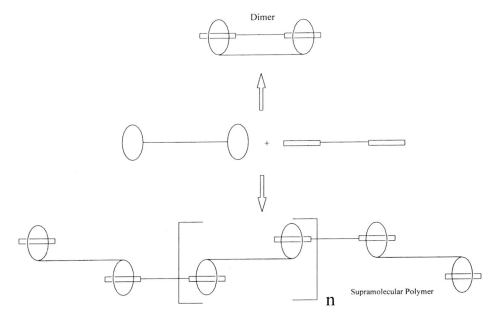

Figure 11.9. Supramolecular polymer resembling a pseudorotaxane.

(Fig. 11.9) [37]. They demonstrated by ^{1}H-NMR that the process of formation of a dimer or a linear supramolecular polymer is concentration dependent.

11.3.3 Networks

Another facet of supramolecular chemistry is *crystal engineering* [38], which includes the study of non-covalent interactions in the solid state for the design of novel solid-state materials. The basic rules (non-covalent interactions) governing complementarity are the same in both the liquid and the solid state. The crystal structure results from a balance of a variety of attractive/repulsive forces. These forces are both directional and non-directional and operate over a range of distances. For the design of organic supramolecules from crystal engineering it is necessary to understand the nature of these interactions [39]. An account of crystal engineering and its comparison with traditional organic synthesis is provided by Desiraju [40]. The eventual goal of these studies is the prediction of the crystal structure given the molecular structure, and eventually of its properties [41].

Recently, the concept of *nanoporous molecular crystals* with voids similar to zeolites engineered in terms of size, shape and chemical environments, has become popular. These voids could potentially be used for separation, catalysis and storage. The concept of 'robust supramolecular modules' and 'synthons' have emerged for rationally designing voids (rather than waiting for their serendipitous formation). Towards achieving this goal, attention is also paid to metal–organic frameworks [42]. Ward and co-workers utilized hydrogen-bonding interactions between guanidinium and sulfonate groups to rationally design a two-dimensional hydrogen-bonded network (Fig. 11.10) with adjustable porosity [43]. Examination of a series of such complexes in the presence of guests led to the rationalization of the resulting inclusion phenomena on the basis of the linkers used to attach the disulfonates. The same motif has recently been used where the linkers between the sulfonate groups and the guest molecule were identical. In this case, packing of the linkers and the guests (between the hydrogen-bonded network sheets) was nearly identical to that observed for the pure guests [44].

The complex between melamine and cyanuric acid (1:1) was reported in the literature in the late 1970s, but it was only in the early 1990s that the contributions from Whitesides and the concept of self-assembly popularized these systems [45]. Whitesides and co-workers reported the formation of tapes (Fig. 11.11), crinkled tapes and cyclic hexamers (rosettes) formed between barbituric acid and *N,N′-bis*(*p*-substituted phenyl)melamine [46]. In this they effectively blocked one face of melamine and, by manipulating substituents at the *para* position, different structures were obtained. Whiteside's putative suggestion that melamine/cyanuric acid formed an extended array (Fig. 11.12) was confirmed recently by Rao *et al.* with the crystal structure [47]. Hamilton and co-workers reported the crystal structure of a 5-substituted isophthalic acid derivative, which forms a cyclic aggregate held together with six pairs of hydrogen bonds, which in a way resembles the trimesic acid (Fig. 11.13) [48].

Lehn reported the synthesis and solid-state architecture of chiral 2,5-diazabicyclo[2.2.2]octane-3,6-dione. The racemic material adopted an infinite chain of alternating (+) and (−) units, whereas the single enantiomer formed a cyclic tetrameric structure (Fig. 11.14) [49].

A tetrameric structure derived from guanosine units (G-quartets) has been implicated in certain biological functions [50]. Metal ions have been known to promote the formation of G-quartets. Sessler *et al.* recently reported the formation of G-quartets both in the solid state and in organic media in the absence of metal ions, from modified guanosine derivatives [51]. Ciuchi *et al.* reasoned that the guanosine unit and the pterin unit present in the folic acid are similar and observed that aqueous solutions of the sodium and the potassium salts of folic acid, with or without NaCl, formed columnar mesophases (Fig. 11.15). These columnar phases consist of tetrameric units held by Hoogsteen-type bonding which arranges in a stacking array [52].

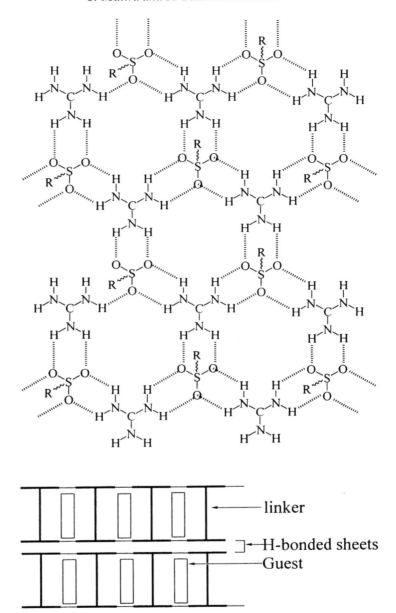

Schematic diagram to show guest inclusion

Figure 11.10. Guanidinium–sulfonate network.

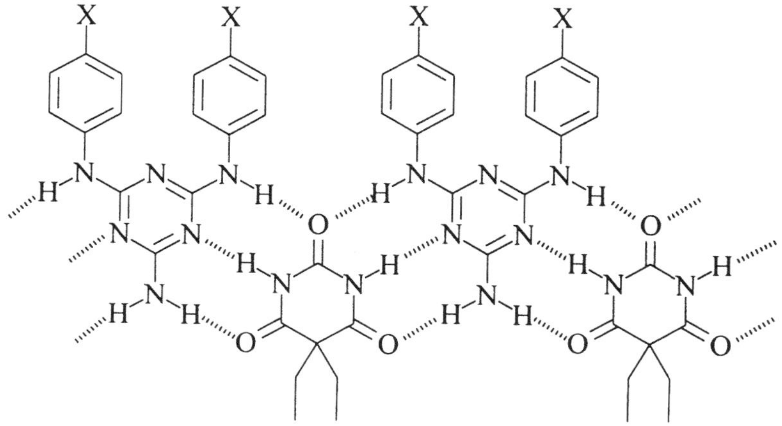

X = F, Cl, Br, I, CH$_3$

Figure 11.11. Tapes dervied from melamine/barbituric acid derivatives.

Zimmerman has reported the synthesis and aggregation behavior of a novel heterocycle having both DDA and (its complementary) AAD sites. The hexameric aggregate formed in organic solvents, apart from the engineered DDA–AAD interactions, may possibly have an extra long range hydrogen-bond to stabilize the assembly [53]. The hexameric aggregate (Fig. 11.16) formed in toluene-d$_8$ was stable even at 90 °C as monitored by ^{1}H-NMR.

11.3.4 Gels as novel materials

Organogelators

Although the applications of polymeric gelators (especially for aqueous systems) in the cosmetics and food processing industries is well known, the systematic study of gelation of organic solvents by low molecular weight organogelators is a relatively recent phenomenon. It is generally believed that three-dimensional networks formed by the gelators trap the solvent molecules through surface tension, thereby preventing its flow. How such gelator molecules form three-dimensional networks is not well understood. However, it seems clear that the individual molecules must utilize various non-covalent interactions like hydrogen bonding, metal coordination, and π–π interactions to interact with one another. Terech and Weiss have summarized the developments up to 1997 in a detailed review [54]. A recent article draws attention to possible applications arising out of these special types of materials [55].

An assorted set of examples of organogelators that have appeared in the

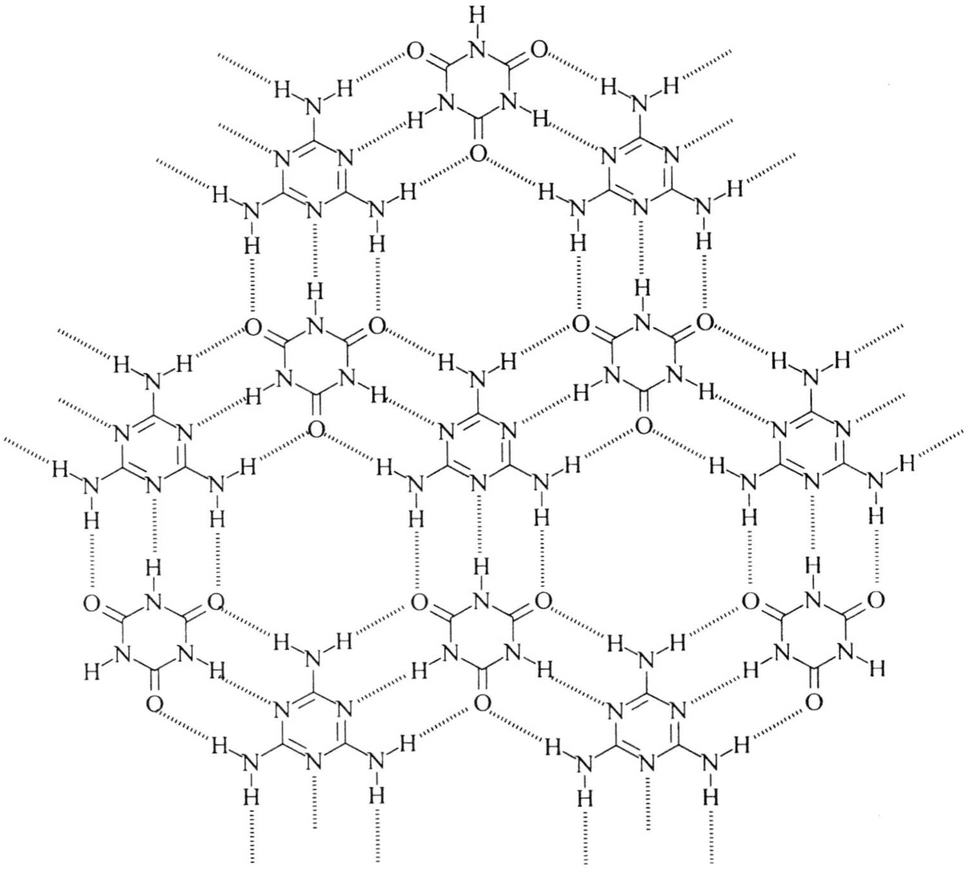

Figure 11.12. Melamine/cyanuric acid crystal structure.

literature recently is illustrated in Fig. 11.17 (see p. 380). The cholesterol moiety is one of the most prominent units present in many of the organogelators. Recently, Shinkai and co-workers reported the synthesis of a *bis*cholesterol, in which two cholesterol units were linked through a scaffold, with provision to bind a barbiturate unit via a hydrogen bond [56]. This compound alone and in the presence of barbiturate could form gels, thus showing gelation promoted by hydrogen-bonded interactions between a host and a guest. Hanabusa *et al.* screened a series of amides derived from cholic acid and observed that only the *i*-Pr derivative was a gelator [57]. Hanabusa's group also reported the synthesis of a chiral gelator derived from 2-amino-2-phenylethanol with multiple hydrogen-bonding sites. Apart from the usual aromatic solvents, mineral oils, silicone oils and edible oils have also been gelled by this gelator [58]. Ferringa's group reported the gelation of organic solvents based on cyclic *bis*-urea compounds that also utilizes hydrogen-bonded interactions [59].

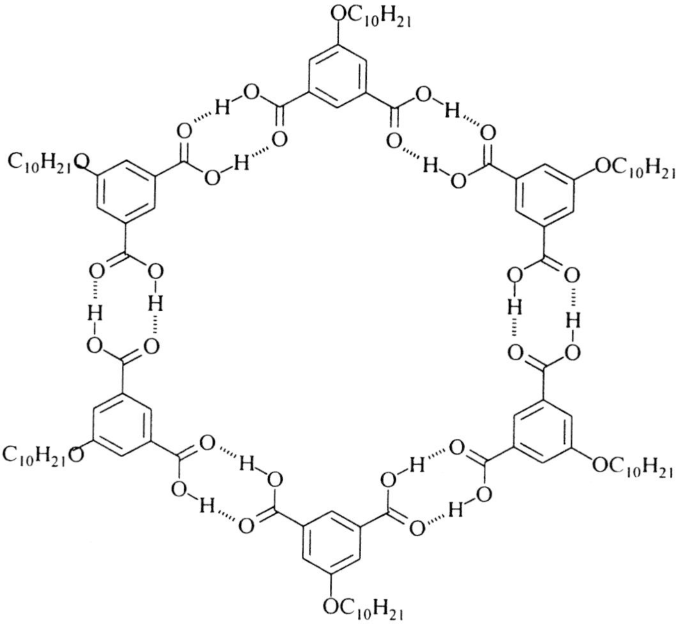

Figure 11.13. A hexameric array derived from an isophthalic acid derivative.

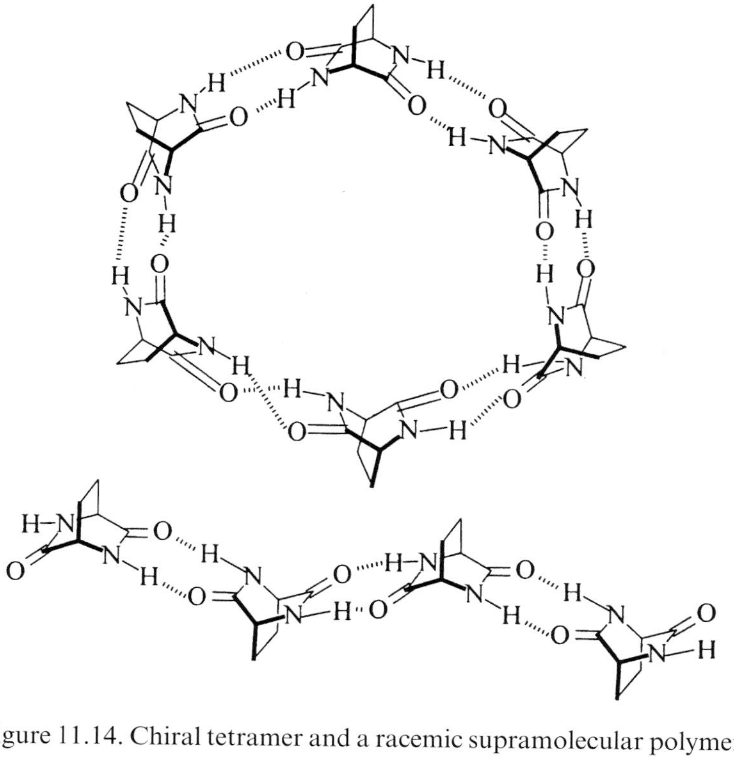

Figure 11.14. Chiral tetramer and a racemic supramolecular polymer.

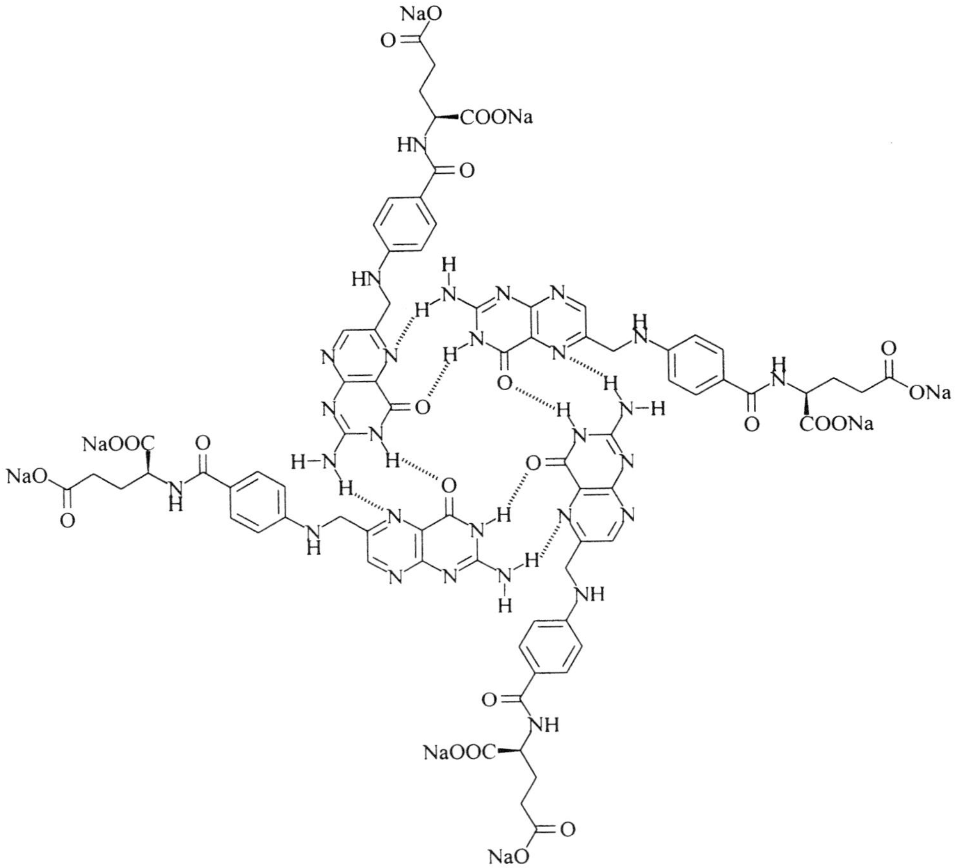

Figure 11.15. A tetrameric array formed by folate that stacks to form columnar mesophases.

During the course of our investigations on bile acid derived molecular tweezers [8c, 8d], we observed that a bile acid appended pyrene unit gelled certain organic solvents *in the presence of* trinitrofluorenone (Fig. 11.18) [60]. No gelation was observed with either of the pure components, and our studies showed that for optimum gelation a 1:1 stoichiometry was needed. The gels were strongly colored because of the charge-transfer band between the electron donor and the electron acceptor (Fig. 11.19). The position of the pyrene unit on the steroid backbone appeared to be important for its gelling ability. These gels were thoroughly characterized by spectroscopic measurements. The phase transition temperatures (T_{gel}) of some selected gels were measured by spectroscopic and other methods. Scanning electron microscopy of the xerogels gave us some insights into the morphology of the gel structures. Most of the

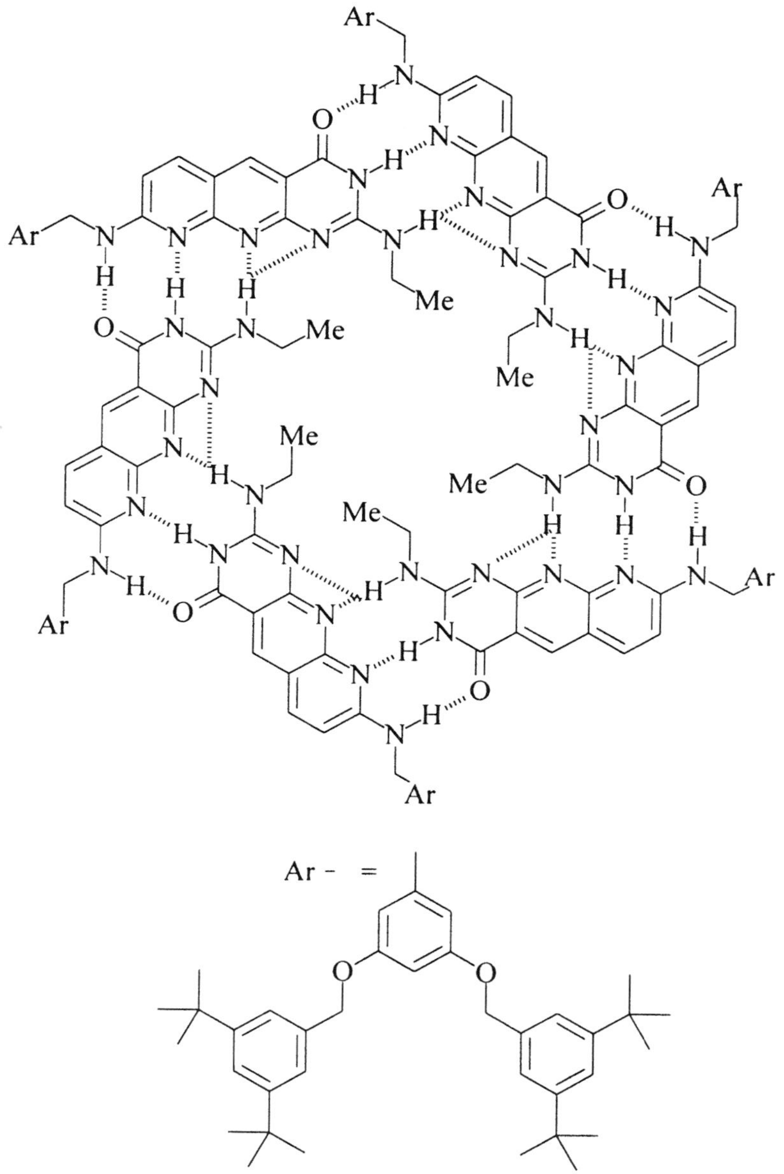

Figure 11.16. DDA–AAD linked hexameric species.

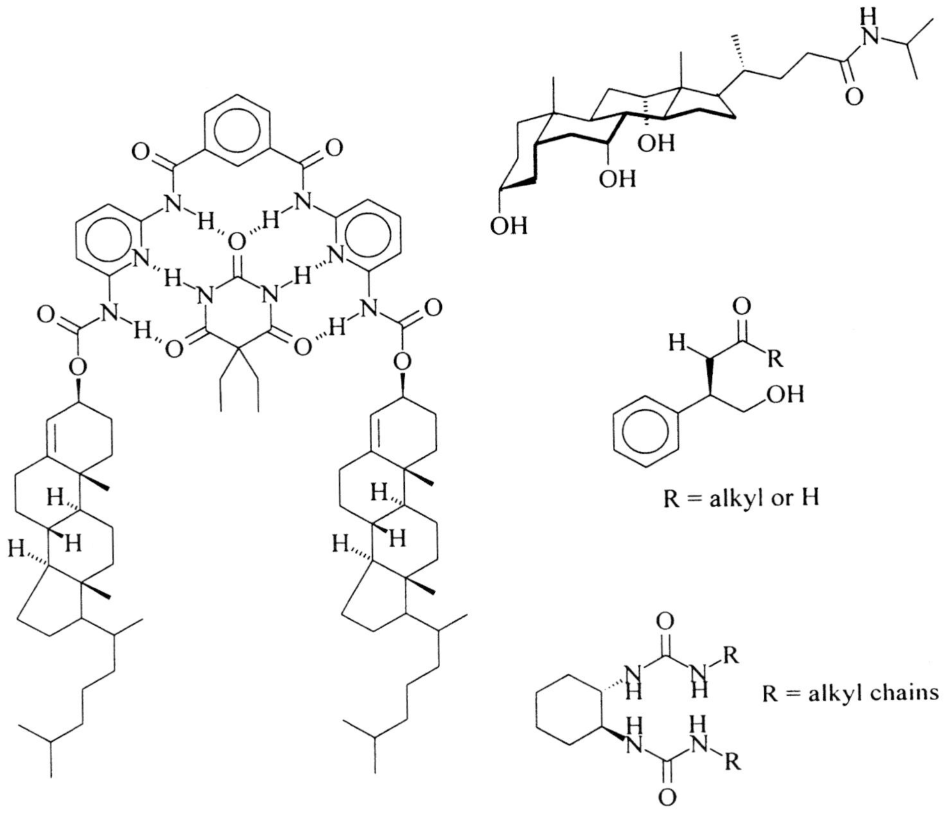

Figure 11.17. Recently reported organogelators from various groups.

xerogels characterized by SEM showed networked fibrous structures, with the fiber diameters of the order of microns (Fig. 11.20).

After the serendipitous discovery of the first donor–acceptor interaction promoted gelation of organic solvents, we decided to probe the role of the bile acid unit. This was important because Weiss and Shinkai have independently shown in their studies that their gelators were very specific in having the cholesterol unit as the key structural component. A generalized structure of the variety of compounds studied by us is shown in Fig. 11.21. It is interesting to note that whereas many of these compounds (without the bile acid backbone!) gelled selected solvents in the presence of TNF, some of these compounds also gelled solvents in the absence of TNF. Efforts are on in the authors' laboratory to study many of these gels in detail in order to generate a structure–function relationship for this class of compounds [61].

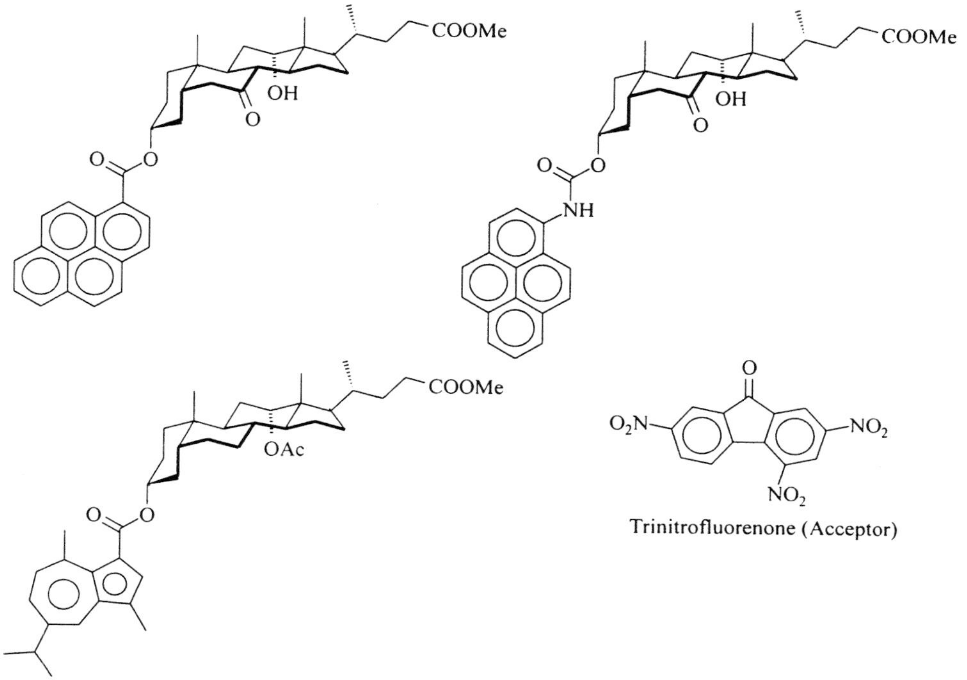

Figure 11.18. Charge-transfer gelators.

Aqueous gels derived from bile salts and their analogues

Bile salts and bile acid conjugates solubilize lecithin and cholesterol in the bile, aid in digestion by removing the products of pancreatic hydrolysis, and probably control certain intracellular events such as the synthesis of cholesterol in the intestinal mucosa [62]. These biologically important surfactants have been extensively studied during the past few decades in order to correlate the pathological implications to molecular structures and interactions. Ueno and Asano have recently reviewed some of the physicochemical properties of bile salts with a detailed discussion of the surface tension, mixed micelle formation and cholesterol solubilization by bile salts and their glycine and taurine conjugates [63]. Cholesterol solubilization is another major property of bile salts, and in this respect the taurine and glycine conjugates of cheno- and ursodeoxycholic acid have been examined in detail [64]. Compared with common surfactants, bile salts present a different three-dimensional structure – with a side chain carboxyl group and up to three hydroxyl groups. The hydroxyl groups attached to the steroid nucleus are all directed towards the concave surface of the molecule. These are therefore amphifacial molecules with a hydrophobic and a hydrophilic surface (Fig. 11.22). Above critical micellar concentration, bile salt

(a)

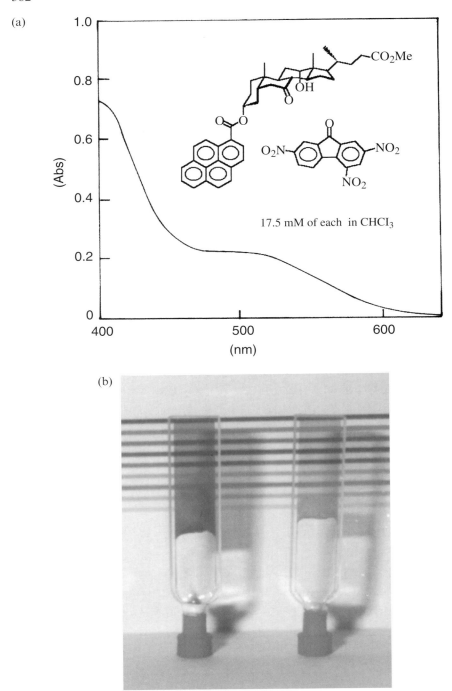

Figure 11.19. (a) Charge-transfer band in the absorption spectrum from ester and TNF. (b) Gels derived from TNF and carbamate (left)/ester (right).

Figure 11.20. SEM of a charge-transfer gel.

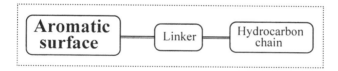

Figure 11.21. Charge-transfer gelators without the bile acid scaffold.

derived *primary micelles* are known to further associate to form *secondary micelles* at higher ionic strength. The aggregation numbers of these salts are known to increase with the increase in the salt concentration. Two models regarding the micellar structure of the bile salts are known [65, 66]. There are numerous review articles on this subject that summarize the relevant variables like temperature, pH, ionic strength etc. which affect the micellar properties [67, 63].

Among bile salts, sodium cholate (NaC) and sodium deoxycholate (NaDC) have been extensively studied by a variety of physical techniques. A few prominent techniques include nuclear magnetic resonance [68], electron spin resonance [69], small angle X-ray scattering [70], and quasi-elastic light scattering [71].

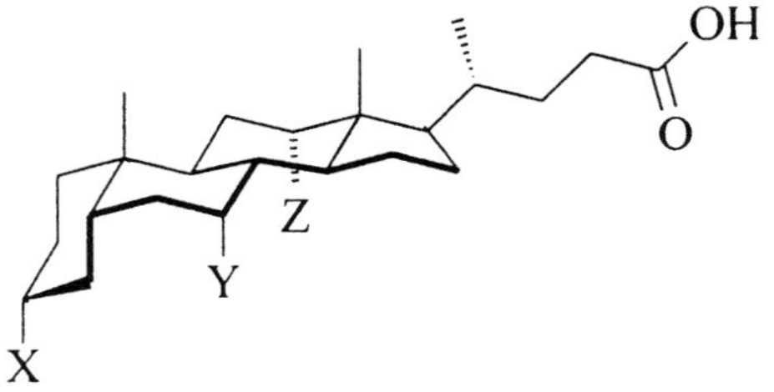

Lithocholic acid   X = OH, Y = Z = H
Deoxycholic acid   X = Z = OH, Y=H
Cholic acid   X = Y = Z = OH

Figure 11.22. The structure of common bile acids.

As early as in 1958, Rich and Blow showed that NaDC forms a helical complex of macromolecular dimensions [72]. They observed that unusual changes occurred in NaDC (0.1 M) in the presence of glycylglycine (0.1 M). There was a large initial increase in the viscosity of the medium, and they were able to draw glassy and brittle fibers, which showed a regular structure by X-ray diffraction. With time the solution reached a gelatinous state and they could lift the cohesive mass, which was more like a gelly.

Ramanathan *et al.* examined various bile acids and suggested that deoxycholic acid formed helical microfibrils, and that lithocholic acid formed bundles [73]. Weiss and co-workers have recently shown that sodium lithocholate (NaLC) formed aqueous gels under certain conditions. NaLC is also capable of gelling ethanol, methanol and ethylene glycol [74].

Cationic analogues of bile salts have attracted the attention of medicinal chemists during the past few decades mainly as anti-microbial [75] and cholesterol lowering agents [76], gall stone dissolution enhancers [77] and DNA transfection agents [78], among others [79]. Most of these analogues have been synthesized by attaching a *bis*(amine) to the bile acid through an amide bond, and converting the other amine to a quaternary ammonium salt. In some cases the amide group was reduced to an amine. Recently we have initiated a program to synthesize new cationic analogues of deoxycholic acid (Fig. 11.23) and study their aggregation and gelation behavior.

The study on the aggregation behavior of these salts showed that, by varying the head group, the critical micellar (aggregation) concentration varied significantly. A pyridinium salt formed stable gels at 0.5–1.0% (w/v) concentration in the presence of NaCl (1.0 M). Preliminary investigations on these molecules revealed that, other than water, only aqueous MeOH (90%) formed a gel. Unlike the gels of NaDC and NaLC, these are not thixotropic. We have also

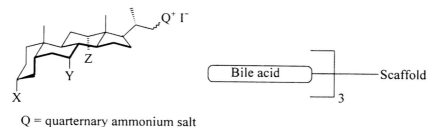

Q = quarternary ammonium salt
X, Y, Z = H/OH

Figure 11.23. General representation of bile acid based aqueous gelators.

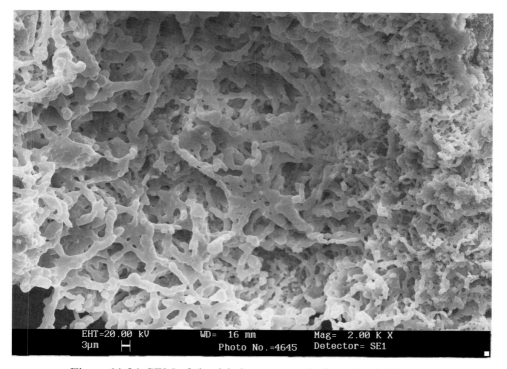

Figure 11.24. SEM of the dried aqueous gel of a cationic bile salt.

observed that sol to gel transition was accompanied by a marked increase in the optical rotation of the sample. The xerogel of one of the cationic surfactants as studied by scanning electron microscopy is shown in Fig. 11.24. Although cationic bile salts were extensively used for other applications, to the best of our knowledge our molecules represent the first example of a cationic bile based gelator [80].

We recently attached three bile acid units to a tripodal scaffold to generate a trimeric analogue (Fig. 11.21) [81]. This compound formed transparent gels in water containing a small amount of an organic co-solvent. We are currently studying these gels by a variety of spectroscopic techniques. Some of these gels might provide novel chiral media for carrying out selective chemical reactions.

11.4 Future perspectives

In this chapter we have attempted to provide an overview of supramolecularly designed materials of different types. Some of the systems shown here will probably find applications as novel materials in the coming years. It is quite clear that organic materials that have potential applications in molecular electronics [82] and other futuristic applications can only be obtained through *rational supramolecular design* [83]. Further research on organogels will lead to a better understanding of gel architecture in terms of molecular structures of gelators. Since organogels as novel materials and media provide flexibility and versatility, the next decade is likely to witness many interesting and useful applications arising out of them [55].

11.4.1 Acknowledgment

The authors thank the Department of Science and Technology, New Delhi, for supporting the research carried out at the Indian Institute of Science, Bangalore.

11.6 References

1. *Chambers 21st Century Dictionary* (1997). Editor in Chief M. Robinson. Allied Chambers (India) Limited, New Delhi.
2. (a) Lehn, J.-M. (1988). *Angew. Chem., Int. Ed. Engl.* **27**, 89. (b) Lehn, J.-M. (1995). *Supramolecular Chemistry, Concepts and Perspectives*. VCH, Weinheim.
3. Lawrence, D. S., Jiang, T. and Levett, M. (1995). *Chem. Rev.* **95**, 2229.
4. Chapman, R. G. and Sherman, J. C. (1997). *Tetrahedron* **53**, 15911.
5. Developments till 1995 have been reviewed in *Comprehensive Supramolecular Chemistry* (1996). (J. L. Atwood, J. E. D. Davies, D. D. MacNicol, F. Vögtle, J.-M. Lehn, Eds.), Vol. 1–10. Pergamon: Tarrytown, NY.
6. Jeffrey, G. A. and Saenger, W. (1991). *Hydrogen Bonding in Biological Structures*. Springer-Verlag, Berlin and Heidelberg.
7. MacGillivray, L. R. and Atwood, J. L. (1997). *Nature* **389**, 469.
8. (a) Hunter, C. A. and Sanders, J. K. M. (1990). *J. Am. Chem. Soc.* **112**, 5525. (b) Zimmerman, S. C. (1993). *Top. Curr. Chem.* **165**, 71. (c) Maitra, U. and D'Souza, L. J. (1994). *J. Chem. Soc., Chem. Commun.*, 2793. (d) D'Souza, L. J. and Maitra, U. (1996). *J. Org. Chem.* **61**, 9494. (e) Otsuki, J., Oya, T., Lee, S.-H. and Araki, K. (1995). *J. Chem. Soc., Chem. Commun.*, 2193.

9. (a) Fredericks, J. R. and Hamilton, A. D. (1996). Metal template control of self-assembly in supramolecular chemistry, in *Perspectives in Supramolecular Chemistry, Vol. 3 (Supramolecular Control of Structure and Reactivity)* (A. D. Hamilton, Ed.), p. 1. John Wiley and Sons, Chichester. (b) For a recent review see: Leininger, S., Olenyuk, B. and Stang, P. J. (2000). *Chem. Rev.* **100**, 853–908.

10. Piguet, C., Bernardinelli, G. and Hopfgartner, G. (1997). *Chem. Rev.* **97**, 2005.

11. Constable, E. C. 'Polynuclear transition metal helicates' in Ref. [5], Vol. 9, p. 213.

12. Zarges, W., Hall, J. and Lehn, J.-M. (1991). *Helv. Chim. Acta.* **74**, 1843.

13. (a) Fujita, M. (1998). *Chem. Soc. Rev.* 417. (b) Fujita, M. 'Self-assembled macrocycles, cages and catenanes containing transition metals in their backbones', in ref. [5], Vol. 9, p. 253.

14. Tanford, C. (1973). *The Hydrophobic Effect: Formation of Micelles and Biological Membranes.* John Wiley and Sons, New York.

15. (a) Fendler, J. H. (1982). *Membrane Mimetic Chemistry.* John Wiley and Sons, New York. (b) Fendler, J. H. and Fendler, E. J. (1975). *Catalysis in Micellar and Macromolecular Systems.* Academic Press, New York. (c) For micellar solutions as reaction media, see Tasicoglu, S. (1996). *Tetrahedron* **52**, 11 113.

16. Kunitake, T. (1992). *Angew. Chem., Int. Ed. Engl.* **31**, 709.

17. Szejtli, J. 'Inclusion of guest molecules, selectivity and molecular recognition by cyclodextrins,' in ref. [5], Vol. 3, p. 189.

18. (a) Tomalia, D. A., Naylor, A. M. and Goddard III, W. A. (1990). *Angew. Chem., Int. Ed. Engl.* **29**, 1038. (b) Tomalia, D. A. and Durst, H. D. (1993). *Top. Curr. Chem.* **165**, 193. (c) Newkome, G. R. and Moorefield, C. N. 'Dendrimers', in ref. [5], Vol. 10, 771. (d) Bosman, A. W., Janssen, H. M. and Meijer, E. W. (1999). *Chem. Rev.* **99**, 1665.

19. Buhleier, E., Wehner, W. and Vögtle, F. (1978). *Synthesis* 155.

20. (a) Knapen, J. W. J., van der Made, A. W., de Wilde, J. C., van Leeuwen, P. W. N. M., Wijkens, P., Grove, D. M. and van Koten, G. (1994). *Nature* **372**, 659. (b) Reetz, M. T., Lohmer, G. and Schwickardi, R. (1997). *Angew. Chem., Int. Ed. Engl.* **36**, 1526.

21. (a) Peerlings, H. W. I. and Meijer, E. W. (1997). *Chem. Eur. J.* **3**, 1563. (b) Thomas, C. W. and Tor, Y. (1998). *Chirality* **10** (1/2), 53. (c) Seebach, D., Rheiner, P. B., Greiveldinger, G., Butz, T. and Sellner, H. (1998). *Top. Curr. Chem.* **197**, 125. (d) Chow, H.-F., Mong, T. K.-K., Wan, C.-W. and Wang, Z.-Y. (1999). 'Chiral dendrimers' in *Advances in Dendritic Macromolecules*, Vol. 4, p. 107. JAI Press Inc.

22. Balasubramanian, R., Rao, P. and Maitra, U. (1999). *J. Chem. Soc., Chem. Commun.*, 2353.

23. (a) Issenberner, J., Moors, R. and Vögtle, F. (1994). *Angew. Chem., Int. Ed. Engl.* **33**, 2413. (b) Smith, D. K. and Diederich, F. (1998). *Chem. Eur. J.* **4**, 1353. (c) Archut, A. and Vögtle, F. (1998). *Chem. Soc. Rev.* **27**, 233. (d) Fischer, M. and Vögtle, F. (1999). *Angew. Chem., Int. Ed. Engl.* **38**, 884.

24. Zeng, F. and Zimmerman, S. C. (1997). *Chem. Rev.* **97**, 1681.

25. Newkome, G. R., He, E. and Moorefield, C. N. (1999). *Chem. Rev.* **99**, 1689.

26. Venturi, M., Serroni, S., Juris, A., Campagna, S. and Balzani, V. (1998). *Top. Curr. Chem.* **197**, 193.

27. (a) Schlüter, A. D. (1998). *Top. Curr. Chem.* **197**, 165. (b) Schlüter, A. D. and Rabe, J. P. (2000). *Angew. Chem, Int. Ed. Engl.* **39**, 864.

28. (a) Zeng, F. and Zimmerman, S. C. (1997). *Chem. Rev.* **97**, 1681. (b) Emrick, T. and Fréchet, J. M. J. (1999). *Curr. Opin. Colloid & Interface Sci.* **4**, 15.

29. Zimmerman, S. C., Zeng, F., Reichert, D. E. C. and Kolotuchin, S. V. (1996). *Science* **271**, 1095.

30. Percec, V., Cho, W.-D., Ungar, C. and Yeardley, D. J. P. (2000). *Angew. Chem., Int. Ed. Engl.* **39**, 1598, and the references cited.

31. Zimmerman, N., Moore, J. S. and Zimmerman, S. C. (1998). *Chem. Ind.*, 604.

32. Fouquey, C., Lehn, J.-M. and Levelut, A.-M. (1990). *Adv. Mater.* **2**, 254.

33. Gulik-Krzywicki, T., Fouquey, C. and Lehn, J.-M. (1993). *Proc. Natl. Acad. Sci. USA* **90**, 163.

34. For a representative example see: St. Pourcain, C.B. and Griffin, A. C. (1995). *Macromolecules* **28**, 4116.

35. Paleos, C. M. and Tsiourvas, D. (1995). *Angew. Chem., Int. Ed. Engl.* **34**, 1696.

36. Sijbesma, R. P., Beijer, F. H, Brunsveld, L., Folmer, B. J. B., Ky Hirschberg, J. H. K., Lange, R. F. M., Lowe, J. K. L. and Meijer, E. W. (1997). *Science* **278**, 1601.

37. Yamaguchi, N. and Gibson, H. W. (1999). *Angew. Chem., Int. Ed. Engl.* **38**, 143.

38. Nangia, A. and Desiraju, G. R. (1998). *Top. Curr. Chem.* **198**, 57.

39. Desiraju, G. R. 'Review of general principles' in ref. [5], Vol. 6, p. 1.

40. Desiraju, G. R. (1995). *Angew. Chem., Int. Ed. Engl.* **34**, 2311.

41. Gavezzotti, A. (1994). *Acc. Chem. Res.* **27**, 309.

42. Li, H., Eddaoudi, M., O'Keeffe, M. and Yaghi, O. M. (1999). *Nature* **402**, 276.

43. Russell, V. A., Evans, C. C., Li, W. and Ward, M. D. (1997). *Science* **276**, 575.

44. Holman, K. T. and Ward, M. D. (2000). *Angew. Chem., Int. Ed. Engl.* **39**, 1653.

45. (a) MacDonald, J. C. and Whitesides, G. M. (1994). *Chem. Rev.* **94**, 2383. (b) Whitesides, G. M., Simanek, E. E., Mathias, J. P., Seto, C. T., Chin, D. N., Mammen, M. and Gordon, D. M. (1995). *Acc. Chem. Res.* **28**, 37.

46. Zerkowski, J. A., Seto, C. T. and Whitesides, G. M. (1992). *J. Am. Chem. Soc.* **114**, 5473.

47. Ranganathan, A., Pedireddi, V .R. and Rao, C. N. R. (1999). *J. Am. Chem. Soc.* **121**, 1752.

48. Yang, J., Marendaz, J.-L., Geib, S. J. and Hamilton, A.D. (1994). *Tetrahedron Lett.* **35**, 3665.

49. Brienne, M.-J., Gabard, J., Leclercq, M., Lehn, J.-M., Cesario, M., Pascard, C., Cheve, M. and Dutruc-Rosset, G. (1994). *Tetrahedron Lett.* **35**, 8157.

50. Borman, S. (1999). *Chem. Eng. News* **77** (27), 36.

51. Sessler, J. L., Sathiosatham, M., Doerr, K., Lynch, V. and Abboud, K. A. (2000). *Angew. Chem., Int. Ed. Engl.* **39**, 1300, and references therein.

52. Ciuchi, F., Di Nicola, G., Franz, H., Gottarelli, G., Mariani, P., Bossi, M. G. P. and Spada, G. P. (1994). *J. Am. Chem. Soc.* **116**, 7064.

53. Kolotuchin, S. V. and Zimmerman, S. C. (1998). *J. Am. Chem. Soc.* **120**, 9092.

54. Terech, P. and Weiss, R. G. (1997). *Chem. Rev.* **97**, 3133.

55. van Esch, J. H. and Feringa, B. L. (2000). *Angew. Chem., Int. Ed. Engl.* **39**, 2263.

56. Inoue, K., Ono, Y., Kanekiyo, Y., Ishi-i, T., Yoshihara, K. and Shinkai, S. (1999). *J. Org. Chem.* **64**, 2933.

57. Hishikawa, Y., Sada, K., Watabale, R., Miyata, M. and Hanabusa, K. (1998). *Chem. Lett.* 795.

58. Hanabusa, K., Maesaka, Y., Kimura, M. and Shirai, H. (1999). *Tetrahedron Lett.* **40**, 2385.

59. Esch, J. V., Schoonbeek, F., de Loos, M., Kooijman, H., Spek, A. L., Kellog, R. M. and Ferringa, B. L. (1999). *Chem. Eur. J.* **5**, 937.
60. Maitra, U., Vijay Kumar, P., Chandra, N., D'Souza, L. J., Prasanna, M. D. and Raju, A. R. (1999). *J. Chem. Soc. Chem. Commun.*, 595.
61. Maitra, U., Vijay Kumar, P., Sangeetha, N. M., Babu, P. and Raju, A. R. (2001) *Tetrahedron: Asymmetry*, **12**, 477.
62. *The Bile Acids: Chemistry, Physiology and Metabolism* (1971). Vol. 1–3 (P. P. Nair and D. Kritchevsky, Eds.), Plenum Press, New York and London.
63. Ueno, M. and Asano, H. 'Physicochemical properties of bile salts', in *Structure–Performance Relationships in Surfactants* (K. Esumi and M. Ueno, Eds.), p. 147. Marcel Dekker Inc., New York, Basel, Hong Kong.
64. (a) Carey, M. C., Montet, J.-C., Phillips, M. C., Armstrong, M. J. and Mazer, N. A. (1981). *Biochemistry* **20**, 3637. (b) Igimi, H. and Carey, M. C. (1981). *J. Lipid. Res.* **22**, 254.
65. Small, D. M. (1968). *Adv. Chem. Ser.* No. 84, 31.
66. Oakenfull, D. G. and Fischer, L. R. (1977). *J. Phys. Chem.* **81**, 1836.
67. (a) Carey, M. C. (1985). *New Comprehensive Biochemistry*, Vol. 12, *Sterols and Bile Acids* (H. Danielsson and J. Sjovall, Eds.), p. 345. Elsevier, Amsterdam. (b) Small, D. M. 'The physical chemistry of cholanic acids', in ref. [62], Vol. 1, p. 249.
68. Murata, Y., Sugihara, G., Fukushima, K., Tanaka, M. and Matsushita, K. (1982). *J. Phys. Chem.* **86**, 4690.
69. Esposito, G., Giglio, E., Pavel, N. V. and Zanobi, A. (1987). *J. Phys. Chem.* **91**, 356.
70. Zakrzewska, J., Marković, V., Vučelić, D., Feigin, L., Dembo, A. and Mogilevsky, L. (1990). *J. Phys. Chem.* **94**, 5078.
71. D'Archivio, A. A., Galantini, L., Giglio, E. and Jover, A. (1998). *Langmuir* **14**, 4776.
72. Rich, A. and Blow, D. M. (1958). *Nature* **182**, 423.
73. Ramanathan, N., Curie, A. L. and Colvin, J. R. (1961). *Nature* **190**, 779.
74. Terech, P., Smith, W. G. and Weiss, R. G. (1996). *J. Chem. Soc., Faraday Trans.* **92** (17), 3157.
75. (a) Bernheim, F. and Lack, L. (1971). *Can. J. Microbiol.* **17**, 323. (b) Bellini, A. M., Quaglio, M. P., Guarneri, M. and Cavazini, G. (1983). *Eur. J. Med. Chem. – Chim. Ther.* **18**, 185, 191.
76. (a) Fears, R., Brown, R., Ferres, H., Grenier, F. and Tyrrell, A. W. (1990). *Biochem. Pharmacol.* **40**, 2029. (b) Firpi, A., Walker, J. T. and Lack, L. (1975). *J. Lipid Res.* **16**, 379.
77. Kwan, K. H., Higuchi, W. I., Molokhia, A. and Hofmann, A.F. (1977). *J. Pharm. Sci.* **66**, 1105.
78. (a) Walker, S., Sofia, M. J., Kakarla, R., Kogan, N. A., Wierichs, L., Longley, C. B., Bruker, K., Axelrod, H. R., Midha, S., Babu, S. and Kahne, D. (1996). *Proc. Natl. Acad. Sci. USA* **93**, 1585. (b) Walker, S., Sofia, M. J. and Axelrod, H. R. (1998). *Adv. Drug Delivery Rev.* **30**, 61.
79. (a) Araki, Y.-I., Lee, A., Sugihara, G., Furuichi, M., Yamashita, S. and Ohseto, F. (1996). *Colloids Surf. B* **8** (1/2), 81. (b) Reid, D. G., Gajjar, K., Robinson, S. P., Hickey, D. M. B., Benson, G. M., Haynes, C., Leeson, P. D. and Whittaker, C. M. (1991). *Chem. Phys. Lipids* **60**, 143. (c) Borgstrom, B. (1977). *Biochim. Biophys. Acta* **488**, 381.

80. Maitra, U., Balsubramanian, R., Ghosh, S. and Raju, A., unpublished results.
81. Maitra, U., Mukhopadhyay, S., Sarkar, A., Rao, P. and Indi, S.S., (2001). *Angew. Chem. Int. Ed. Engl.* **40**, 2281.
82. Reed, M. A. and Tour, J. M. (2000). *Sci. Am.*, June 2000, 69.
83. Brinker, C. J., Lu, Y., Sellinger, A. and Fan, H. (1999). *Adv. Mater.* **11**, 579.

12

Controlling crystal architecture in molecular solids: the supramolecular approach

ANDREW D. BOND AND WILLIAM JONES

12.1 Introduction

The properties of molecular solids are well known to depend not only on the structure of the molecule itself, but also on the three-dimensional architecture adopted during crystallization. Such is the variation in properties with crystal structure that the existence of different polymorphs can, for example, switch on and off photoinduced reactions as well as profoundly affect the pharmaceutical activity of drugs [1–3].

The structure adopted by a molecule in the crystalline state is determined by a complex balance of forces, each possessing varying degrees of strength, distance dependence and directionality. The ability to control these forces to create predetermined three-dimensional architectures is a formidable problem. Given the current situation regarding the possible *ab initio* prediction of crystal structures (even for simple high-symmetry systems), it is unlikely that a rigorous and effective methodology for structure prediction will soon emerge [4]. The solid-state chemist, therefore, approaches the problem by attempting to identify particular interactions within a crystal which occur regularly and which seem to play a significant and repetitive role in directing molecules during crystallization. This approach has been compared with the synthon approach in classical organic chemistry in which the target is a complex molecule – the synthon approach identifies structural units which can be linked by known synthetic methods to give the desired molecule [5]. In designing a molecular solid, the equivalent position is to identify robust, predictable intermolecular interactions which can be used to create supermolecules [6, 7]. Whilst this methodology will not obviate the need for structure determination, it does reduce the number of possible motifs which might be present in the crystal. Crystal structure prediction then reduces (only slightly, but perhaps sufficiently in some cases that we will see) to a viable objective.

12.2 Intermolecular forces in molecular solids

In order to rationalize and subsequently predict the structures of molecular solids, it is necessary to understand the forces that hold the molecules together in the crystalline state. Crystallization of a pure substance from the liquid or gas phase is an entropically unfavorable process and must, therefore, result in an enthalpy gain sufficient to overcome the entropy decrease. This enthalpy gain will arise from a balance of attractive and repulsive forces between molecules within the crystal. The attractive forces may be divided into two types: dispersion interactions and electrostatic interactions. As a result of their transient nature, dispersion interactions are weak and non-directional, but are always attractive. Electrostatic interactions are stronger and more directional, and may be either attractive or repulsive. Hydrogen bonds may be considered as special electrostatic interactions; the absence of electrons on the hydrogen atom (other than that involved in the covalent bond) reduces the repulsive forces, making hydrogen bonds particularly strong and directional [8]. The attractive and repulsive intermolecular forces sum to give the total binding enthalpy (usually referred to as the lattice enthalpy) of the crystal.

The various distance dependencies and directionalities of the intermolecular forces give rise to numerous overall stabilizing arrangements – the only prediction that can be made with absolute confidence is that no overall destabilizing arrangement is permitted. As an initial simplification, molecules may be considered as impenetrable systems whose packing is governed solely by shape and size. Molecular recognition on the basis of shape requires complementary disposition of molecular features – 'bumps must fit into hollows'. This approach was formally described by Kitaigorodskii as the close-packing principle [9]. The need for close packing introduces certain limitations on the symmetry relationships between molecules in the crystal. Inversion centers and space symmetry elements (translations, screw axes and glide planes) necessarily lead to close packing, while rotation axes and mirror planes introduce voids into the structure. These observations provide an effective means to rationalize the observed space group statistics for organic molecules. Close packing may be considered to optimize dispersive interactions between molecules such that 'empty space is an energetic waste' [4]. In many cases, the symmetry requirements that give rise to close packing also give rise to favorable intermolecular electrostatic interactions. The common carboxylic acid dimer, for example, is formed across an inversion center. The overall three-dimensional arrangement of molecules in the crystalline state results from a complex interplay between non-directional dispersion interactions and electrostatic interactions with varying degrees of strength, directionality and distance dependence. In some cases, given interactions will act in concert while in other cases they may be in conflict. This complex balance of intermolecular forces ultimately gives rise to numerous possible equilibrium arrangements, making *ab initio* crystal structure prediction a formidable task.

12.3 Supramolecular chemistry in crystal structure design

Clearly, the task of predicting crystal structures solely on the basis of energetic requirements is not straightforward, and an alternative approach has emerged which utilizes the molecular recognition concepts of supramolecular chemistry. Following Lehn's initial statement that 'supermolecules are to molecules and the intermolecular bond what molecules are to atoms and the covalent bond' [10]. chemists were quick to recognize that a crystal of an organic compound is the ultimate supermolecule [11]. With this realization, the analogies between crystal formation and traditional organic synthesis followed rapidly. Targets in organic synthesis are molecules assembled via covalent bonds. In a similar manner, crystals may be considered as targets assembled via intermolecular interactions. A unifying concept within traditional organic synthesis, introduced by Corey in 1967, was the definition of synthons as 'structural units within molecules which can be formed and/or assembled by known or conceivable synthetic operations' [5]. Desiraju has extended this concept into the area of crystal design with the realization that 'supramolecular synthons are structural units within supermolecules which can be formed and/or assembled by known or conceivable synthetic operations involving intermolecular interactions' [6, 12]. The task of crystal structure design (or prediction) then reduces considerably to the identification and utilization of supramolecular synthons which are robust and transferable between structures. To build a desired supramolecular array simply requires that the molecules carry the appropriate substituents (the appropriate 'supramolecular functionality') to create the synthon.

The identification of synthons is effectively equivalent to identifying relative arrangements of supramolecular functionalities that will give rise to the greatest enthalpy of interaction. For organic molecules containing heteroatoms, the charge distribution over the molecule is non-uniform and the relative arrangements of neighboring molecules will influence the electrostatic interactions, thus affecting the lattice enthalpy. The overall packing arrangement must be one which optimizes attractive interactions and minimizes repulsive ones. Clearly, the repeated observation of a given structural motif within crystal structures must mean that this arrangement goes some way towards satisfying the enthalpy requirements. The synthons that will occur most commonly and

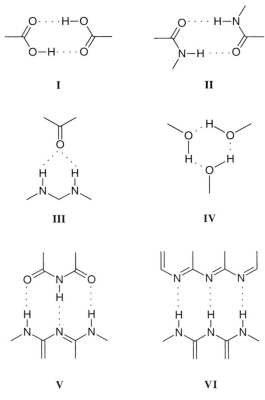

Scheme 12.1. Examples of synthons based on strong hydrogen bonds.

reliably, therefore, are those which give rise to large attractive enthalpies of interaction, i.e. those based upon strong intermolecular interactions. As the strengths of the intermolecular interactions comprising the synthon decrease, its robustness and predictability decreases accordingly.

12.3.1 Design strategies based on strong hydrogen bonds

The strongest and most directional of the intermolecular interactions which might comprise a supramolecular synthon are 'classical' hydrogen bonds, i.e. where both the hydrogen bond acceptor and the donor atom to which the hydrogen atom is bound are highly electronegative atoms such as O and N. The distinctive geometrical attributes of strong hydrogen bonds have been discussed extensively and their predictable directionality may be utilized in crystal design [8, 13, 14]. Some of the frequently observed synthons based upon such hydrogen bonds are shown in Scheme 12.1. The concept of structure design by utilization of such synthons is conveniently illustrated by the often-quoted

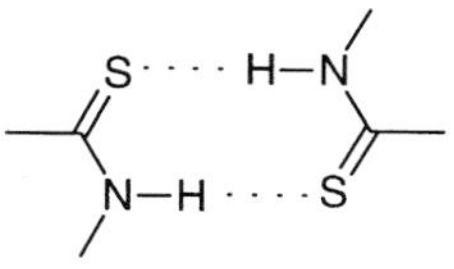

VII

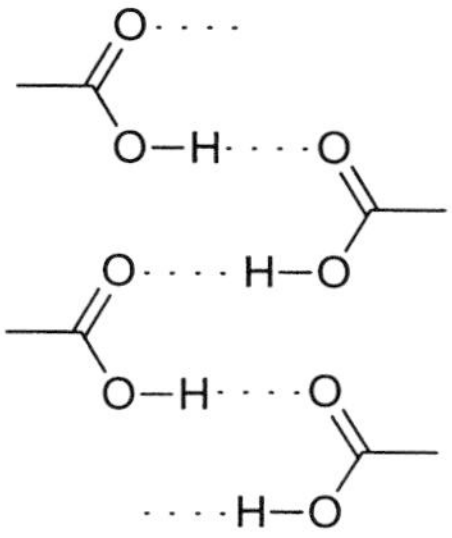

VIII

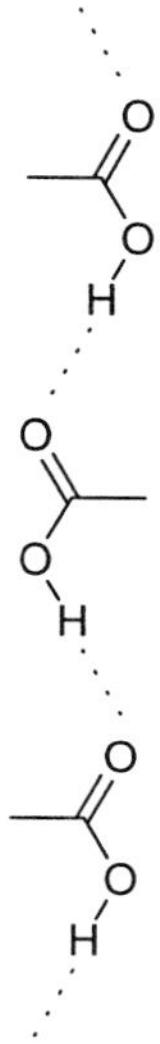

IX

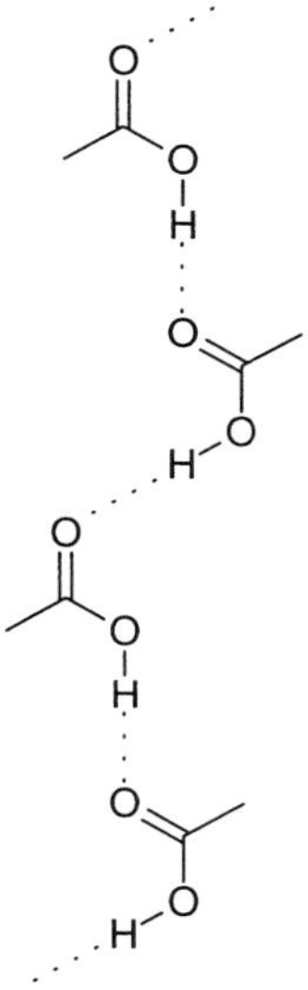

X

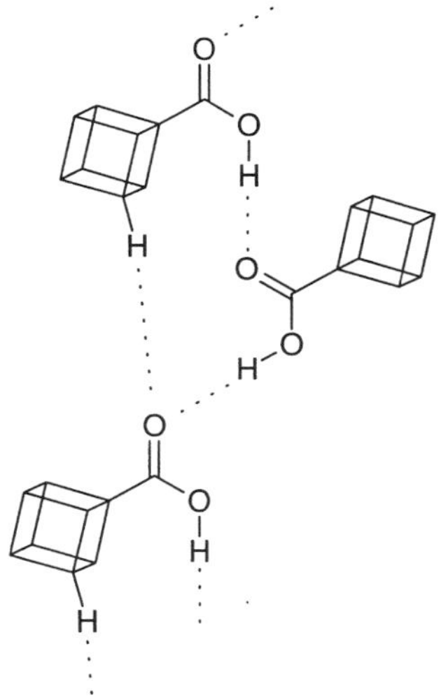

XI

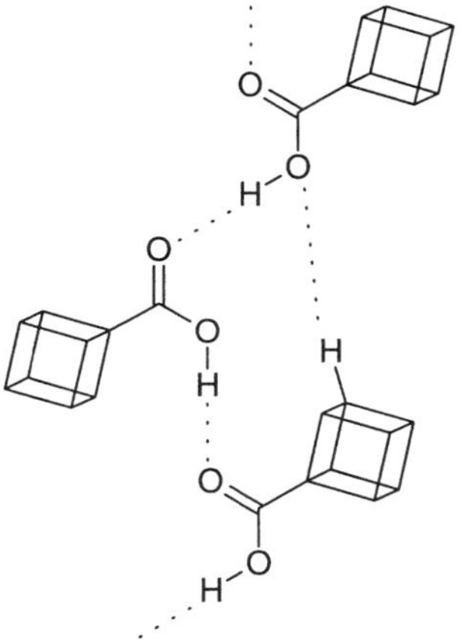

XII

Scheme 12.1. (*cont.*)

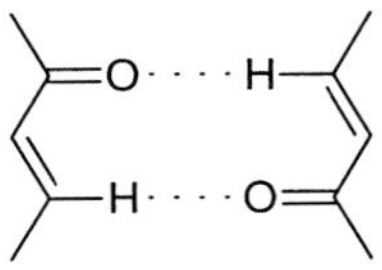

XIII 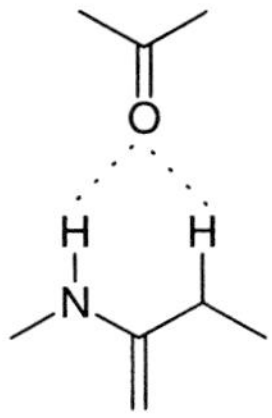   **XIV** 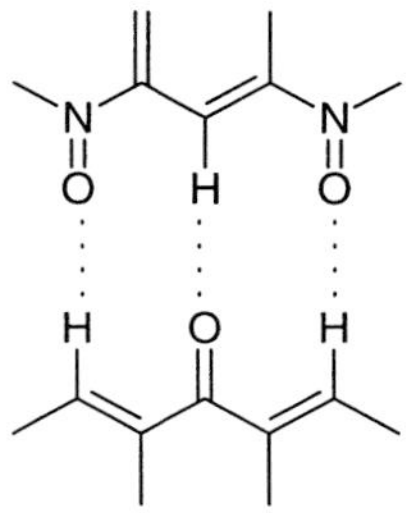   **XV**

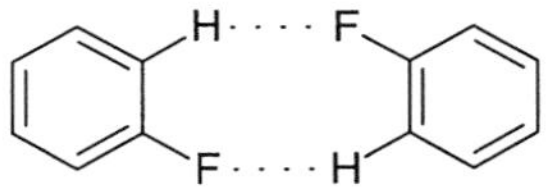

XVI 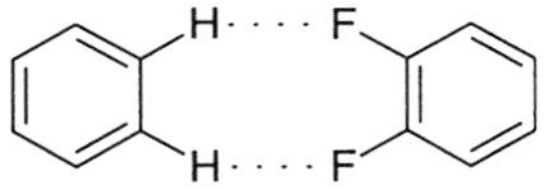   **XVII**

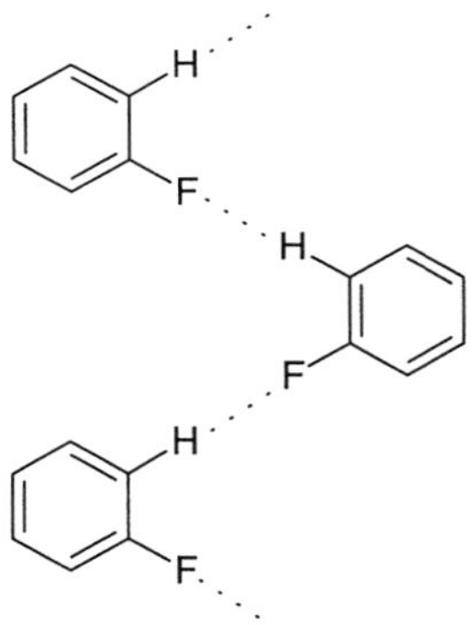

XVIII 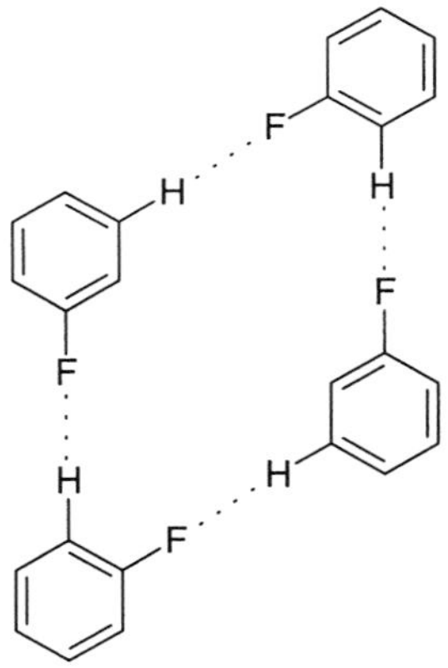   **XIX**

Scheme 12.1. (*cont.*)

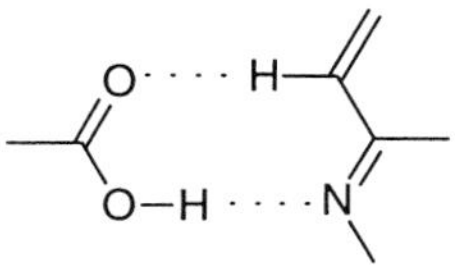

XX

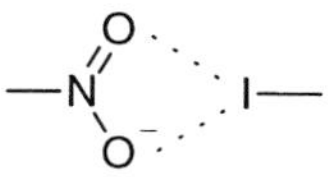

XXI

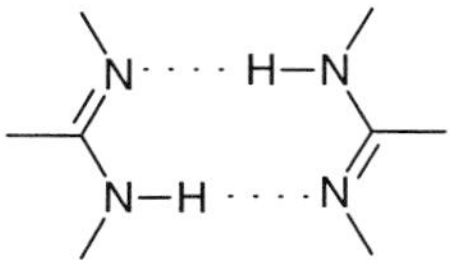

XXII

Scheme 12.1. (*cont.*)

crystal structures of the aromatic carboxylic acids: benzoic **1** [15], terephthalic **2** [16], isophthalic **3** [17], trimesic **4** [18] and adamantane-1,3,5,7-tetracarboxylic **5** [19] acids link via synthon **I** in a logical progression to produce isolated dimers, linear tapes, crinkled tapes, sheets and a three-dimensional diamondoid structure, respectively (Fig. 12.1). For **4** and **5**, the primary structures illustrated are interpenetrated in the crystals to improve packing efficiency. Several general reviews concerned with structure design strategies based on strong hydrogen bonds have been undertaken previously [20–22]. Further comprehensive review of the vast literature is beyond the scope of this chapter and the discussion is restricted accordingly to several illustrative examples.

Successful crystal design strategies depend upon the identification and utilization of predictable structure-directing functionalities. The most extensively studied group to date is undoubtedly the carboxylic acid unit, with pioneering contributions from Leiserowitz and Etter [23, 24]. Design strategies incorporating this functionality have been illustrated by the aromatic carboxylic acids (Fig. 12.1). A functionality comparable to the carboxylic acid unit in its directionality and predictability is the amide unit. Primary and secondary amides regularly form hydrogen-bonded dimers via synthon **II** [25]. In addition, the possibility exists for the formation of one-dimensional α-networks via N$-$H$\cdots$O hydrogen bonds [26]. Thus, the crystal structure of benzamide **6**, for example, contains centrosymmetric dimers linked into α-networks (Fig. 12.2) [27]. Lauher and

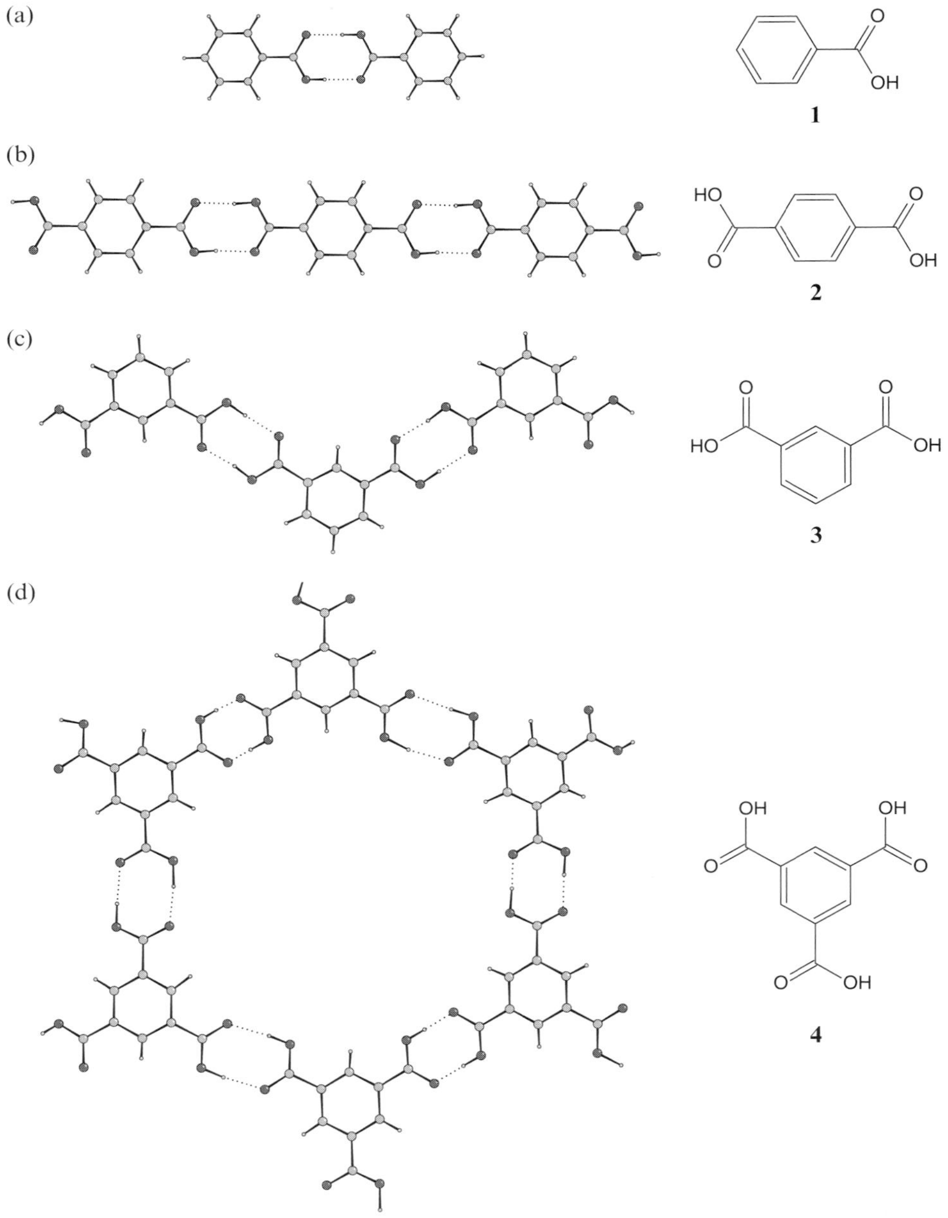

(a)

(b)

(c)

(d)

Figure 12.1. Aromatic carboxylic acids linked via synthon **I**: (a) **1** forms centro-symmetric dimers, (b) **2** forms linear tapes, (c) **3** forms crinkled tapes and (d) **4** forms planar sheets.

(e)

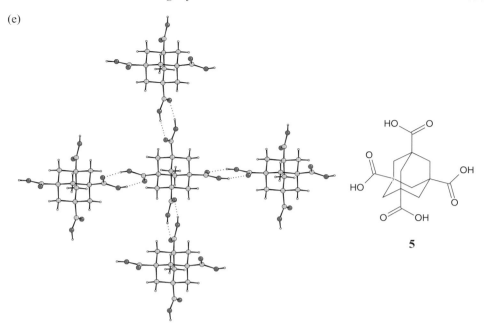

Figure 12.1. (cont.) Aromatic carboxylic acids linked via synthon **I**: (e) **5** forms a three-dimensional diamondoid structure.

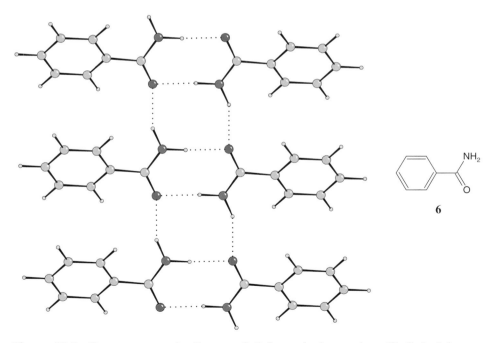

Figure 12.2. Centrosymmetric dimers of **6** formed via synthon **II**, linked into an α-network by additional N−H···O hydrogen bonds.

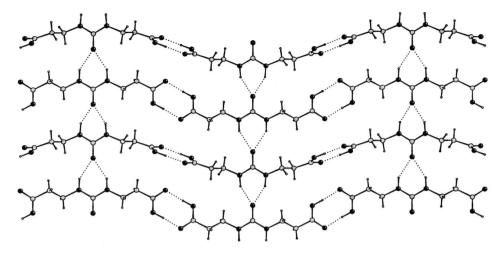

Figure 12.3. The β-network formed by 3,3′-ureylenedipropionic acid.

co-workers have attempted to combine into a single molecule functionalities such as the carboxylic acid and amide groups to produce predictable supramolecular features. Orthogonal combination of two sets of α-network-generating functionalities, for example, can be used to construct a two-dimensional β-network. Initial studies with substituted ureas showed that the urea functionality will reliably form an α-network via synthon **III** [28]. To produce a β-network, additional carboxylic acid functionality is introduced (Fig. 12.3). Such strategies have been extended successfully to include urea diamide derivatives [29], oxalamides [30], and urea analogues such as substituted pyridones and pyrimidones [31, 32].

A further, potentially even more powerful, approach for structure design involves the combination of different molecules. A strategy based on a single molecule is limited by the fact that it becomes progressively harder to incorporate all of the desired supramolecular functionality required to generate a specific structure. A binary system can provide complementary molecules with a greater degree of reliable recognition and greater overall structural predictability. Hollingsworth and co-workers have utilized the urea functionality in conjunction with α,ω-dinitriles **7** to design layered structures mediated by synthon **III** and N−H ··· N interactions (a typical sheet structure with $n = 3$ is shown in Fig. 12.4) [33]. An interesting, and at first sight somewhat surprising, result from this study is that layered 1:1 complexes are only formed for **7** with $n = 3, 4$ and 5. For $n = 6$ and higher analogues, the crystal system changes from a 1:1 complex to a channel inclusion compound. This transition has been attributed to the balance between directional hydrogen bonding and packing efficiency. Within the 1:1 complex, synthon **III** dictates a 4.58 Å repeat within the

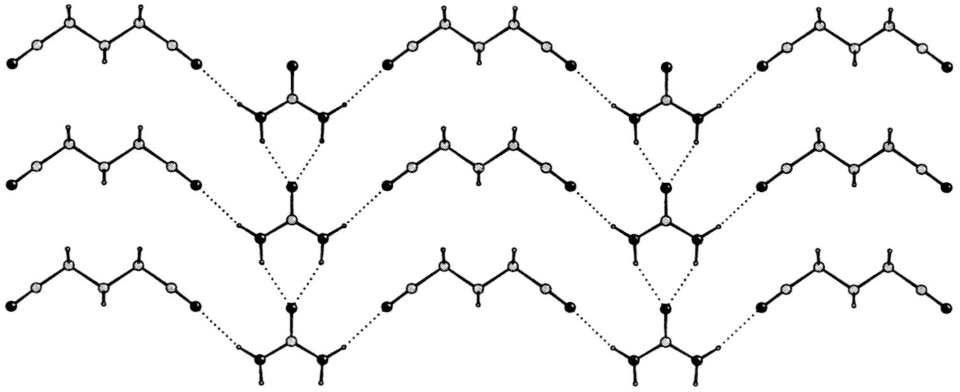

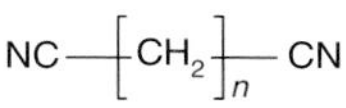

$$NC-[CH_2]_n-CN$$

7

Figure 12.4. The β-network formed between urea and glutaronitrile, **7** ($n=3$).

α-network, imposing the same separation between adjacent alkyl chains of the α,ω-dinitrile in a given layer. The separation *between* layers, however, is in the range 3.25–3.45 Å, suggesting that the alkyl chains *within* the layers are packed inefficiently. As indicated in Section 12.2, empty space is an energetic waste since it reduces the stabilizing contribution from dispersion interactions. This inefficiency can be tolerated up to the $n=5$ limit, above which the structure switches to maximize the dispersion interactions (at the expense of the $N-H\cdots N$ hydrogen-bond interactions) in the inclusion compound. This series of compounds provides a cautionary note, expedient at this stage of the discussion: although the supramolecular synthon approach is undoubtedly elegant and provides significant predictive power, it should not be forgotten that crystal structures ultimately result from a delicate balance of directional forces *and* isotropic dispersion interactions.

Nguyen *et al.* have recently examined the structure-directing properties of the hydroxyl group, which would appear to be a useful supramolecular functionality capable of forming strong, predictable $O-H\cdots O$ networks such as that illustrated in synthon **IV** [34]. The structures of a number of new urea and oxalamide derivatives (**8**–**10**) were found, however, to exhibit considerable variation

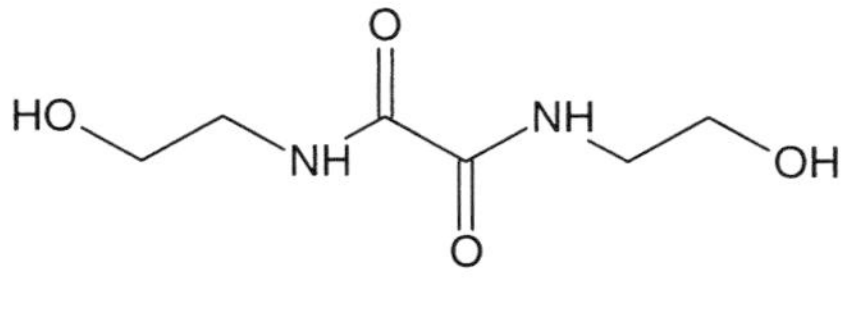

8

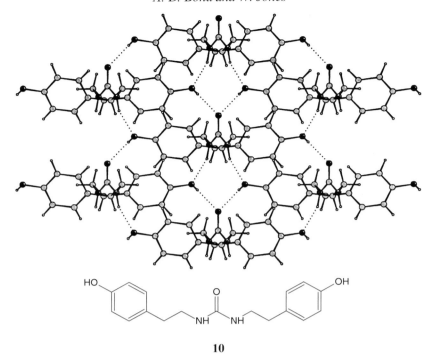

10

Figure 12.5. Chain structure of **10** in which hydrogen bonding between the hydroxyl
and carbonyl groups disrupts the expected urea α-network.

in an unpredictable manner. While **8** and **9** both form the expected β-network
(albeit with unexpected inversion symmetry for the hydroxyl–hydroxyl interactions in **9**), **10** forms a unique chain structure in which the normal urea α-network
is disrupted by hydrogen bonding between the carbonyl group and the hydroxyl
groups of **10** (Fig. 12.5). Co-crystallization of **9** with the bipyridine **11** does
produce the expected α-network of **9** linked into β-networks by O−H⋯N interactions between the hydroxyl groups and the nitrogen atom of the pyridine components (Fig. 12.6). The predicted β-network is also observed in a co-crystal of **8**
with the oxalamide of 4-aminomethylpyridine **12**. The pyridine nitrogen atom in
11 and **12** is a much better hydrogen-bond acceptor than the hydroxyl oxygen
atom and consequently the O−H⋯N hydrogen bond is a more reliable structure-
directing element. Thus, while the usefulness of the hydroxyl group appears to be
limited by its lack of predictability, the hydroxyl–pyridine pairwise interaction is
far more robust and useful for crystal design strategies.

A potential difficulty for the design of binary systems is maintaining the
independence of the two supramolecular functionalities. In the case of the **8:12**
co-crystal, two possible arrangements may be envisaged since both molecules
possess the appropriate oxalamide functionality to generate α-networks. The

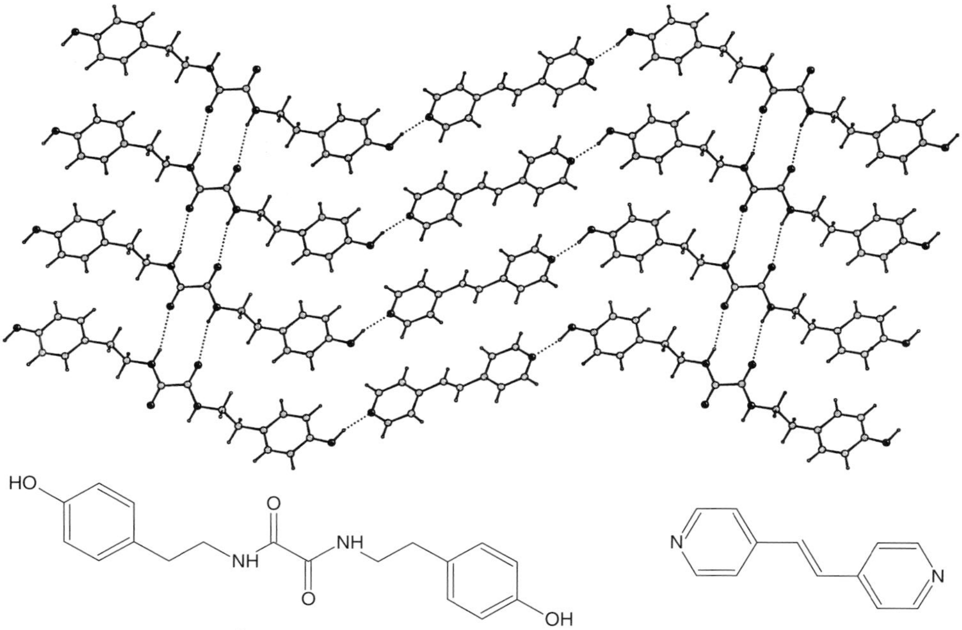

9

11

Figure 12.6. The β-network formed in the **9:11** co-crystal.

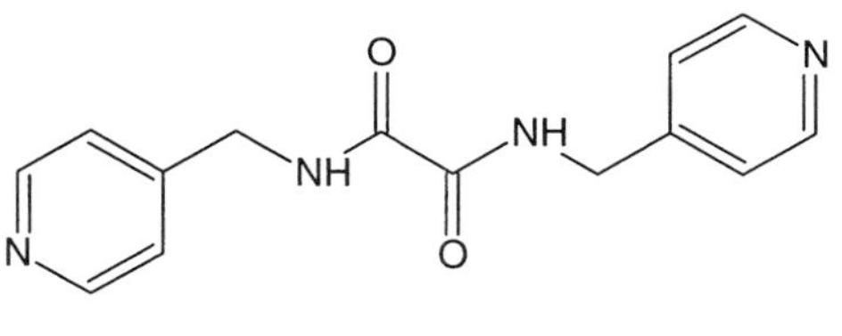

12

first possibility, the 'checkerboard' arrangement, would have the two components alternating in a single α-network (Fig. 12.7a). The second arrangement,
the 'stripes' pattern, would contain two independent α-networks, one formed
from **8** and one formed from **12** (Fig. 12.7b). In both arrangements, the
α-networks are brought together by O−H · · · N interactions to yield
β-networks. The observed arrangement in the **8:12** co-crystal is the 'checkerboard', and the same arrangement has been observed in a co-crystal of **8** with
the dicarboxylic acid **13**. This preference may be attributed to the exact

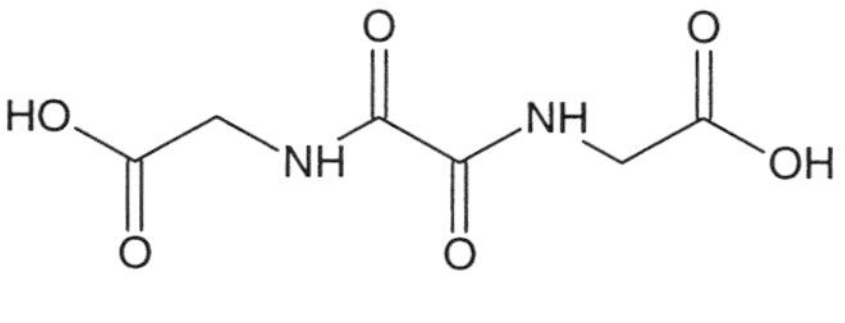

13

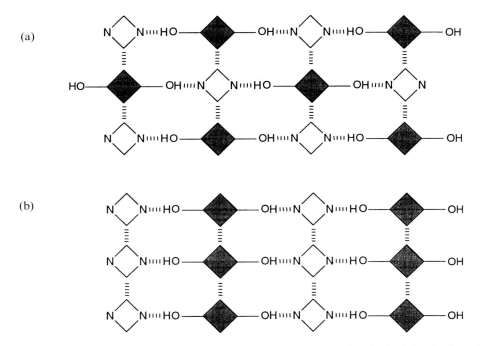

Figure 12.7. Alternative arrangements for the **8:12** co-crystal: (a) the 'checkerboard' arrangement, and (b) the 'stripes' arrangement.

distance requirements of the 'stripes' pattern – the two independent α-networks must be exactly commensurate for a continuous β-network to be formed, necessitating alternate **8** and **12** in the α-networks.

Whitesides *et al.* have studied extensively the 1:1 complex of cyanuric acid **14** and melamine **15** (denoted **CA·M**) as a model two-component hydrogen-bonded system [35]. Consideration of the geometrical disposition of the supramolecular functionality in the two components led to a proposed sheet structure for **CA·M** containing hexagonal rosettes (Fig. 12.8a) [36]. Verification of the structure was hindered for a number of years by a difficulty in obtaining single crystals suitable for X-ray analysis (owing to the limited solubility of **14** and **15** in most organic solvents), but has recently been reported by Ranganathan *et al.* with the utilization of hydrothermal synthesis methods [37]. This spectacular example of noncovalent synthesis represents a notable success for the supramolecular approach to crystal design. Whitesides *et al.* have also investigated the robustness of the hydrogen bonding in **CA·M** on introduction of a range of substituents in **14** and **15** [35, 38]. By interrupting the hydrogen bonds (either by replacement of the NH groups in **14** with CR_2 or NR groups, or by substitution of two of the NH_2 groups in **15** with NHR groups) it is

(a)

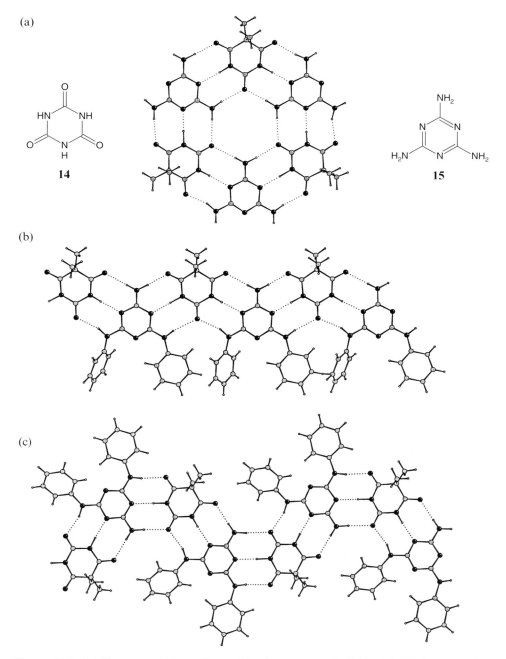

14

15

(b)

(c)

Figure 12.8. (a) Hexagonal sheets formed in the co-crystal of **14** with **15**. Interrupting the hydrogen bonding by substitution of **14** or **15** produces (b) linear tapes and (c) crinkled tapes.

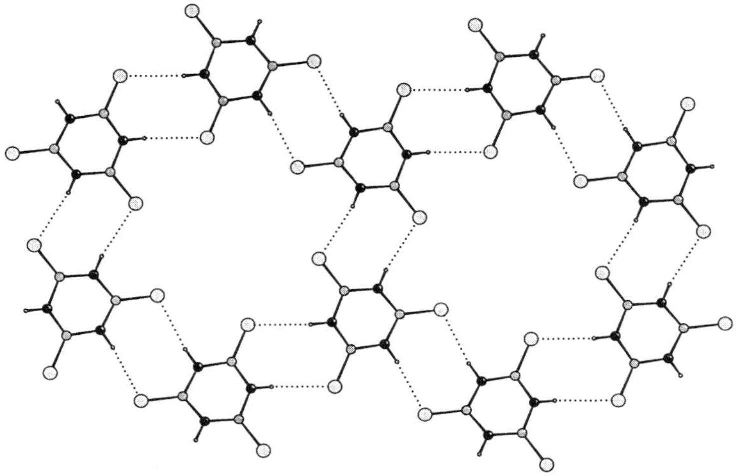

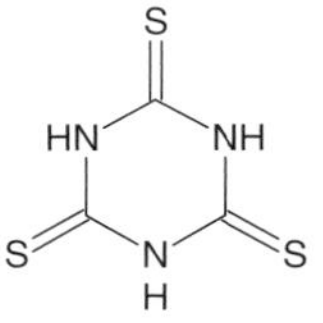

Figure 12.9. Hexagonal sheets in the crystal structure of the acetone solvate of **16**, mediated by synthon **VII**.

possible to construct supramolecular tapes rather than two-dimensional sheets. Both linear and crinkled tapes may be formed, both of which preserve the hydrogen-bond motif between the two components (Figs. 12.8b, c).

It is interesting to note that a co-crystal of trithiocyanuric acid **16** with **15** (denoted **TCA·M**) has also been synthesized recently and shown to have a structure identical to that of **CA·M** [37]. These two structures suggest an apparently equivalent structure-directing role for N−H···O and N−H···S hydrogen bonds. Indeed, **14** and **16** crystallized independently from highly polar solvents such as dimethylsulfoxide (dmso) adopt very similar solvate structures incorporating tapes mediated by the analogous synthons **II** and **VII**, respectively [39]. Crystallization of **14** and **16** from less polar solvents, however, produces vastly different structures. While **14** forms the same non-solvated structure from both methanol and acetone, **16** forms 1:1 adducts in both cases. Crystallization of **16** from methanol produces dimers mediated by synthon **VII**, linked into tapes by N−H···O interactions with the solvent molecules. From acetone, **16** crystallizes to produce essentially planar hexagons constructed from six molecules mediated by synthon **VII** (Fig. 12.9). This supramolecular assembly is one of very few examples constructed entirely by sulfur-mediated hydrogen bonds. The difference in structure between **TCA·M** and **CA·M** undoubtedly arises largely from the differences in the strengths of the N−H···O and N–H···S hydrogen bonds, but may also indicate differences in the directionality of the two interactions. Clearly, caution must be exercised in assuming equivalence in the structure-directing properties of apparently similar hydrogen-bond interactions.

As a final example illustrative of design strategies based on strong hydrogen bonds, a recent re-examination of the carboxylic acid group warrants mention. The carboxylic acid group itself exists in two conformations, *syn*planar **17(a)** and *anti*planar **17(b)** [23]. The *syn* conformation is more stable by

17(a)   17(b)

c. 4.5 kJ mol^{-1} and the most common motif adopted by carboxylic acids is the ubiquitous *syn–syn* dimer (synthon **I**). In addition, catemers of types **VIII** (*syn–syn*), **IX** (*anti–anti*) and **X** (*syn–anti*) have been observed, of which **VIII** is most common, and **IX** and **X** are extremely rare. One example of a structure containing pattern **X** is that of 1,4-cubanecarboxylic acid **18** [40]. Kuduva *et al.*

18

have subsequently investigated the crystal chemistry of a number of cubane-carboxylic acid derivatives and found that the majority of these also contain pattern **X** [41]. This disproportionately high occurrence of **X** must be attributable to some other feature of the cubanecarboxylic acid molecule. Indeed, it has been suggested by the authors that the supramolecular behaviour of cubanes in the solid state arises from the acidic nature of the cubyl-H atom. The acidity of an unactivated cubyl-H is comparable to NH_3 (p$K_a \sim 38$), at least 10^5–10^6 times more acidic than vinyl and phenyl hydrogens [42]. The formation of pattern **X** is attributed to its stabilization from C–H···O type interactions between the cubyl C–H donors and the carboxyl O acceptor atoms. Consideration of these interactions leads to the identification of patterns **XI** and **XII**, which contain within them the smaller **X**. These two motifs are constructed with a combination of strong O–H···O hydrogen bonds and weaker C–H···O interactions. The active role of weak C–H···O hydrogen bonds in determining O–H···O networks has been reported for numerous other crystal structures, including fumaric acid [43, 44], citraconic acid [45], and bis-formamides [46]. It is clear that successful supramolecular design strategies must also take account of weak interactions such as these.

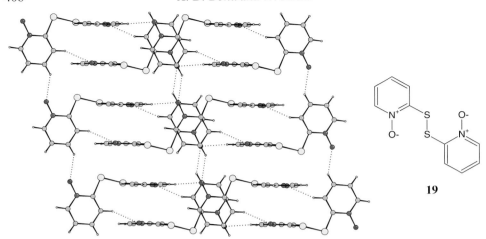

19

Figure 12.10. Projection of the crystal structure of **19**. Each molecule forms eight C—H⋯O hydrogen bonds (four illustrated and four to molecules in adjacent layers).

12.3.2 Design strategies based on weak hydrogen bonds

Where the hydrogen atom involved in an intermolecular hydrogen bond is bound to a less electronegative atom, the strength of the resulting interaction will be less than that of the classical hydrogen bonds discussed previously. Indeed, there has been considerable debate as to whether such interactions may reasonably be termed 'hydrogen bonds' [47, 48]. From a crystal design perspective, the distinction between hydrogen bonds, however weak, and dispersion interactions lies in their directionality; hydrogen bonds are inherently directional while dispersion interactions are isotropic. A recent survey of the Cambridge Structural Database (CSD) by Steiner and Desiraju identified significant directionality in C—H ⋯ O contacts, with linear contact geometries being favored [49]. In addition, there is a gradual decrease in directionality for C—H ⋯ O interactions with decreasing C—H polarization. This is exactly the picture expected for hydrogen bond interactions. Similar conclusions have been drawn from crystallographic data regarding the directional nature of C—H⋯N, C—H⋯Cl and C—H⋯F interactions [50, 51].

An interesting example of a crystal structure mediated entirely by weak C—H ⋯ O hydrogen bonds is that of 2,2′-dithiobis(pyridine-*N*-oxide) (DTPO) **19** [52]. The crystal structure of **19** consists of interpenetrated homochiral networks in which each molecule forms eight C—H⋯O bonds (Fig. 12.10). The open network character of the crystal structure suggests that it may be suitable for guest inclusion. Indeed, it is possible to incorporate isolated molecules of tetracyanobenzene (TCNB) **20** to form the hydrate [DTPO]$_2$[TCNB](H$_2$O)$_4$, which contains ladders of DTPO molecules clathrated by TCNB guests (Fig. 12.11a).

(a)

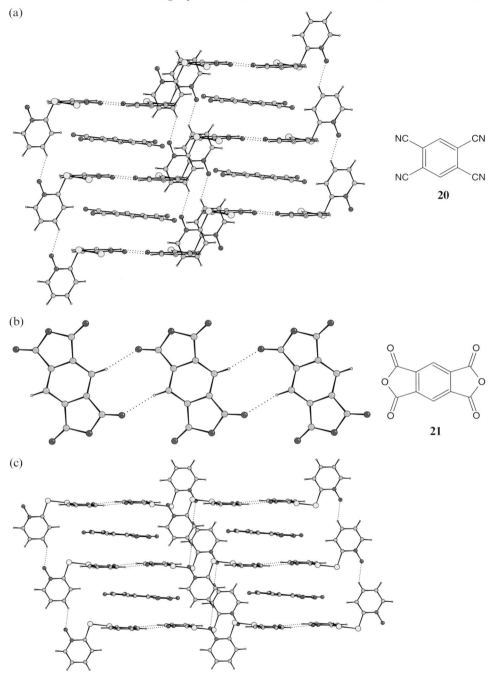

20

(b)

21

(c)

Figure 12.11. (a) Lattice of **19** incorporating isolated molecules of **20**. Molecules of **21** associate via synthon **XIII** to form chains (b) which thread the lattice in one direction, giving rise to a solid-state polypseudorotaxane structure (c). For clarity, water molecules are omitted from (a).

Adjacent ladders are cross-linked by strong O−H···O hydrogen bonds via the water molecules. Incorporation into **19** of pyromellitic dianhydride (PMDA) **21** produces [DTPO]$_2$[PMDA] which has a similar ladder structure, but in this case the PMDA molecules are also able to associate via C−H···O hydrogen bonds (synthon **XIII**) to form tapes (Fig. 12.11b). In the absence of water molecules, adjacent ladders are bound via additional C−H···O bonds. The overall structure, therefore, consists of a three-dimensional, C−H···O bound DTPO network, threaded by PMDA tapes also bound via C−H···O interactions (Fig. 12.11c). This fascinating structure represents a solid-state polypseudorotaxane mediated entirely by weak C−H···O hydrogen bonds.

Anthony *et al.* have recently demonstrated an equivalent structure-directing role for strong and weak hydrogen bonds [53]. Two oxa-androgens, **22** and **23**, were found to have similar crystal structures, despite the substitution of a C−O−H···O interaction in **22** for a C−H···O interaction in **23**. Further-

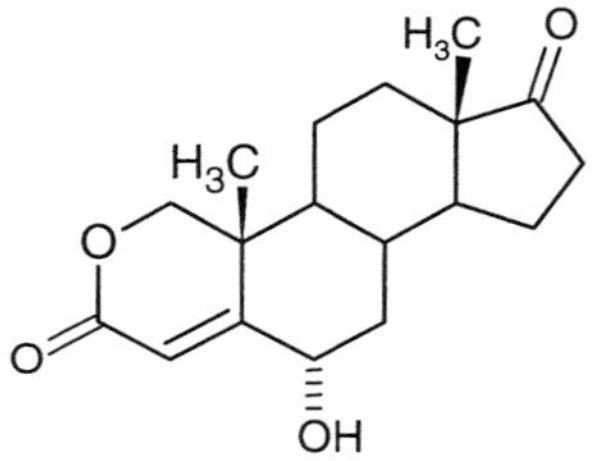

22

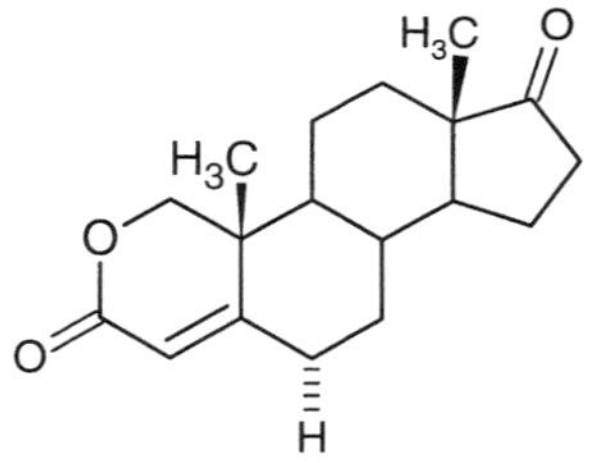

23

more, it was possible to produce a binary solid solution of the two molecules – this is often considered as the most stringent test of isostructurality and is only possible where two molecules have similar overall shapes and similar supramolecular recognition properties. Similar substitutions of weak hydrogen bonds for strong ones may be envisaged for all existing synthons based upon O−H···O and N−H···O interactions, etc., facilitating the use of weak hydrogen bonds in crystal design strategies. A classic example is given by the

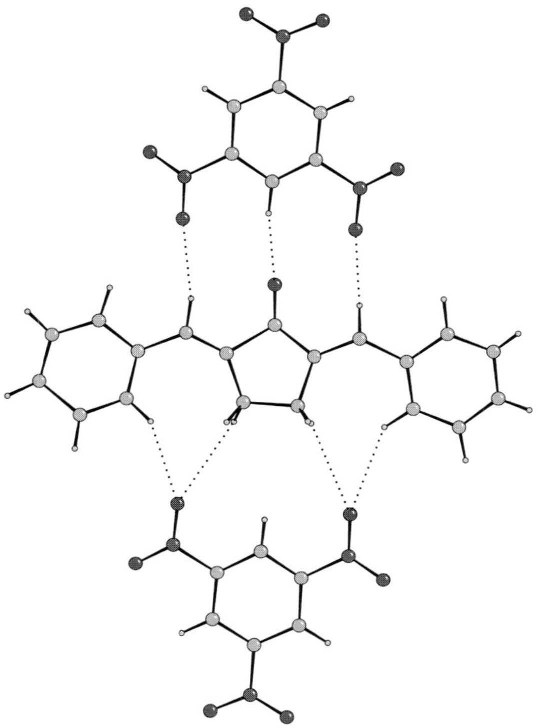

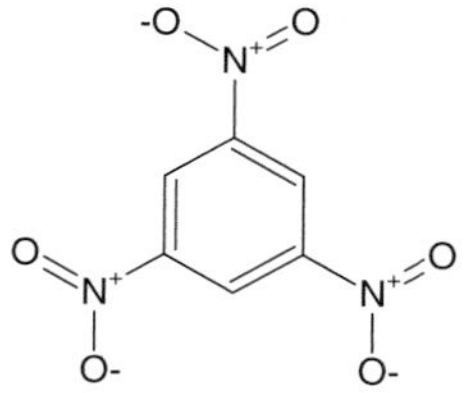

Figure 12.12. Trimer formed from **24** and **25** incorporating synthon **XV**.

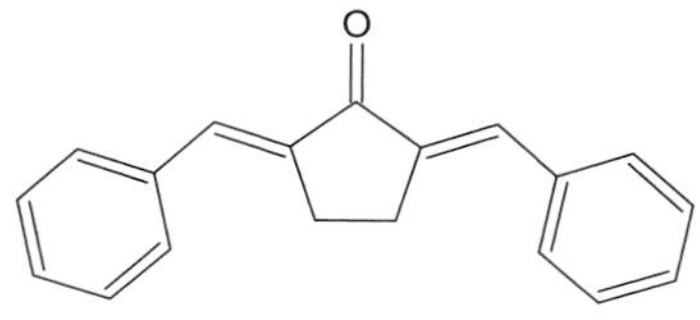

isomorphous structures of the urea:barbital and acetamide:barbital complexes in which a single $N-H\cdots O$ bond is replaced by a $C-H\cdots O$ bond, producing the modified synthon **XIV** from **III** [54]. An excellent example where strong hydrogen bonds are replaced entirely by $C-H\cdots O$ bonds is given by synthon **XV**, observed in the 1:2 molecular complex of 1,3,5-trinitrobenzene **24** and 2,5-dibenzylidenecyclopentanone **25** (Fig. 12.12) [55]; this may be compared with the strongly hydrogen-bonded synthon **V**. In both synthons **XIV** and **XV**, the $C-H\cdots O$ interactions play structure-directing roles equivalent to those of strong $N-H\cdots O$ hydrogen bonds. Owing to the inherent weakness of these interactions, however, a design strategy based upon multipoint recognition rather than a single interaction provides a greater degree of predictability.

Evidence for the weak hydrogen bond nature of $C-H\cdots\pi$ interactions has been reported recently by Madhavi *et al.* in a study of the complexes of 2,3,7,8-tetraphenyl-1,9,10-anthyridine **26** with toluene and chlorobenzene [56]. It was found that diffraction quality crystals of **26** could not readily be obtained from common solvents, with the exception of toluene, which produced a 1:1 solvate,

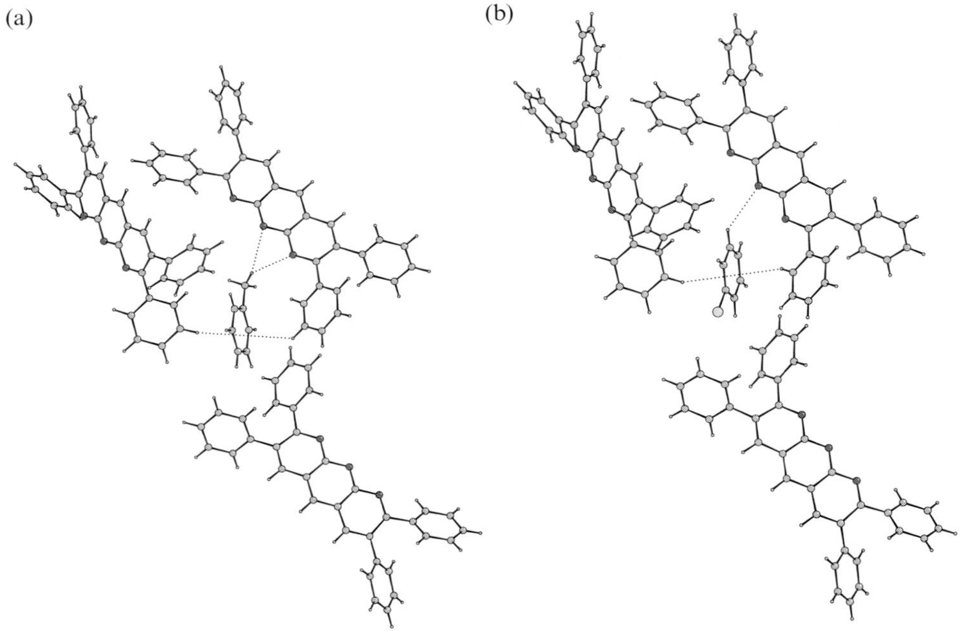

(a)   (b)

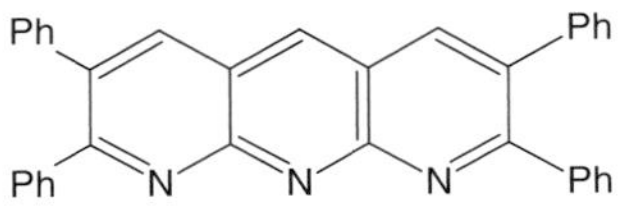

26

Figure 12.13. Molecules of (a) toluene and (b) chlorobenzene incorporated with different orientations into a lattice of **26**.

27. Invoking the well-known chloro-methyl interchange hypothesis, recrystallization of **26** from chlorobenzene was attempted, successfully producing a 1:1 solvate **28**. The two solvates have essentially identical structures, but with crucial differences in the mode of solvent inclusion. Both the toluene molecule in **27** and the chlorobenzene molecule in **28** are situated in a pocket formed by three molecules of **26**, but with the methyl (Fig. 12.13a) and chloro (Fig. 12.13b) substituents pointing in almost opposite directions. Where chloro–methyl interchange operates, the two substituents behave in an identical manner with only shape and size factors controlling the packing. The different positions of the substituents in **27** and **28** indicates that some other factors are involved in determining the orientation of the solvent molecules in one or both cases. In **27**, two $C-H \cdots \pi$ interactions may be identified, one short (2.54 Å), and one considerably longer (3.01 Å), such that the toluene molecule is positioned at unequal distances from adjacent glide-related molecules of **26**. The two corresponding contacts in **28** are equidistant (2.62 Å) such that the chlorobenzene molecule is disposed halfway between adjacent molecules of **26**. The presence of the particularly short C−H ⋯ π contact in **27** correlates with the electron-rich nature of the aromatic ring in toluene (compared with chlorobenzene), consistent with increasing hydrogen-bond strength as the basicity of the acceptor group increases. Additional interactions of C−H ⋯ N type are present in **27** between the methyl substituent and the molecules of **26**, along with one further C−H ⋯ π interaction. In **28**, only a single additional C−H ⋯ N interaction is present. Thus, although interactions of this type make only small contributions to the overall crystal binding energy, they appear to be crucial in determining the orientation of the solvent molecules in this example. In general, they play a significant role in determining the secondary structures of crystals.

12.3.3 *The interplay between strong and weak hydrogen bonds*

We have seen that crystal structures result from an optimum balance of both strong and weak intermolecular interactions, both of which may be considered as structure-directing to some extent. For weak interactions such as C−H ⋯ O and C−H ⋯ π hydrogen bonds to find major utility in crystal design strategies, however, their effect must be significant and predictable in the presence of other stronger interactions. For a hydrogen bond of the type C−H ⋯ X, a strong interaction (indeed, the *strongest* interaction) might be expected to arise where X is the most electronegative atom, F. The role of C−H ⋯ F interactions in the crystal structures of fluorobenzenes has been examined by Thalladi *et al.* [51]. It was found that C−H ⋯ F interactions do indeed have a significant structure directing effect, and synthons **XVI–XIX** have been identified, analogous to previously identified C−H ⋯ O and C−H ⋯ N containing synthons. A comparison with the corresponding chloro, bromo and iodo structures highlights that F behaves distinctly differently from the other halogens, which prefer the formation of halogen–halogen interactions (Section 12.3.4). It has been shown, however, that the C−F group is actually a very poor hydrogen-bond acceptor and is unable to compete favorably in the presence of O and N atom acceptors [57]. Only when the C−H group acidity is enhanced to the levels of the fluorobenzenes and only in the absence of competing acceptors is the hydrogen-bond nature of the C−H ⋯ F interaction revealed. Thus, although interactions such as C−H ⋯ F may be shown to play a structure-directing role in some cases, it is unlikely that they may find general utility in crystal design strategies since they are not effective in the presence of other stronger interactions.

While design strategies based entirely on strong hydrogen bonds may be somewhat more predictable, consideration of supramolecular synthons

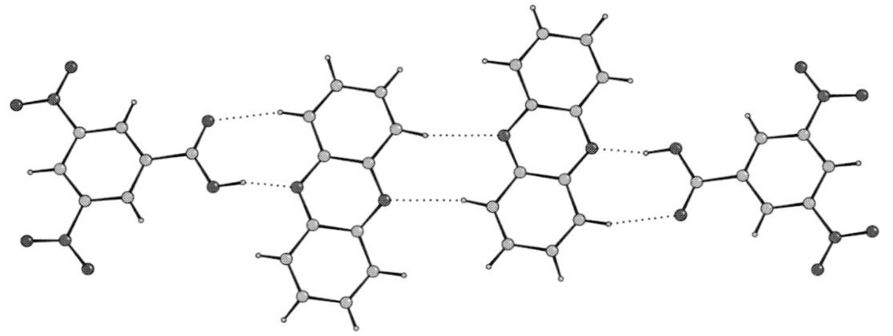

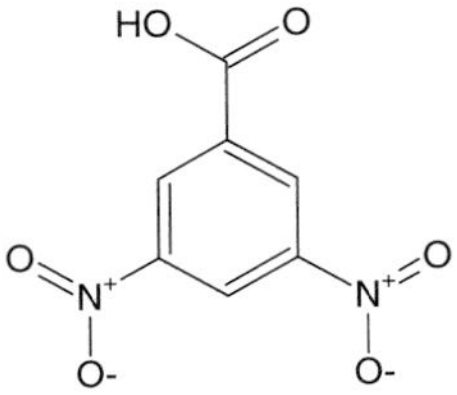

Figure 12.14. Four-membered unit formed between **29** and **30** via synthon **XX**.

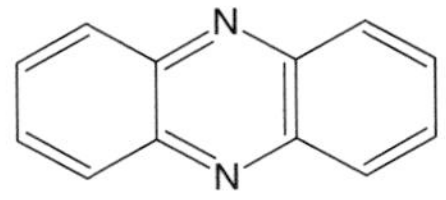

constructed from both strong and weak interactions is likely to provide a more complete understanding of crystal packing. Pedireddi *et al.* have illustrated the use of a strong and weak hydrogen bond in the same synthon to create various supramolecular arrays [58]. Co-crystallization of molecules incorporating −COOH units with *N*-heterocycles were found to produce structures incorporating the synthon **XX**, consisting of strong O−H $\cdots$ N and weak C−H $\cdots$ O hydrogen bonds. Initially, the synthon was observed in a 2:1 complex of 3,5-dinitrobenzoic acid **29** with phenazine **30**, where four-membered units are assembled via **XX** and additional C−H $\cdots$ N interactions between the phenazine molecules (Fig. 12.14). In order to produce molecular tapes assembled only via **XX**, co-crystallization of **30** with terephthalic acid **2** was envisaged, but this was hindered by solubility differences between the two molecules. It was noted, however, that a non-covalent analogue of terephthalic acid would be expected to result from the coupling of two aliphatic acid molecules via the extremely robust carboxylic acid dimer, synthon **I**. This approach proved to be successful with the production of a 2:1 co-crystal of phenazine and malonic acid **31**, incorporating tapes in which the malonic acid dimers were linked to the phenazine molecules by **XX** (Fig. 12.15). This complex provides a particularly elegant example of the interchange between covalent and non-covalent linkages in the formation of supramolecular arrays.

Further investigations of the robustness of synthon **XX** have recently been reported by Batchelor *et al.* [59]. Co-crystallization of phenazine with the *trans*-olefinic acids, fumaric **32** and mesaconic **33**, produces 1:1 supramolecular tapes incorporating only synthon **XX** (Figs. 12.16a, b). With the *cis*-olefinic acids, maleic **34** and citraconic **35**, however, 2:1 tapes are produced in which the

31 $n = 1$
36 $n = 2$
37 $n = 3$
40 $n = 4$

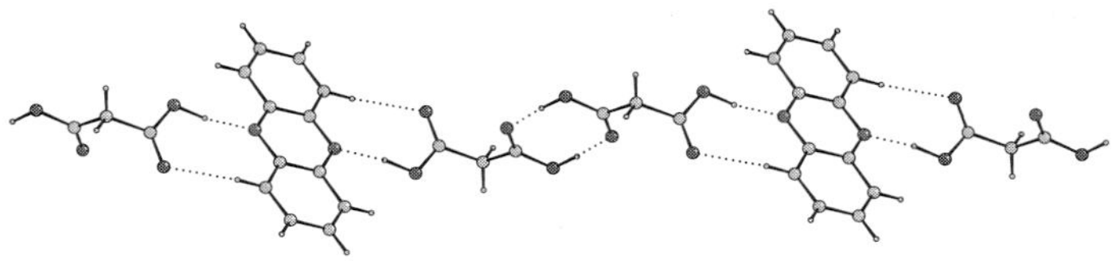

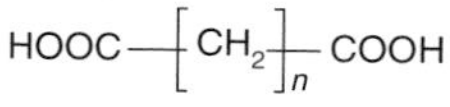

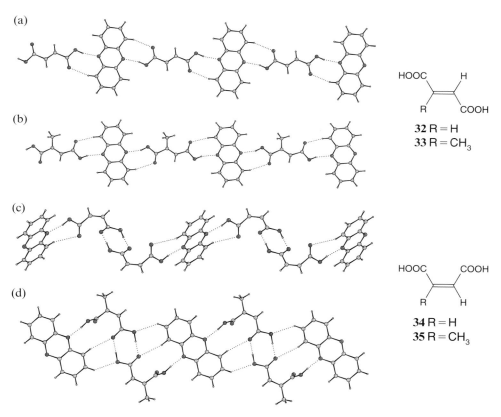

Figure 12.15. Extended tapes incorporating synthon **XX** formed between **30** and **31**. Two molecules of **31** are linked via synthon **I** into supramolecular analogue of terephthalic acid (**2**).

32 R = H
33 R = CH_3

34 R = H
35 R = CH_3

Figure 12.16. Tapes formed between **30** and the olefinic acids: (a) **32** and (b) **33** form 1:1 tapes incorporating only synthon **XX**, whereas (c) **34** and (d) **35** form 2:1 tapes incorporating both synthon **XX** and synthon **I**.

acid molecules dimerize via synthon **I** in the same manner as that observed by Pedireddi for the complex **30:31** (Figs. 12.16c, d). Co-crystallization with the aliphatic acids, succinic **36** and glutaric **37**, also give rise to 2:1 and 1:1 complexes respectively. Thus, 1:1 tapes incorporating only synthon **XX** are formed where the acid groups point in the same direction (*cis*-olefinic acids and aliphatic acids with *n* odd), and 2:1 tapes incorporating both synthons **XX** and **I** are formed where the acid groups point in opposite directions (*trans*-olefinic acids and aliphatic acids with *n* even). Synthon **XX** has also been observed recently in a co-crystal of phenazine with 2,2′-biphenyl carboxylic acid **38**, in

38

which rotation about the C—C single bond in **38** allows for the —COOH groups to point in opposite directions, giving a 1:1 stoichiometry [60]. In the case of the complex **30:35**, rotation of one —COOH group with respect to the plane of the acid molecule results in disruption of synthon **XX** such that the C—H···O interaction is not observed. Such a rotation, however, allows for two additional C—H···O interactions between the phenazine molecule and the acid dimers (Fig. 12.16d). This complex provides an excellent example of the subtle balance between particular supramolecular arrangements where both strong and weak intermolecular interactions are present.

Similar co-crystallization studies have been performed with 4,4′-bipyridyl **39** and the aliphatic acids, malonic **31**, glutaric **37**, and adipic **40**, again giving rise to molecular tapes mediated primarily by synthon **XX** [61]. In this case, however, the study was extended to include co-crystallization of **39** with the thiodicarboxylic acids, thiodiglycolic **41** and thiodipropionic **42**. In the case of **42**, tapes were formed via synthon **XX** in the same manner as for the aliphatic acids (Fig. 12.17a). For **41**, however, an entirely new structure was produced incorporating cyclic supermolecules composed of two molecules of **39** and two molecules of **41**, linked via synthon **XX** (Fig. 12.17b). Formation of the cyclic supermolecule for **41** compared with the formation of tapes for the analogous glutaric acid **37** is attributed to the differences in molecular geometry arising from the geometric requirements of the sulfur atom – the smaller angle about the central sulfur atom in **41** facilitates a geometry suitable for stabilizing interaction between the aromatic bipyridyl molecules [62]. This

(a)

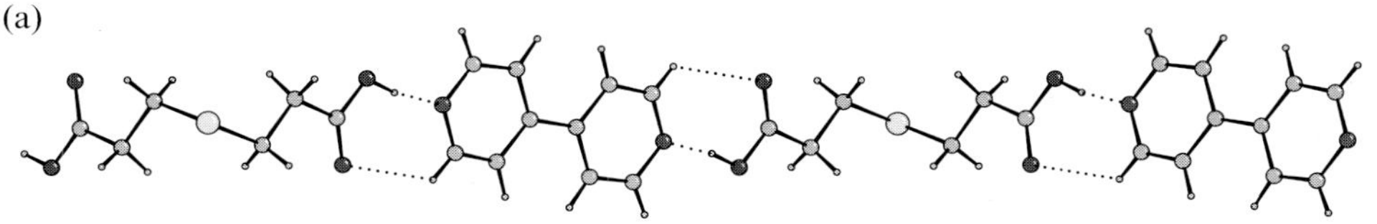

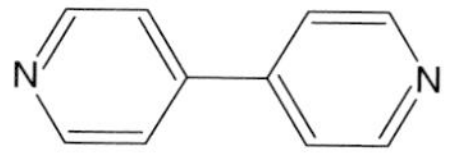

39

(b)

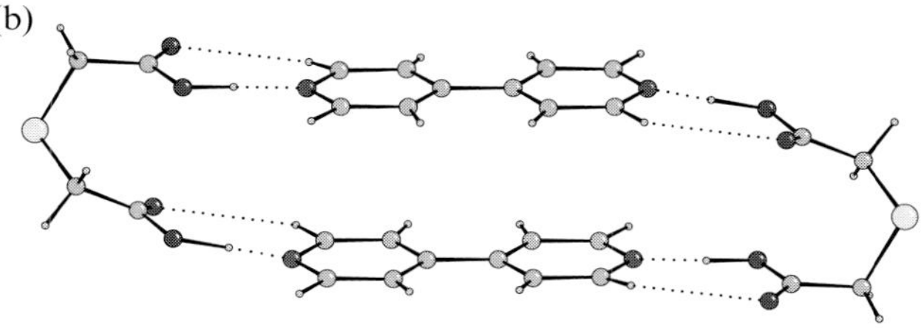

HOOC—[ ]$_n$—S—[ ]$_n$—COOH

41 $n=1$
36 $n=2$

Figure 12.17. Supramolecular features in the co-crystals **39:41** and **39:42** formed via synthon **XX**: (a) **39:42** forms extended tapes, whereas (b) **39:41** forms cyclic supermolecules.

additional interaction, not of the hydrogen-bond type considered so far, is sufficient to stabilize the cyclic supermolecule, disrupting the predicted tape formation.

12.3.4 Other types of intermolecular interaction

As the previous example illustrates, in addition to hydrogen bonds, there exists a symphony of other intermolecular interactions which may be crucial in determining the crystal structure adopted by a molecule. Typical interactions frequently cited as weak, but with a significant structure-directing effect include halogen–halogen, nitro–halogen, nitro–nitro, sulfur–cyano and aromatic–aromatic (π–π) interactions. Halogen–halogen interactions have been used in crystal design studies since the early work of Schmidt [63], and manifest themselves clearly in the crystal structures of the elements themselves. The nature of these interactions is subject to some debate, but their undoubted directionality has been shown in a crystallographic study by Pedireddi *et al.* [64]. The anomalous behaviour of fluorine, where C—H···F interactions are favored over F···F, has been noted previously (Section 12.3.3). Several examples of crystal design based upon the predictable interaction of halogen substituents with nitro groups have also been reported, including a 1:1 complex of 1,4-diiodobenzene **43** with 1,4-dinitrobenzene **44** which forms tapes via synthon **XXI** (Fig. 12.18) [65].

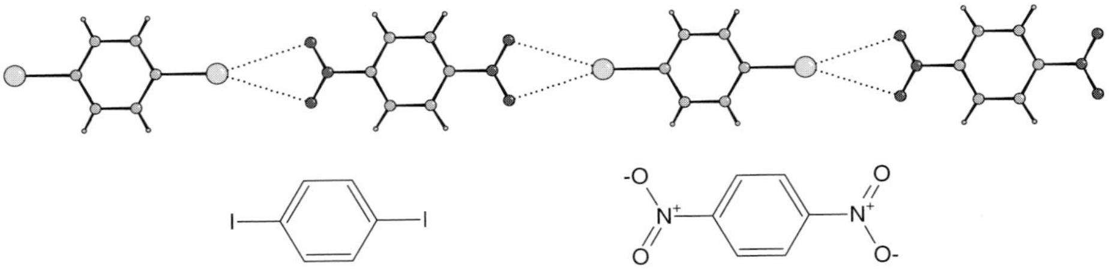

A. D. Bond and W. Jones

Figure 12.18. Extended tape formed between **43** and **44** via synthon **XXI**.

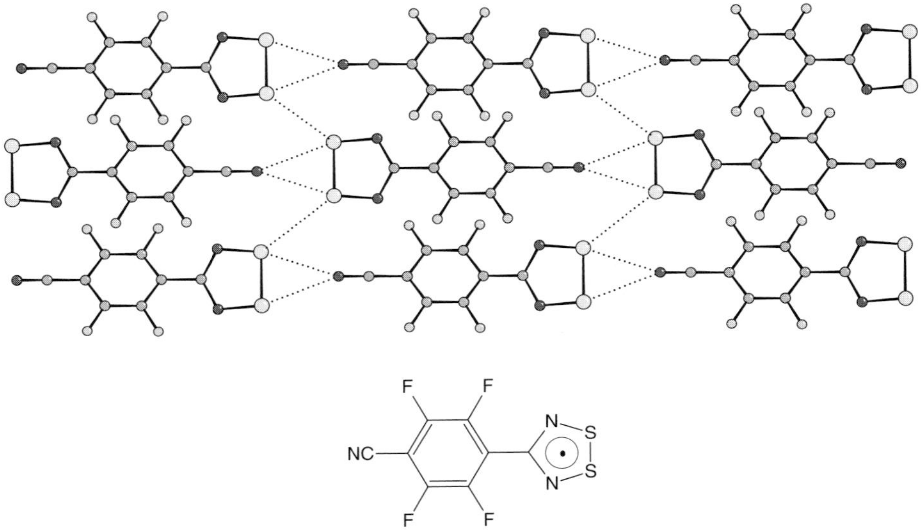

Figure 12.19. Tapes of **45** mediated by C−N···S interactions.

An interesting recent example of a structure moderated via sulfur–cyano interactions is that of the 1,2,3,5-dithiazolyl radical, [p-NC−C$_6$F$_4$−CNSSN]· **45**, which exists in two polymorphic forms, denoted the α- and β-phase, respectively [66, 67]. Radicals of this type generally form dimers in the solid state, rendering them diamagnetic. The α-phase of **45**, however, forms tapes of monomeric radicals held together by C−N ··· S interactions (Fig. 12.19), aligned in an anti-parallel manner. The monomeric nature of this structure allows **45** to retain its paramagnetic character in the solid state. Identical tapes are present in the β-phase, but these align in a parallel manner. This arrangement results in weak ferromagnetic ordering at the unprecedentedly high temperature of 36 K.

12.4 Design strategies in metallo-organic and organometallic solids

The formal development of supramolecular chemistry and crystal engineering has occurred predominantly through organic rather than inorganic chemistry. While this organic bias is likely to persist for some time, appreciation of supramolecular concepts within organometallic and metallo-organic solids is increasing rapidly. For organic solids, we can generally be confident that the covalent bonds holding together the molecule are significantly stronger than the intermolecular forces, such that the molecular connectivity is entirely predictable. In organometallic systems, however, the presence of metal atoms introduces additional structural freedom such that possibilities exist for significant structural variability and flexibility on a molecular level. The availability of multiple, almost isoenergetic binding modes for ligands such as CO, CN^-, NCS^-, etc., give rise to several molecular isomers of very similar energy. Indeed, it is often possible to compensate for less-stable molecular isomers with enhanced intermolecular interactions in the solid state [68]. Thus, in organometallic systems it becomes especially difficult to differentiate between intramolecular and intermolecular interactions, and the overall solid-state structure ultimately depends upon the interplay between both. This additional level of complexity compared with organic systems renders *ab initio* structure prediction even more difficult for organometallics. The organometallic crystal engineer, therefore, is particularly reliant on the simplifications available through the supramolecular approach.

12.4.1 Strong hydrogen bonds in organometallic solids

In many instances, the metal atom in an organometallic or metallo-organic molecule may be considered to be encased completely by organic ligands and, therefore, unable to participate directly in intermolecular interactions. In such cases, the molecular peripheries are entirely organic in nature and the intermolecular forces which govern the crystal structure would be expected to be very similar to those in purely organic systems. Braga *et al.* have examined the extent to which the synthons adopted by common organic supramolecular functionalities (carboxylic acid, hydroxyl, primary and secondary amide groups) are transferable to organometallic systems when the functionality is incorporated into a metal-coordinated ligand, and it is found that there is indeed a certain degree of similarity [69, 70]. Transition metal complexes carrying one supramolecular −COOH functionality, such as $[(\eta^6\text{-}^tBuC_6H_4COOH)Cr(CO)_3]$ **46** (Fig. 12.20) [71], regularly form dimers via synthon **I**. In $(\eta^2\text{-maleic acid})Fe(CO)_4$ **47**, where two −COOH functionalities are present, each maleic acid molecule

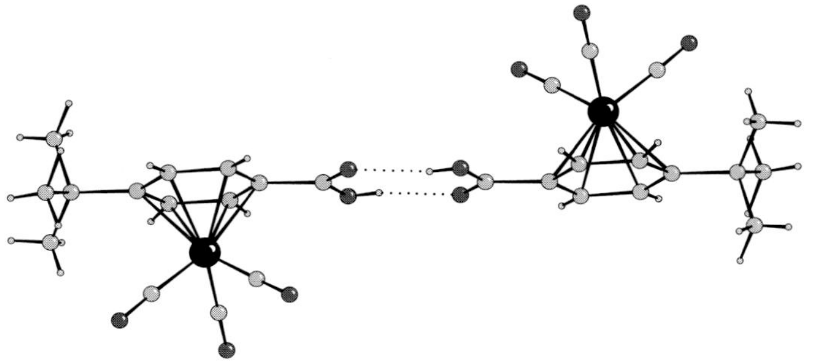

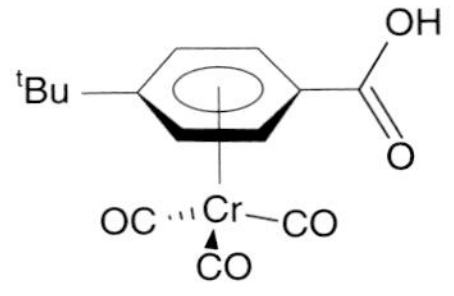

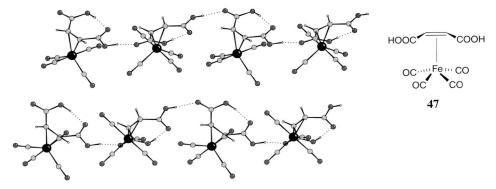

Figure 12.20. Centrosymmetric dimers in **46** formed via synthon **I**.

Figure 12.21. Chains of **47** incorporating both intra- and intermolecular O—H···O hydrogen bonds.

forms one intramolecular and one intermolecular hydrogen bond to yield chains (Fig. 12.21) [72]; the same arrangement is observed in maleic acid itself [73]. [(η^2-fumaric acid)Fe(CO)$_4$] **48** exists in two polymorphic forms, one containing chains mediated by synthon **I** (Fig. 12.22a), and one in which synthon **I** coexists with a catemer pattern similar to **VIII** (Fig. 12.22b) [72, 74]. Clearly, the robust synthons based upon the carboxylic acid functionality are entirely transferable between organic and organometallic systems.

Synthons based upon the amide and hydroxyl functionality also display some degree of transferability to organometallic systems. The cyclic synthon **II** is observed in the crystal structures of the primary amide complexes [(η^5-C$_5$H$_5$) Mn(CO)(NO)(CONH$_2$)] **49** (Fig. 12.23a) [75] and the solvate [(C$_4$H$_4$N$_2$O$_6$S$_3$) Cu(II)·CHO(NMe$_2$)] **50** (Fig. 12.23b) [76]. Hydrogen-bonded tetrameric and hexameric motifs, similar to synthon **IV**, exist within the structures of [(η^3-C$_7$H$_{10}$OH) (η^5-C$_5$H$_5$)Mo(CO)$_2$] **51** (Fig. 12.24a) and [(η^5-C$_5$H$_4$C$_2$H$_4$OH)$_2$Fe$_2$(CO)$_4$] **52** (Fig. 12.24b), respectively [77, 78]; similar motifs are observed frequently in organic

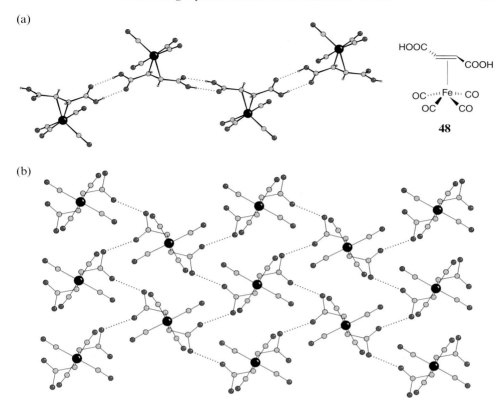

Figure 12.22. Polymorphs of **48**: (a) contains chains mediated by synthon **I**, whereas (b) contains a catemer pattern similar to **VIII**. Hydrogen atoms are omitted from (b).

monoalcohols [79]. Frequently, however, additional hydrogen bonding of $N-H\cdots$ $O, N-H\cdots N, N-H\cdots Cl$, and $C-H\cdots O$ types may be identified, and these interactions play a significant, and somewhat less predictable, role in determining the crystal structures.

As with organic systems, a greater degree of predictability may be achieved by utilizing multipoint recognition strategies, rather than those based upon single interactions. Mingos and co-workers have employed triple hydrogen-bond patterns for the design of aggregates based on transition-metal complexes and organic molecules [80]. For example, co-crystals of the metal complex $Ni(dithiobiuret)_2$ **53** with the complementary organic molecules 1,8-naphthalimide **54**, bemigride **55** and uracil **56** display self-assembly via triple complementary hydrogen-bond interactions (synthon **V**) [81, 82]. In the case of the **53:54** complex, **54** bears no further supramolecular functionality and acts as a terminating group, resulting in isolated supramolecular assemblies within the crystal structure (Fig. 12.25a). In **53:55**, a similar arrangement is observed,

(a)

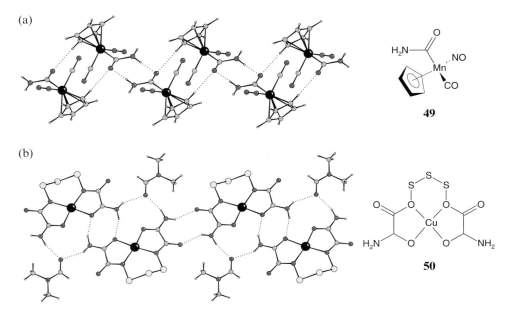

(b)

Figure 12.23. Synthon **II** observed in the crystal structures of the amide complexes (a) **49** and (b) **50**.

but the smaller size of **55** allows for additional N−H···O interactions between the assemblies, forming tapes (Fig. 12.25b). Tapes are also observed in **53:56**, but additional supramolecular functionality is introduced in the form of the ring NH group in **56** – this facilitates cross-linking of the tapes via incorporated water molecules to yield two-dimensional sheets (Fig. 12.25c). A spectacular example of 'one-pot' self-assembly based upon simultaneous coordinate- and hydrogen-bond interactions is given by the complex of Cu(5-(2-pyridylmethylene)hydantoin)$_2$ **57** with melamine **15** [83]. Microwave irradiation of a mixture of 5-(2-pyridylmethylene)hydantoin, melamine and CuCl$_2$·2H$_2$O in methanol produces single crystals of the complex **57:15**, containing supramolecular assemblies based on synthon **V**, linked into sheets via synthon **XXII** (Fig. 12.26). The full three-dimensional architecture of this complex may be rationalized by considering C−H···O hydrogen bonds between the sheets – these weak interactions are shown again to be important in determining crystal structures.

12.4.2 Weak hydrogen bonds in organometallic systems

Many organometallic complexes carry ligands such as cyclopentadienyl, phenyl, methylene and methylidene groups, all of which are capable of

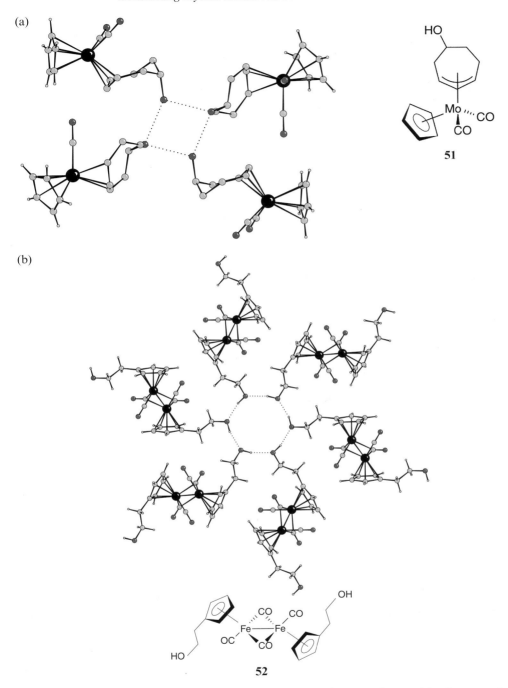

Figure 12.24. Hydrogen-bonded motifs similar to synthon **IV** observed in the crystal structures of (a) **51** and (b) **52**. For clarity, hydrogen atoms are omitted from the C₇ ring in (a).

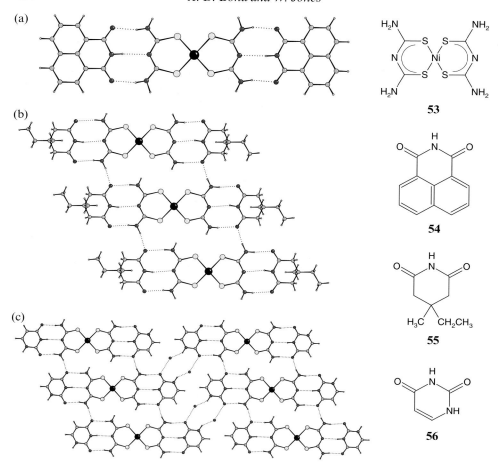

Figure 12.25. Supramolecular assemblies based on synthon **V** observed in complexes of **53** with **54**, **55** and **56**: (a) **53**:**54** forms isolated supramolecular assemblies, (b) **53**:**55** forms tapes and (c) **53**:**56** forms two-dimensional sheets.

hydrogen-bond donation via acidic C−H groups. In addition, ligands such as CO and CN⁻, good hydrogen bond acceptors, are common for the stabilization of low transition-metal oxidation states. As a result, the potential for C−H···O and C−H···N interactions in organometallic solids is great. A crystallographic study of C−H···O interactions in bridged and terminal metal carbonyls has been undertaken by Braga *et al.*, and there is a definite manifestation of directional C−H···O hydrogen bonds [84]. The crystal structure of $[(\eta^5\text{-}C_5H_5)_2Fe_2(CO)_2(\mu\text{-}CO)(\mu\text{-}CHCH_3)]$ **58**, for example, contains chains of molecules linked by bifurcated C−H···O interactions (Fig. 12.27), clearly reminiscent of the α-network in urea [85]; it has been noted that bifurcated

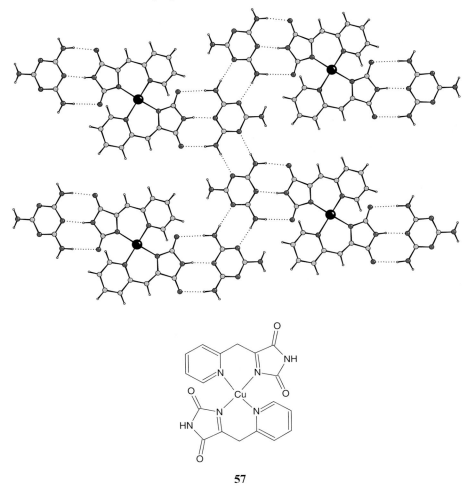

57

Figure 12.26. An extended sheet in the crystal structure of **57:15** incorporating synthons **V** and **XXII**.

interactions are particularly common in organometallic crystals [86]. Examples of methylidene and methylene groups involved in intermolecular C−H···O interactions include [Co$_3$(μ$_3$-CH)(CO)$_9$] **59** (Fig. 12.28a) and [(η^5-C$_5$H$_5$)$_2$Mn$_2$ (CO)$_4$(μ-CH$_2$)] **60** (Fig. 12.28b) [87, 88]. Clearly, C−H···O interactions have considerable influence in determining the crystal structures adopted by organometallic compounds, perhaps to an even greater extent than in pure organic systems, and control of these interactions is vital for structure design strategies.

As for organic systems, it is likely that multipoint recognition strategies will provide greater predictability for weak hydrogen bonds in organometallic crystals – systematic studies incorporating such strategies have yet to be reported,

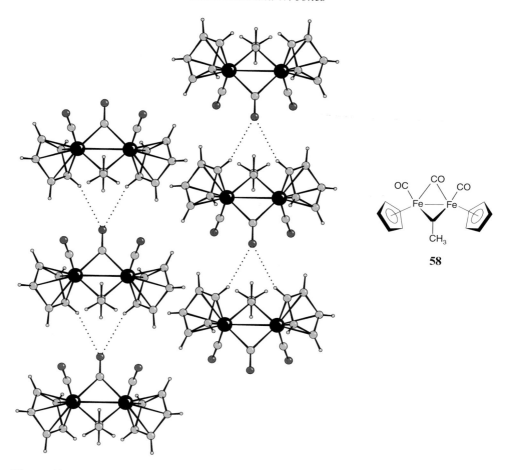

Figure 12.27. Molecules of **58** linked into chains by bifurcated C—H···O interactions, similar to the α-network formed by urea.

although there are undoubtedly numerous structures in the literature in which such interactions are important. For example, a host–guest complex of DTPO **19** (considered in Section 12.3.2) with $Mg(C_5H_4NOS)_2(H_2O)_2$ **61** has been reported which displays multipoint recognition through C—H···S interactions [89]. Molecules of **61** are linked into chains (Fig. 12.29a) which are inserted into **19** in two perpendicular directions (Fig. 12.29b). This may be compared with similar chains formed by pyromellitic dianhydride **21** via the synthon **XIII**, inserted in one direction in the **19:21** co-crystal (Figs. 12.11b, c). In **19:61**, additional supramolecular functionality is introduced above and below the plane of the chains in the form of water molecules from the diaqua-magnesium unit. These facilitate strong O—H···O hydrogen bonds with the N—O groups of **19** such that the C—H···O network observed in **19:21** and in

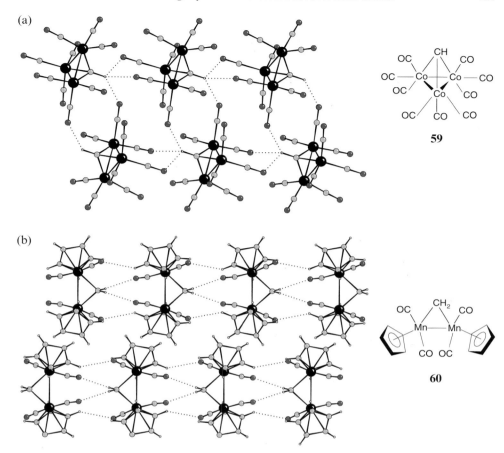

Figure 12.28. C−H···O interactions involving methylidene and methylene groups in the crystal structures of (a) **59** and (b) **60**.

pure **19** is completely disrupted. The coordination geometry of the magnesium center allows for arrangement of the supramolecular functionality in a suitable manner for this novel three-dimensional architecture – it is difficult to envisage a purely organic molecule that could perform the same function.

12.4.3 *The role of metal atoms in organometallic solids*

As we have seen, the role of metal atoms in organometallic and metallo-organic solids may be simply to act as coordination centers, organizing the organic ligands (and the supramolecular functionality that they possess) in three dimensions. It is possible, however, for the metal atom itself to become involved in intermolecular interactions – it is in these cases that the system might be

(a)

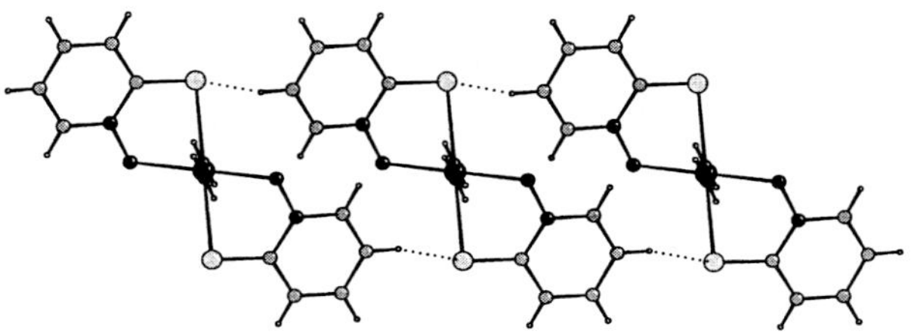

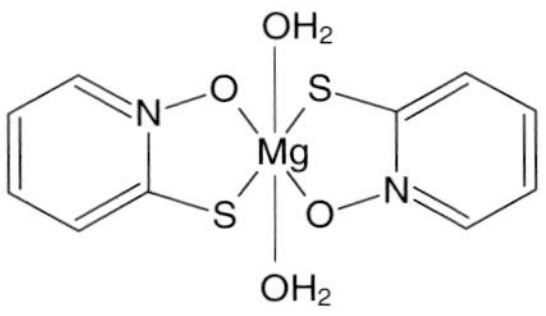

61

(b)

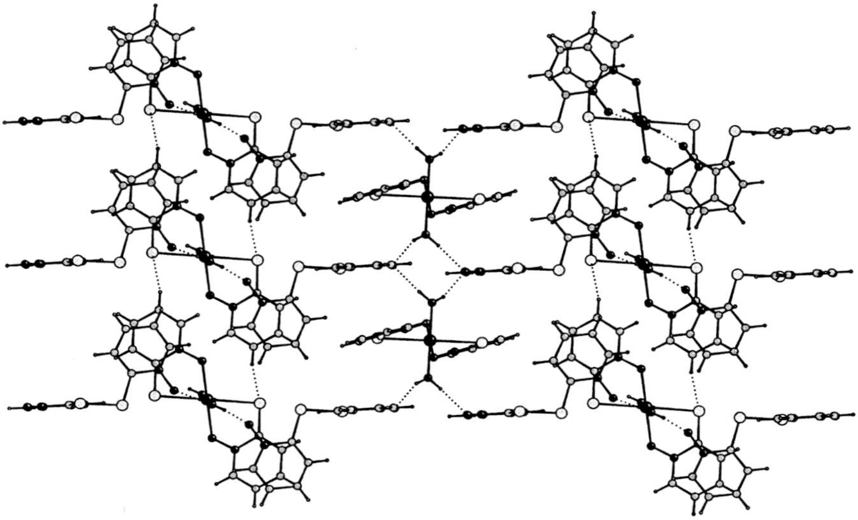

Figure 12.29. (a) Molecules of **61** are linked into chains via C−H⋯S interactions and (b) inserted in two perpendicular directions into a lattice of **19**.

expected to differ most significantly from the pure organic case. Metal-bound hydrogen atoms in polymetallic complexes can form hydrogen bonds with suitable acceptors (usually CO), provided that the hydrogen is not sterically hindered [90]. The directionality and distance preferences for M−H⋯O interactions are found to be comparable with those of C−H⋯O hydrogen bonds and it has been suggested that the two interactions are also comparable in strength. Thus, it should be possible to relate M−H⋯O interactions in organometallic systems with C−H⋯O interactions in pure organics, and perhaps derive synthons incorporating M−H⋯O hydrogen bonds. The number of complexes in which M−H⋯O interactions are actually observed is limited by the fact that hydrogen atoms bound to a metal center are commonly encased by other organic ligands, and also by the difficulty in locating hydride ligands in an

(a)

(b)

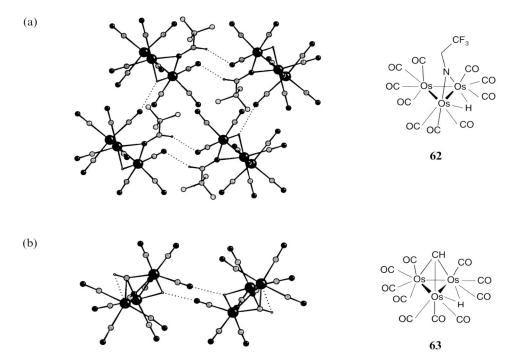

Figure 12.30. M−H⋯O interactions involving bridging hydride ligands in the crystal structures of (a) **62** and (b) **63**.

X-ray diffraction experiment. From the available structures in the CSD, it is observed that hydride ligands forming M−H⋯O bonds are present generally as μ_2-bridges spanning M−M bonds [90]. For example, the structure of $[(\mu\text{-H})(\mu\text{-NCHCF}_3)Os_3(CO)_{10}]$ **62**, as determined by neutron diffraction, contains an M−H⋯O type hydrogen bond between the bridging hydride ligand and a terminally bound CO (Fig. 12.30a) [91]. In $[(\mu\text{-H})Os_3(CO)_{10}(\mu_3\text{-CH})]$ **63**, centrosymmetric rings are formed via a similar interaction (Fig. 12.30b) [92]. A rare example of an M−H⋯O interaction involving a hydride ligand not bound in a bridging manner is the crystal structure of $[(\eta^5\text{-C}_5H_5)_2Mo(H)(CO)]$ $[(\eta^5\text{-C}_5H_5)Mo(CO)_3]$ **64**, in which M−H⋯O interactions link two cations into centrosymmetric dimers (Fig. 12.31) [93]. The ability of the hydride ligand to become involved in intermolecular interactions depends upon the characteristics of the metal center and the other ligands present – the acidity of transition metal hydrides increases across the d-block elements [94–96].

Intermolecular hydrogen bond interactions in which the metal atom acts as an acceptor are also observed. The strengths of X−H⋯M interactions are

A. D. Bond and W. Jones

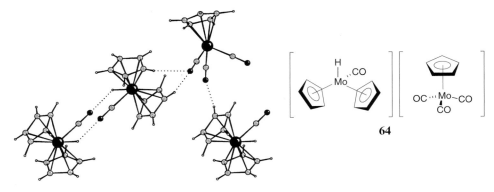

Figures 12.31. The crystal structure of **64**, incorporating centrosymmetric dimers linked by M−H···O interactions involving non-bridging hydride ligands.

estimated to be comparable with those of other strong hydrogen bonds [97], and they should, therefore, be sufficiently robust to find utility in crystal design strategies. Crystallographic studies are again inhibited by the uncertainty in the location of hydrogen atoms, and the number of relevant neutron diffraction studies performed to date is small. Nonetheless, a systematic survey of the CSD reveals a definite existence of X−H··· M (X = C, N, O) interactions where the metal atom is sterically available and electron rich – the latter requirement is usually observed with late transition metals in low oxidation states [98, 99]. For example, O−H··· M interactions are seen with water molecules incorporated in the crystal structures of Cu(N-acetyl-α-alinato)$_2$ (N-methylimidazole)$_2$·2H$_2$O **65** (Fig. 12.32a) and Cu$_2$(μ-2-pyridylamino) phthalazine-N^1,N^3,N^4,N^6)Cl$_2$(μ-Cl) (μ-OH)·1.5H$_2$O **66** (Fig. 12.32b), and also with a methanol molecule incorporated in the crystal structure of Cu$_2$(N,N'-bis(2-(o-hydroxybenzhydrylidene)amino)ethyl-1,2-ethanediamine) (CH$_3$COO)$_2$·2CH$_3$OH **67** (Fig. 12.32c) [100–102]. In cases where the ligands bound to the metal acceptor contain electronegative atoms (mainly O, N, S, Cl), the hydrogen atom is commonly observed to participate in polyfurcated interactions with the expanded electron-rich system. Examples of intermolecular C−H ··· M interactions are observed in Cu(cyclodecane-1,3-dionato)$_2$ **68**, and the solvate Pt(2,2′-bis(5,6-dihydro1,3-thiazine))(C$_6$H$_5$)$_2$·CHCl$_3$ **69** [103, 104]. In the first case, the C−H vector points toward the Cu atom and the two nearby oxygens of the ligand (Fig. 12.33a), while in the second, the C−H bond of the chloroform molecule points directly at the electron-rich Pt center (Fig. 12.33b).

Clearly, identification of metal-mediated intermolecular interactions in organometallic solids is the first step towards understanding their role in

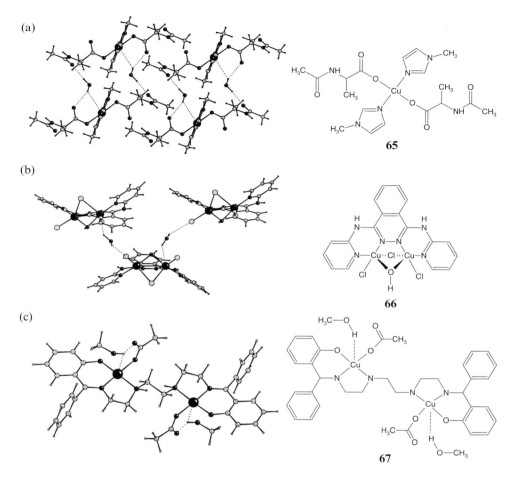

Figure 12.32. O−H···M interactions in the crystal structures of (a) **65**, (b) **66** and (c) **67**.

controlling crystal architectures. Their apparent strength and directionality would suggest that such interactions may eventually find utility in crystal design strategies. A complete understanding, however, is still far from being realized.

12.4.4 Other interactions in organometallic systems

Other notable interactions unique to organometallic systems include metal–metal interactions such as those between gold atoms – the 'aurophilic effect'. The strength of these Au···Au interactions has been estimated to be in

(a)

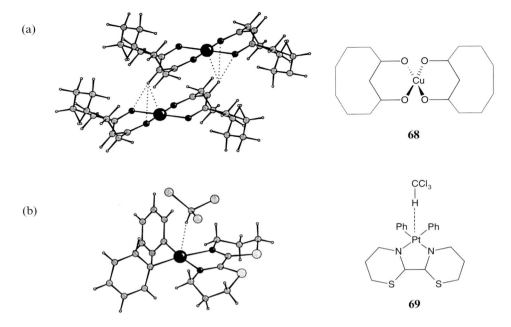

68

(b)

69

Figure 12.33. C−H···M interactions in the crystal structures of (a) **68** and (b) **69**.

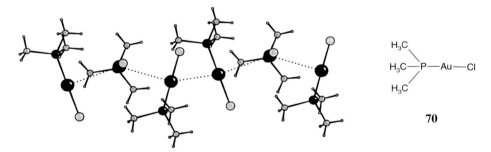

70

Figure 12.34. Molecules of **70** linked via Au···Au interactions into one-dimensional helices.

the range 20–40 kJ mol^{-1}, comparable to medium-strength hydrogen bonds [105]. The tendency for gold to form Au ··· Au interactions may be likened to the formation of short halogen ··· halogen contacts (Section 12.3.4). The crystal structure of chloro(trimethylphosphine)gold(I) **70** provides an illustrative example of supramolecular association via Au ··· Au interactions – the molecules associate into extended one-dimensional helices (Fig. 12.34) [106]. An interesting structure reported recently is that of bis(3-bromopyridine)gold(I) dichloroaurate(I) **71**, which contains intermolecular Au ··· Au

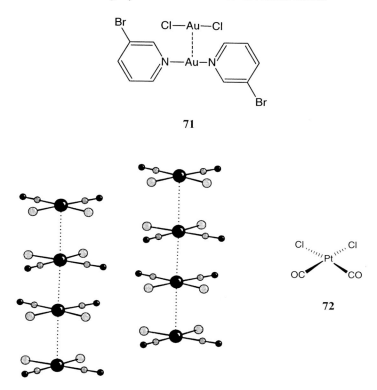

71

72

Figure 12.35. The crystal structure of **72** incorporating stacks mediated by Pt···Pt interactions.

interactions, C−H···Cl hydrogen bonds and Br···Cl interactions, all of which are considered to be structure directing [107]. Similar metal···metal interactions are observed in *cis*-[PtCl$_2$(CO)$_2$] **72**, where Pt···Pt interactions are observed in stacks of molecules (Fig. 12.35) [108].

The proliferation of aromatic rings in organometallic systems provides ample opportunity for aromatic–aromatic interactions, in both an edge-to-face and face-to-face manner. For example, in the structures of the isomeric complexes, [Ru$_6$C(CO)$_{11}$(η^6-C$_6$H$_6$)(μ_3-η^2:η^2:η^2-C$_6$H$_6$)] **73(a)** and *trans*-[Ru$_6$C(CO)$_{11}$ (η^6-C$_6$H$_6$)$_2$] **73(b)**, face-to-face interactions link the molecules into crinkled and linear tapes, respectively (Figs. 12.36a, b) [109]. In **73(a)**, the benzene rings adopt a planar offset arrangement with interplanar separations of 3.29 and 3.56 Å for the η^6:η^6 and the μ_3:μ_3 interactions, respectively – an arrangement similar to that in graphite. In **73(b)**, a nearly staggered orientation exists with an interplanar separation of 3.52 Å. The triphenylphosphine ligand, common in organometallic complexes, has been shown to adopt a

(a)   (b)

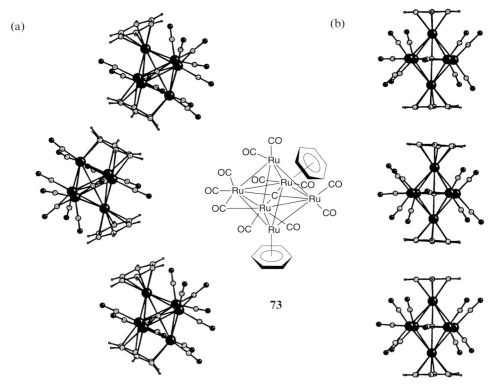

Figure 12.36. Molecular isomers of **73** form (a) crinkled and (b) linear tapes mediated by aromatic–aromatic interactions.

common supramolecular arrangement, referred to as the 'sextuple phenyl embrace'. Similar arrangements are observed between Ph_4P^+ cations, a common choice of counterion for synthetic organometallic chemists (Fig. 12.37) [110]. This motif arises from concerted edge-to-face aromatic interactions between neighboring PPh_3 groups, and the total energy of the interaction has been estimated as 60–85 kJ mol^{-1} [111]. Here, the authors recognize the underlying relationship between supramolecular recognition and the inherent enthalpic requirements of crystal packing: 'phenylated cations provide quality crystals with favorable lattice energies … tacit recognition of advantageous supramolecularity in their molecular crystals'.

12.6 Concluding remarks

In order to utilize the desired properties of molecules it is equally important that crystal structure as well as molecular structure be controlled or designed.

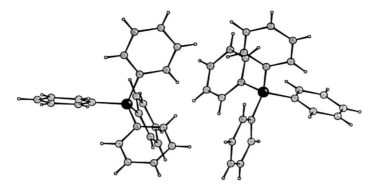

Figure 12.37. The 'sextuple phenyl embrace' illustrated between two PPh_4^+ cations.

The reasons for this are numerous: crystal structure is known to affect reactivity, photochemistry, optical properties, magnetic structures and ferroelastic properties, to give but a few examples. In this chapter, we have described some of the more explicit examples of the supramolecular approach to crystal design via utilization of what are variously called synthons, couplings, etc. Without doubt, the hydrogen bond remains the intermolecular interaction of choice. Its robust and directional nature leads to the observation of targeted primary and secondary supramolecular motifs in a satisfyingly high proportion of cases. In some materials, albeit relatively few, control of the whole three-dimensional architecture is achieved. As we have seen, however, crystal structures are not determined simply by a small number of strong interactions, but by the cooperative effect of a large number of strong and weak interactions. Progress is being made towards understanding the nature of these forces and their influence on crystal packing, but *ab initio* structure prediction for general organic molecules remains a faraway goal.

The recent elegant work of Lauher, Fowler and co-workers in the systematic design of a polymerizable triacetylene exemplifies the predictive use rather than simple retrospective analysis of crystal-engineering principles [112] – an impressive example of the approach foreseen by Schmidt in 1971 [63]. We are also seeing the systematic analysis of organometallic and metallo-organic crystals with useful synthons being identified in these systems. In many cases, a satisfying similarity with organic systems is observed, although additional factors associated with the role of the metal atom and the effect of the crystal field on intramolecular structure add an extra dimension. Not all synthons are equally robust and persistent, of course. In some cases, they may be countered by the presence of stronger interactions, more effective in satisfying the overall enthalpy requirements of crystal packing.

Practical considerations may also arise. Supramolecular functionality introduced to control crystal architecture must be added without degrading the fundamental molecular properties of primary interest. Co-crystals offer an alternative approach for controlling crystal architecture without necessarily modifying the primary molecule of interest. In addition, we must recognize the role of the solvent from which the crystal grows. The occurrence of solvates and polymorphs, particularly relevant in the pharmaceutical industry, is still a relatively poorly understood aspect of crystal chemistry. The manner in which synthons are modified from normal geometries in non-crystalline organic structures is also yet to be explored fully.

In this chapter, we have aimed to present illustrative examples – the success stories. Some of these successes undoubtedly arise by serendipity, essentially by post-mortem analysis of observed crystal structures. Others are indeed the result of inspired design. The continuing challenge for crystal engineers is to tip the balance away from the fortuitous towards the true manipulation of the three-dimensional architecture of molecular crystals.

12.6.1 Acknowledgement

All crystal packing diagrams in this chapter were produced using the software package *CAMERON* [113].

12.7 References

1. Brittain, H. G. (1999). In *Polymorphism in Pharmaceutical Solids*. Dekker, New York.
2. Jones, W. (1997). In *Organic Molecular Solids: Properties and Applications*. CRC Press, New York.
3. Haleblian, J. and McCrone, W. (1969). Pharmaceutical applications of polymorphism. *J. Pharm. Sci.* **58**, 911.
4. Gavezzotti, A. (Ed.). (1997). In *Crystal Symmetry and Molecular Recognition*. John Wiley and Sons, Chichester.
5. Corey, E. J. (1967). General methods for the construction of complex molecules. *Pure Appl. Chem.* **14**, 19.
6. Desiraju, G. R. (1995). Supramolecular synthons in crystal engineering – a new organic synthesis. *Angew. Chem., Int. Ed. Engl.* **34**, 2311.
7. Lehn, J.-M. *Supramolecular Chemistry: Concepts and Perspectives*. VCH, Weinheim.
8. Jeffrey, G. A. (1997). *An Introduction to Hydrogen Bonding*. Oxford University Press.
9. Kitaigorodskii, A. I. (1961). *Organic Chemical Crystallography*. Consultants Bureau.
10. Lehn, J.-M. (1988). Supramolecular chemistry – scope and perspectives. Molecules, supermolecules and molecular devices. *Angew. Chem., Int. Ed. Engl.* **27**, 89.

11. Dunitz, J. D. (1991). Phase transitions in molecular crystals from a chemical viewpoint. *Pure Appl. Chem.* **63**, 177.
12. Desiraju, G. R. (1989). *Crystal Engineering: the Design of Organic Solids.* Elsevier, Amsterdam.
13. Aakeroy, C. B. and Seddon, K. R. (1993). The hydrogen bond and crystal engineering. *Chem. Soc. Rev.* **22**, 397.
14. Bernstein, J., Etter, M. C. and Leiserowitz, L. (1994). In *The Role of Hydrogen Bonding in Molecular Assemblies* (H.-B. Burgi and J. D. Dunitz, Eds.). VCH, Weinhein.
15. Sim, G. A., Robertson, J. M. and Goodwin, T. H. (1955). The crystal and molecular structure of benzoic acid, *Acta Crystallogr.* **8**, 157.
16. Bailey, M. and Brown. C. J. (1967). The crystal structure of terephthalic acid. *Acta Crystallogr.* **22**, 387.
17. Alcala, R. and Martinez Carrera, S. (1972). The crystal structure of isophthalic acid. *Acta Crystallogr.* **B28**, 1671.
18. Duchamp, D. J. and Marsh, R. E. (1969). The crystal structure of trimesic acid (benzene-1,3,5-tricarboxylic acid). *Acta Crystallogr.* **B25**, 5.
19. Ermer, O. (1988). Fivefold-diamond structure of adamantane-1,3,5,7-tetracarboxylic acid. *J. Am. Chem. Soc.* **110**, 3747.
20. Fyfe, M. C. T. and Stoddart, J. F. (1997). Synthetic supramolecular chemistry. *Acc. Chem. Res.* **30**, 393.
21. Aakeroy, C. B. (1997). Crystal engineering: strategies and architectures. *Acta. Crystallogr.* **B53**, 569.
22. Nangia, A. and Desiraju, G. R. (1998). Supramolecular structures – reason and imagination. *Acta Crystallogr.* **A54**, 934.
23. Leiserowitz, L. (1976). Molecular packing modes: carboxylic acids. *Acta Crystallogr.* **B32**, 775.
24. Etter, M. C. (1990). Encoding and decoding hydrogen-bond patterns of organic compounds. *Acc. Chem. Res.* **23**, 120.
25. Leiserowitz, L. and Schmidt, G. M. J. (1969). Molecular packing modes, part III: primary amides. *J. Chem. Soc. A.* p. 2372.
26. Leiserowitz, L. and Tuval, M. (1978). Molecular packing modes: *N*-methylamides. *Acta Crystallogr.* **B34**, 1230.
27. Penfold, B. R. and White, J. C. B. (1959). The crystal and molecular structure of benzamide. *Acta Crystallogr.* **12**, 130.
28. Zhao, X., Chang, Y.-L., Fowler, F. W. and Lauher, J. W. (1990). An approach to the design of molecular solids. The ureylenedicarboxylic acids. *J. Am. Chem. Soc.* **112**, 6627.
29. Chang, Y.-L., West, M.-A., Fowler, F. W. and Lauher, J. W. (1993). An approach to the design of molecular solids. Strategies for controlling the assembly of molecules into two-dimensional layered structures. *J. Am. Chem. Soc.* **115**, 5991.
30. Coe, S., Kane, J. J., Nguyen, T. L., Toledo, L. M., Wininger, E., Fowler, F. W. and Lauher, J. W. (1997). Molecular symmetry and the design of molecular solids: The oxalamide functionality as a persistent hydrogen bonding unit. *J. Am. Chem. Soc.* **119**, 86.
31. Toledo, L. M., Lauher, J. W. and Fowler, F. W. (1994). Design of molecular solids. Application of 2-amino-4(1*H*)-pyridones to the preparation of hydrogen-bonded α- and β-networks. *Chem. Mater.* **6**, 1222.
32. Liao, R.-F., Lauher, J. W. and Fowler, F. W. (1996). The application of 2-amino-4-pyrimidones to supramolecular synthesis. *Tetrahedron* **52**, 3153.

33. Hollingsworth, M. D., Brown, M. E., Santarsiero, B. D., Huffman, J. C. and Goss, C. R. (1994). Template-directed synthesis of 1:1 layered complexes of α,ω-dinitriles with urea: packing efficiency versus specific functional group interactions. *Chem. Mater.* **6**, 1227.

34. Nguyen, T. L., Scott, A., Dinkelmeyer, B., Fowler, F. W. and Lauher, J. W. (1998). Design of molecular solids: utility of the hydroxyl functionality as a predictable design element. *New J. Chem.* **22**, 129.

35. Zerkowski, J. A., Seto, C. T. and Whitesides, G. M. (1992). Solid-state structures of 'rosette' and 'crinkled tape' motifs derived from the cyanuric acid-melamine lattice. *J. Am. Chem. Soc.* **114**, 5473.

36. Seto, C. T. and Whitesides, G. M. (1990). Self-assembly based on the cyanuric acid melamine lattice. *J. Am. Chem. Soc.* **112**, 6409.

37. Ranganathan, A., Pedireddi, V. R. and Rao, C. N. R. (1999). Hydrothermal synthesis of organic channel structures: 1:1 hydrogen-bonded adducts of melamine with cyanuric and trithiocyanuric acids. *J. Am. Chem. Soc.* **121**, 1752.

38. Zerkowski, J. A., MacDonald, J. C., Seto, C. T., Wierda, D. A. and Whitesides, G. M. (1994). Design of organic structures in the solid state: molecular tapes based on the network of hydrogen bonds present in the cyanuric acid: melamine complex. *J. Am. Chem. Soc.* **116**, 2382.

39. Pedireddi, V. R., Chatterjee, S., Ranganathan, A. and Rao, C. N. R. (1997). Noncovalent synthesis of layered and channel structures involving sulfur-mediated hydrogen bonds. *J. Am. Chem. Soc.* **119**, 10867.

40. Ermer, O. and Lex, J. (1987). Shortened C-C bonds and antiplanar O=C−O−H torsion angles in 1,4-cubancarboxylic acid. *Angew. Chem., Int. Ed. Engl.* **26**, 447.

41. Kuduva, S. S., Craig, D. C., Nangia, A. and Desiraju, G. R. (1999). Cubanecarboxylic acids. Crystal engineering considerations and the role of C−H···O hydrogen bonds in determining O−H···O networks. *J. Am. Chem. Soc.* **121**, 1936.

42. Hare, M., Emrick, T., Eaton, P. E. and Kass, S. R. (1997). Cubyl anion formation and an experimental determination of the acidity and C−H bond dissociation energy of cubane. *J. Am. Chem. Soc.* **119**, 237.

43. Brown, C. J. (1966). The crystal structure of fumaric acid. *Acta Crystallogr.* **21**, 1.

44. Bednowitz, A. L. and Post, B. (1966). Direct determination of the crystal structure of β-fumaric acid. *Acta Crystallogr.* **21**, 566.

45. Batchelor, E. and Jones, W. (1998). Citraconic acid. *Acta Crystallogr.* **C54**, 238.

46. Chaney, J. D., Goss, C. R., Folting, K., Santarsiero, B. D. and Hollingsworth, M. D. (1996). Formyl C−H···O hydrogen bonding in crystalline bis-formamides? *J. Am. Chem. Soc.* **118**, 9432.

47. Donohue, J. (1968). In *Selected Topics in Hydrogen Bonding* (A. Rich and N. Davidson, Eds.). Freeman, San Francisco.

48. Cotton, F. A., Daniels, L. M., Jordan, G. T. and Murillo, C. A. (1997). The crystal packing of bis(2,2'-dipyridylamido)cobalt(II), Co(dpa)$_2$, is stabilised by C−H···N bonds: are there any real precedents? *J. Chem. Soc., Chem. Commun.*, 1673.

49. Steiner, T. and Desiraju, G. R. (1998). Distinction between the weak hydrogen bond and the van der Waals interaction. *J. Chem. Soc., Chem. Commun.*, 891.

50. Taylor, R. and Kennard, O. (1982). Crystallographic evidence for the existence of C−H⋯O, C−H⋯N and C−H⋯Cl interactions. *J. Am. Chem. Soc.* **104**, 5063.

51. Thalladi, V. R., Weiss, H. C., Blaser, D., Boese, R., Nangia, A. and Desiraju, G. R. (1998). C−H⋯F interactions in the crystal structures of some fluorobenzenes. *J. Am. Chem. Soc.* **120**, 8702.

52. Bodige, S. G., Rogers, R. D. and Blackstock, S. C. (1997). Supramolecular networks via pyridine *N*-oxide C−H⋯O hydrogen bonding in the crystal structures of 2,2′-dithiobis(pyridine *N*-oxide) and its complexes with 1,2,4,5-tetracyanobenzene and pyromellitic dianhydride. *J. Chem. Soc., Chem. Commun.*, 1669.

53. Anthony, A., Jaskolski, M., Nangia, A. and Desiraju, G. R. (1998). Isostructurality in crystalline oxa-androgens: a case of C−O−H⋯O and C−H⋯O interaction mimicry and solid solution formation. *J. Chem. Soc., Chem. Commun.*, 2537.

54. Berkovitch-Yellin, Z. and Leiserowitz, L. (1980). The role of Coulomb forces in the crystal packing of amides: a study based on experimental electron densities. *J. Am. Chem. Soc.* **102**, 7677.

55. Biradha, K., Sharma, C. V. K., Panneerselvam, K., Shimoni, L., Carrell, H. L., Zacharias, D. E. and Desiraju, G. R. (1993). Solid-state supramolecular assembly via C−H⋯O hydrogen bonds – crystal structures of the complexes of 1,3,5-trinitrobenzene with dibenzylideneacetone and 2,5-dibenzylidenecyclopentanone. *J. Chem. Soc., Chem. Commun.*, 1473.

56. Madhavi, N. N. L., Katz, A. K., Carrell, H. L., Nangia, A. and Desiraju, G. R. (1997). Evidence for the characterisation of the C−H⋯π interaction as a weak hydrogen bond: toluene and chlorobenzene solvates of 2,3,7,8-tetraphenyl-1,9,10-anthyridine. *J. Chem. Soc., Chem. Commun.*, 1953.

57. Shimoni, L. and Glusker, J. P. (1994). The geometry of intermolecular interactions in some crystalline fluorine-containing organic compounds. *Struct. Chem.* **5**, 383.

58. Pedireddi, V. R., Jones, W., Chorlton, A. and Docherty, R. (1996). Creation of crystalline supramolecular assemblies using C−H⋯O/O−H⋯N pair-wise hydrogen bond coupling. *J. Chem. Soc., Chem. Commun.*, 997.

59. Batchelor, E., Klinowski, J. and Jones, W. (2000). Crystal engineering using co-crystallisation of phenazine with dicarboxylic acids. *J. Mater. Chem.* **10**, 839.

60. Shaameri, Z. and Jones, W. (2001). Molecular complexes between 2,2′-biphenyldicarboxylic acid and phenazine: anhydrous and hydrated forms. *Mol. Cryst. Liq. Cryst.* **356**, 131.

61. Pedireddi, V. R., Chatterjee, S., Ranganathan, A. and Rao, C. N. R. (1998). A study of supramolecular hydrogen bonded complexes formed by aliphatic dicarboxylic acids with azaaromatic donors. *Tetrahedron* **54**, 9457.

62. Hunter, C. A. and Sanders, J. K. M. (1990). The nature of π–π interactions. *J. Am. Chem. Soc.* **112**, 5525.

63. Schmidt, G. M. J. (1971). Photodimerisation in the solid state. *Pure Appl. Chem.* **27**, 647.

64. Pedireddi, V. R., Reddy, D. S., Goud, B. S., Craig, D. C., Rae, A. D. and Desiraju, G. R. (1994). The nature of halogen–halogen interactions and the crystal structure of 1,3,5,7-tetraiodoadamantane. *J. Chem. Soc., Perkin Trans. 2* 2353.

65. Allen, F. H., Goud, B. S., Hoy, V. J., Howard, J. A. K. and Desiraju, G. R. (1994). Molecular recognition via iodo···nitro and iodo···cyano interactions: crystal structures of the 1:1 complexes of 1,4-diiodobenzene with 1,4-dinitrobenzene and 7,7,8,8-tetracyanoquinodimethane (TCNQ). *J. Chem. Soc., Chem. Commun.*, 2729.

66. Banister, A. J., Bricklebank, N., Clegg, W., Elsegood, M. R. J., Gregory, C. I., Lavender, I., Rawson, J. M. and Tanner, B. K. (1995). The first solid-state paramagnetic 1,2,3,5-dithiadiazoyl radical; X-ray crystal structure of [p-NCC$_6$F$_4$CNSSN]·. *J. Chem. Soc., Chem. Commun.*, 679.

67. Banister, A. J., Bricklebank, N., Lavender, I., Rawson, J. M., Gregory, C. I., Tanner, B. K., Clegg, W., Elsegood, M. R. J. and Palacio, F. (1996). Spontaneous magnetization in a sulfur-nitrogen radical at 36 K. *Angew. Chem., Int. Ed. Engl.* **35**, 2533.

68. Braga, D., Grepioni, F., Wadepohl, H., Gebert, S., Calhorda, M. J. and Veiros, L. F. (1995). Intramolecular and intermolecular bonding in crystalline clusters of the type (Cp^R)$_3$M$_3$(CO)$_3$ [M = Co, Rh, Ir; Cp^R = C$_5$H$_5$, C$_5$Me$_5$, C$_5$H$_4$Me]. *Organometallics* **14**, 5350.

69. Braga, D., Grepioni, F., Sabatino, P. and Desiraju, G. R. (1994). Hydrogen bonding in organometallic crystals 1: from carboxylic acids and alcohols to carbonyl complexes. *Organometallics* **13**, 3532.

70. Biradha, K., Desiraju, G. R., Braga, D. and Grepioni, F. (1996). Hydrogen bonding in organometallic crystals 3: transition-metal complexes containing amido groups. *Organometallics* **15**, 1284.

71. van Meurs, F. and van Koningsveld, H. (1974). The molecular structure of 4-tert-butyl-π-(tricarbonyl-chromium)benzoic acid. *J. Organomet. Chem.* **78**, 229.

72. Hsiou, Y., Wang, Y. and Liu, L.-K. (1989). Structures of tetracarbonyl (2-3-η-maleic acid)iron, *cis*-[Fe(C$_4$H$_4$O$_4$)(CO)$_4$], and tetracarbonyl (2-3-η-fumaric acid)iron, *trans*-[Fe(C$_4$H$_4$O$_4$)(CO)$_4$]. *Acta Crystallogr.* **C45**, 721.

73. James, M. N. G. and Williams, G. B. J. (1974). A refinement of the crystal structure of maleic acid. *Acta Crystallogr.* **B30**, 1249.

74. Pedone, C. and Sirigu, A. (1968). Crystal structure and absolute configuration of (−)-tetracarbonyl(fumaric acid)iron. *Inorg. Chem.* **7**, 2614.

75. Messer, D., Landgraf, G. and Behrens, H. (1979). η^5-cyclopentadienyl-carbonyl-nitrosyl-carbamoyl-mangan(I), die erste rontgenstrukturanalyse eines carbarmoylkomplexes mit einer unsubstituierten NH$_2$-gruppe. *J. Organomet. Chem.* **172**, 349.

76. Brunn, K., Endres, H. and Weiss, J. (1988). Reactions of the organic donor tetra(methylthio)tetrathiafulvalene with halides of copper and gold: (TTM-TTF)(CuI)$_2$, (TTM-TTF)$^+$[CuCl$_2$]$^-$, (TTM-TTF)$^{2+}$[AuCl$_4$]$^{2-}$ and C$_4$H$_6$S$_4$AuCl$_2$. *Z. Naturforsch, Teil B.* **43**, 113.

77. Pearson, A. J., Mallik, S., Mortezaei, R., Perry, M. W. D., Shively, R. J. and Youngs, W. J. (1990). Stereocontrol during the alkylation of enolates attached to π-allyl-Mo(CO)$_2$Cp systems. *J. Am. Chem. Soc.* **112**, 8034.

78. Tenhaeff, S. C., Tyler, D. R. and Weakley, T. J. R. (1992). Di-μ-carbonyl-dicabonylbis[(2-hydroxyethyl)-η-5-cyclopentadienyl]diiron(*Fe–Fe*). *Acta Crystallogr.* **C48**, 162.

79. Brock, C. P. and Duncan, L. L. (1994). Anomalous space-group frequencies for monoalcohols C$_n$H$_m$OH. *Chem. Mater.* **6**, 1307.

80. Burrows, A. D., Chan, C.-W., Chowdry, M. M., McGrady, J. E. and Mingos, D. M. P. (1995). Multidimensional crystal engineering of bifunctional metal complexes containing complementary triple hydrogen bonds. *Chem. Soc. Rev.* **24**, 329.

81. Houlton, X., Mingos, D. M. P. and Williams, D. J. (1994). Multi-dimensional crystal engineering of metal complexes based on complementary hydrogen bonding. *Transition Met. Chem.* **19**, 653.

82. Houlton, A., Mingos, D. M. P. and Williams, D. J. (1994). Molecular recognition of nucleoside bases by a metal complex. A novel bifunctionality with implications for metallo-drug design. *J. Chem. Soc., Chem. Commun.*, 503.

83. Chowdry, M. M., Mingos, D. M. P., White, A. J. P. and Williams, D. J. (1996). Novel supramolecular self-assembly of a transition-metal-organo network based on simultaneous coordinate- and hydrogen-bond interactions. *J. Chem. Soc., Chem. Commun.*, 899.

84. Braga, D., Grepioni, F., Biradha, K., Pedireddi, V. R. and Desiraju, G. R. (1995). Hydrogen bonding in organometallic crystals 2: $C-H \cdots O$ hydrogen bonds in bridged and terminal first-row metal carbonyls. *J. Am. Chem. Soc.* **117**, 3156.

85. Orpen, A. G. (1983). Structural chemistry of binuclear metal centres – crystal and molecular structures of the μ-vinyl and μ-methylcarbene complexes $[Fe_2(CO)_2(\mu\text{-}CO)(\mu\text{-}CHCH_2)(\eta\text{-}C_5H_2)_2][BF_4]$ and $[Fe_2(CO)_2(\mu\text{-}CO)(\mu\text{-}CHMe)(\eta\text{-}C_5H_5)_2]$. *J. Chem. Soc., Dalton Trans.*, 1427.

86. Braga, D., Grepioni, F. and Desiraju, G. R. (1998). Crystal engineering and organometallic architecture. *Chem. Rev.* **98**, 1375.

87. Leung, P., Coppens, P., McMullan, R. K. and Koetzle, T. F. (1981). The structure of nonacarbonyl-μ_3-methylidene-triangulo-tricobalt: X-ray and neutron diffraction studies. *Acta Crystallogr.* **B37**, 1347.

88. Clemente, D. A., Biagini, M. C., Rees, B. and Herrmann, W. A. (1982). Molecular structure and experimental electron density of (μ-methylene)bis[dicarbonyl(η^5-cyclopentadienyl)manganese] at 130 K. *Inorg. Chem.* **21**, 3741.

89. Bond, A. D. and Jones, W. (2000). A host–guest complex of diaquabis(1-hydroxy-2(1*H*)-pyridinethionato-*O,S*]-magnesium(II) and 2,2′-dithiobispyridine-*N*-oxide. *Acta Crystallogr.* **C56**, 436.

90. Braga, D., Grepioni, F., Tedesco, E., Biradha, K. and Desiraju, G. R. (1996). Hydrogen bonding in organometallic crystals 4: $M-H \cdots O$ hydrogen-bonding interactions. *Organometallics* **15**, 2692.

91. Dawoodi, Z., Mays, M. J. and Orpen, A. G. (1981). Neutron diffraction structure analysis of $[(\mu\text{-}H)(\mu\text{-}NCHCF_3)Os_3(CO)_{10}]$ at 20 K. *J. Organomet. Chem.* **219**, 251.

92. Shapley, J. R., Cree-Uhciyama, M. E. and St. George, G. M. (1983). Synthesis and reactivity of a 'semi' triply bridging methylidene complex. Crystal structure of $HOs_3(CO)_{10}(CH)$. *J. Am. Chem. Soc.* **105**, 140.

93. Adams, M. A., Folting, K., Huffman, J. C. and Caulton, K. G. (1979). Structural determination of $[(C_5H_5)_2Mo(H)CO][(C_5H_5)Mo(CO)_3]$. *Inorg. Chem.* **18**, 3020.

94. Shriver, D. F. (1970). Transition metal basicity. *Acc. Chem. Res.* **3**, 231.

95. Bursten, B. E. and Gatter, M. G. (1984). Molecular orbital studies of organometallic hydride complexes 2. The correlation of hydrogen atom reactivity with valence orbital energetics. *Organometallics* **3**, 895.

96. Pearson, R. G. (1985). The transition-metal-hydrogen bond. *Chem. Rev.* **85**, 41.

97. Zhang, K., Gonzalez, A. A., Mukerjee, S. L., Chou, S.-J., Hoff, C. D., Kubat-Martin, K. A., Barnhar, D. and Kubas, G. J. (1991). Solution calorimetric and stopped-flow kinetic study of ligand substitution for the complexes $M(CO)_3(PCy)_2(L)$ (M = Cr, Mo, W). Comparison of first-, second- and third-row transition-metal-ligand bonds at sterically-crowded metal centres, *J. Am. Chem. Soc.* **113**, 9170.

98. Braga, D., Grepioni, F., Biradha, K. and Desiraju, G. R. (1996). Agostic interactions in organometallic compounds. A Cambridge Structural Database study. *J. Chem. Soc., Dalton Trans.*, 3925.

99. Braga, D., Grepioni, F., Tedesco, E., Biradha, K. and Desiraju, G. R. (1997). Hydrogen bonding in organometallic crystals 6: X−H · · · M hydrogen bonds and M · · · (H−X) pseudo-agostic bonds. *Organometallics* **16**, 1846.

100. Battaglia, L. P., Corradi, A. B., Menabue, L., Pellacani, G. C., Prompolini, P. and Saladani, M. (1982). Ternary complexes of copper(II) with *N*-protected amino acids and *N*-methlyimidazole. Crystal and molecular structures of bis(*N*-acetyl-α-alaninato)bis(*N*-methlyimidazole)copper(II) dihydrate. *J. Chem. Soc., Dalton Trans.*, 781.

101. Marongiu, G. and Lingafelter, E. C. (1982). μ-{1,4-bis[(2-pyridyl)amino]phthalazine-N^1,N^3,N^4,N^6}-μ-chloro-dichloro-μ-hydroxodicopper(II) sesquihydrate. *Acta Crystallogr.* **B38**, 620.

102. Chiari, B., Hatfield, W. E., Piovesana, O., Tarantelli, T., ter Haar, L. W. and Zanazzi, P. F. (1983). Exchange interaction in multinuclear transition-metal complexes 3. Synthesis, X-ray structure and magnetic properties of $Cu_2L(CH_3COO)_2·2CH_3OH$ (L^{2-} = anion of *N,N′*-bis(2-((*o*-hydroxybenzhydrylidene)amino)ethyl)-1,2-ethanediamine), a one-dimensional Heisenberg antiferromagnet having through-bond coupled copper(II) ions. *Inorg. Chem.* **22**, 1468.

103. Modenbach, C. L., Fronczek, F. R., Berg, E. W. and Taylor, T. C. (1983). Crystal and molecular structures of bis(cyclodecane-1,3-dionato)copper(II) and bis(cyclotridecane-1,3-dionato)copper(II). *Inorg. Chem.* **22**, 4083.

104. Bruno, G., Lanza, S. and Nicolo, F. (1990). Structure of [Pt(C$_6$H$_6$)$_2$(btz-*N,N'*)]·CHCl$_3$, btz = 2,2'-bi-5,6–5,6-dihydro-4*H*-1,3-thiazine. *Acta Crystallogr.* **C46**, 765.

105. Schmidbaur, H. (1995). High-carat gold compounds. *Chem. Soc. Rev.* **24**, 391.

106. Angermaier, K., Zeller, E. and Schmidbauer, H. (1994). Crystal structures of chloro(trimethylphosphine)gold(I), chloro(tri-^{i}propylphosphine)gold(I) and bis(trimethylphosphine)gold(I) chloride. *J. Organomet. Chem.* **472**, 371.

107. Freytag, M. and Jones, P. G. (2000). Hydrogen bonds C−H · · · Cl as a structure-determining factor in the gold(I) complex bis(3-bromopyridine)gold(I) dichloroaurate(I). *J. Chem. Soc., Chem. Commun.*, 277.

108. Bagnoli, F., Belli dell'Amico, D., Calderazzo, F., Englert, U., Marchetti, F., Herberich, G. E., Pasqualetti, N. and Ramello, S. (1996). New preparative access to the crystal and molecular structures of *cis*-[PtCl$_2$(CO)$_2$]. *J. Chem. Soc., Dalton Trans.* 4317.

109. Braga, D., Grepioni, F., Johnson, B. F. G., Chen, H. and Lewis, J. (1991). Molecular organisation and dynamic behaviour of arene cluster complexes in the solid state. Crystal structures of [Os$_4$H$_2$(CO)$_{10}$(η^6-C$_6$H$_5$Me)], [Os$_4$H$_2$(CO)$_{10}$(η^6-C$_6$H$_4$Me$_2$)] and [Ru$_6$C(CO)$_{14}$(η^6-C$_6$H$_3$Me$_3$-1,3,5)]. *J. Chem. Soc., Dalton Trans.*, 2559.

110. Dance, I. and Scudder, M. (1996). Supramolecular motifs: concerted multiple phenyl embraces between Ph_4P^+ cations are attractive and ubiquitous.. *Chem. Eur. J.* **2**, 481.

111. Dance, I. and Scudder, M. (1995). The sextuple embrace, a ubiquitous concerted supramolecular motif. *J. Chem. Soc. Chem., Commun.*, 1039.

112. Xiao, J., Yang, M., Lauher, J. W. and Fowler, F. W. (2000). A supramolecular solution to a long-standing problem: the 1,6-polymerization of a triacetylene. *Angew. Chem., Int. Ed. Engl.* **39**, 2132.

113. Watkin, D. J., Prout, C. K. and Pearce, L. J. (1996). *CAMERON*. Chemical Crystallography Laboratory, University of Oxford, Oxford.

Index